This report contains the collective views of an international group of experts and does not necessarily represent the decisions or the stated policy of the United Nations Environment Programme, the International Labour Organisation, or the World Health Organization

Environmental Health Criteria 80

PYRROLIZIDINE ALKALOIDS

Published under the joint sponsorship of the United Nations Environment Programme, the International Labour Organisation, and the World Health Organization

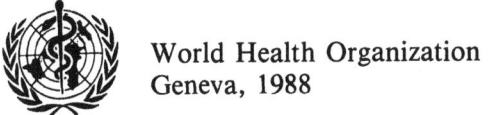

World Health Organization
Geneva, 1988

The **International Programme on Chemical Safety (IPCS)** is a joint venture of the United Nations Environment Programme, the International Labour Organisation, and the World Health Organization. The main objective of the IPCS is to carry out and disseminate evaluations of the effects of chemicals on human health and the quality of the environment. Supporting activities include the development of epidemiological, experimental laboratory, and risk-assessment methods that could produce internationally comparable results, and the development of manpower in the field of toxicology. Other activities carried out by IPCS include the development of know-how for coping with chemical accidents, coordination of laboratory testing and epidemiological studies, and promotion of research on the mechanisms of the biological action of chemicals.

ISBN 92 4 154280 2

©World Health Organization 1988

Publications of the World Health Organization enjoy copyright protection in accordance with the provisions of Protocol 2 of the Universal Copyright Convention. For rights of reproduction or translation of WHO publications, in part or *in toto,* application should be made to the Office of Publications, World Health Organization, Geneva, Switzerland. The World Health Organization welcomes such applications.

The designations employed and the presentation of the material in this publication do not imply the expression of any opinion whatsoever on the part of the Secretariat of the World Health Organization concerning the legal status of any country, territory, city or area or of its authorities, or concerning the delimitation of its frontiers or boundaries.

The mention of specific companies or of certain manufacturers' products does not imply that they are endorsed or recommended by the World Health Organization in preference to others of a similar nature that are not mentioned. Errors and omissions excepted, the names of proprietary products are distinguished by initial capital letters.

ISSN 0250-863X
PRINTED IN FINLAND
DASS — VAMMALA — 5500

CONTENTS

	Page
ENVIRONMENTAL HEALTH CRITERIA FOR PYRROLIZIDINE ALKALOIDS	
PREFACE	12
INTRODUCTION - PYRROLIZIDINE ALKALOIDS AND HUMAN HEALTH	13
1. SUMMARY AND RECOMMENDATIONS	19
1.1 Summary	19
1.2 Sources and chemical structure	19
1.3 Mechanisms and features of toxicity	20
1.4 Effects on man	22
1.4.1 Nature and extent of health risks	22
1.5 Methods for prevention	24
1.6 Recommendations	24
1.6.1 General recommendations	24
1.6.2 Recommendations for research	25
2. PROPERTIES AND ANALYTICAL METHODS	27
2.1 Chemical structure and properties	27
2.2 Analytical methods	32
2.2.1 Extraction	33
2.2.1.1 Plant tissue	33
2.2.1.2 Biological fluids and tissues	33
2.2.2 Analysis for pyrrolizidine alkaloids	33
2.2.2.1 Thin-layer chromatography (TLC)	33
2.2.2.2 High-performance liquid chromatography (HPLC)	34
2.2.2.3 Gas chromatography (GC) and mass spectrometry (MS)	34
2.2.2.4 Nuclear magnetic resonance (NMR) spectrometry	35
2.2.2.5 The Ehrlich reaction	35
2.2.2.6 Indicator dyes	36
2.2.2.7 Direct weighing	36
2.3 Determination of metabolites in animal tissues	37
3. SOURCES AND PATHWAYS OF EXPOSURE	38
3.1 Hepatotoxic pyrrolizidine alkaloids and their sources	38
3.2 Pneumotoxic and other toxic pyrrolizidine alkaloids	40

			Page
	3.3	Pathways of exposure	42
		3.3.1 Contamination of staple food crops	43
		3.3.2 Herbal infusions	43
		3.3.3 Use of PA-containing plants as food	47
		3.3.4 Contaminated honey	50
		3.3.5 Milk	51
		3.3.6 Meat	54
		3.3.7 Use of PAs as chemotherapeutic agents for cancer	54
4.	METABOLISM		55
	4.1	Absorption, excretion, and tissue distribution	55
		4.1.1 Absorption	55
		4.1.2 Excretion and distribution	55
	4.2	Metabolic routes	58
		4.2.1 Hydrolysis	58
		4.2.2 $\underline{N}$-oxidation	59
		4.2.3 Conversion to pyrrolic metabolites	59
	4.3	Effects of treatments affecting metabolism	61
	4.4	Other factors affecting metabolism	63
	4.5	Other metabolic routes	63
	4.6	Metabolism of pyrrolizidine $\underline{N}$-oxides	64
	4.7	Metabolism in man	64
5.	MECHANISMS OF TOXICITY AND OTHER BIOLOGICAL ACTIONS		65
	5.1	Metabolites responsible for toxicity	65
		5.1.1 Metabolic basis of toxicity	65
		5.1.2 Isolation of pyrrolic metabolites	66
		5.1.3 Chemical aspects of pyrrolic metabolites	66
		5.1.3.1 Preparation	66
		5.1.3.2 Chemistry associated with toxic actions	67
		5.1.4 Possible further metabolites	68
	5.2	Toxic actions of pyrrolic metabolites	69
		5.2.1 Animals	69
		5.2.1.1 Pyrrolic esters (dehydroalkaloids)	69
		5.2.1.2 Pyrrolic alcohols (dehydronecines)	70
		5.2.2 Cell cultures	71
		5.2.3 Possible participation of membrane lipid peroxidation	72
	5.3	Chemical and metabolic factors affecting toxicity	72
		5.3.1 Structural features of a toxic alkaloid	72

Page

	5.3.2 Activation and detoxication	73
	5.3.3 Factors affecting the toxicity of active metabolites	74
	5.3.3.1 Reactivity of the metabolite	74
	5.3.3.2 The number of reactive groups	74
5.4	Metabolites associated with the biological actions of pyrrolizidine alkaloids	75
	5.4.1 Acute hepatotoxicity	75
	5.4.2 Chronic hepatotoxicity	75
	5.4.3 Pneumotoxicity	76
	5.4.4 Toxicity in other tissues	76
	5.4.5 Carcinogenicity	77
	5.4.6 Antitumour activity	77
5.5	Prevention and treatment of pyrrolizidine poisoning	78
	5.5.1 Modified diets	78
	5.5.2 Pre-treatment to enhance the detoxication of active metabolites	79
	5.5.3 Other treatments	80

6. EFFECTS ON ANIMALS 81

   6.1 Patterns of disease caused by different plant genera and of organ involvement in different species 81
   6.2 Field observations - outbreaks in farm animals . . 81
   6.3 Studies on farm animals 84
   6.4 Experimental animal studies 87
      6.4.1 Effects on the liver 87
          6.4.1.1 Relative hepatotoxicity of different PAs and their $\overline{N}$-oxides . 87
          6.4.1.2 Factors affecting hepatotoxicity . 87
          6.4.1.3 Acute effects 92
          6.4.1.4 Mechanism of toxic action 97
          6.4.1.5 Chronic effects 98
      6.4.2 Effects on the lungs 103
          6.4.2.1 Acute effects 105
          6.4.2.2 Chronic effects 110
          6.4.2.3 Mechanisms of toxic action 114
      6.4.3 Effects on the central nervous system . . . 121
      6.4.4 Effects on other organs 121
      6.4.5 Teratogenicity 124
      6.4.6 Fetotoxicity 124
      6.4.7 Mutagenicity 126
          6.4.7.1 Chromosome damage 129
      6.4.8 Carcinogenesis 131
          6.4.8.1 Purified alkaloids 162

Page

    6.4.8.2 Plant materials 167
    6.4.8.3 Pyrrolizidine alkaloid metabolites
       and analogous synthetic
       compounds 171
    6.4.8.4 Molecular structure and
       carcinogenic activity 173
  6.4.9 Antimitotic activity 174
  6.4.10 Immunosuppression 175
  6.4.11 Effects on mineral metabolism 175
  6.4.12 Methods for the assessment of chronic
      hepatotoxicity and pneumotoxicity 176
 6.5 Effects on wild-life 177
  6.5.1 Deer 177
  6.5.2 Fish. 177
  6.5.3 Insects 178

7. EFFECTS ON MAN 179

 7.1 Clinical features of veno-occlusive disease
   (VOD) 179
 7.2 Salient pathological features of veno-
   occlusive disease 181
 7.3 Human case reports of veno-occlusive disease . . . 183
 7.4 VOD and cirrhosis of the liver 201
 7.5 Differences between VOD and
   Indian childhood cirrhosis (ICC) 203
 7.6 Chronic lung disease 204
 7.7 <u>Trichodesma</u> poisoning 205
 7.8 Relationship between dose level and toxic
   effects 206
 7.9 Pyrrolizidine alkaloids as a chemotherapeutic
   agent for cancer 213
 7.10 Prevention of poisoning in man 214

8. BIOLOGICAL CONTROL 216

9. EVALUATION OF HUMAN HEALTH RISKS AND EFFECTS
  ON THE ENVIRONMENT 217

 9.1 Human exposure conditions 217
  9.1.1 Reported sources of human exposure 217
  9.1.2 Plant species involved 217
  9.1.3 Modes and pathways of exposure 218
    9.1.3.1 Contamination of grain crops . . . 218
    9.1.3.2 Herbal medicines 218
    9.1.3.3 PA-containing plants used as food
       and beverages 219
    9.1.3.4 Other food contaminated by PAs . . 219

		Page
	9.1.4 Levels of intake	220
9.2	Acute effects of exposure	222
	9.2.1 Acute liver disease	222
9.3	Chronic effects of exposure	223
	9.3.1 Cirrhosis of the liver	223
	9.3.2 Mutagenicity and teratogenicity	223
	9.3.3 Cancer of the liver	223
	9.3.4 Effects on other organs	224
9.4	Effects on the environment	225
	9.4.1 Agriculture	225
	9.4.2 Wild-life	226
	9.4.3 Insects	226
	9.4.4 Soil and water	226

REFERENCES 227

APPENDIX I. PYRROLIZIDINE ALKALOIDS AND THEIR
PLANT SOURCES 275

APPENDIX II.
TABLE 1. PLANTS CONTAINING HEPATOTOXIC PYRROLIZIDINE
ALKALOIDS 303

TABLE 2. PLANTS CONTAINING KNOWN ALKALOIDS THAT ARE
NON-HEPATOTOXIC (AMINOALCOHOLS AND ESTERS) . 337

WHO TASK GROUP ON PYRROLIZIDINE ALKALOIDS

Members

Professor M.S. Abdullahodjaeva, Uzbek Republican Centre for Pathological Anatomy, Tashkent State Medical Institute, Tashkent, USSR

Dr C.C.J. Culvenor, Commonwealth Scientific and Industrial Research Organization, Division of Animal Health, Parkville, Victoria, Australia (Chairman)

Professor P.P. Dykun, Department of Biophysics, Petrov Research Institute of Oncology, Leningrad, USSR

Dr H.N.B. Gopalan, University of Nairobi, Department of Botany, Nairobi, Kenya

Dr R.J. Huxtable, Department of Pharmacology, University of Arizona, Tucson, Arizona, USA

Dr A.R. Mattocks, MRC Toxicology Unit, Medical Research Council Laboratories, Carshalton, Surrey, United Kingdom

Dr V. Murray, National Poisons Information Service, New Cross Hospital, London, United Kingdom

Dr B. Smith, Division of Food Regulatory Affairs, Food Directorate, Health Protection Branch, Tunney's Pasture, Ottawa, Canada

Professor H.D. Tandon, National Academy of Medical Sciences (India), Ansari Nagar, New Delhi, India (Chairman)

Academician F.YU. Yunusov, Department of Chemical Sciences, Uzbek SSR Academy of Sciences, Tashkent, USSR

Secretariat

Dr R. Montesano, Unit of Mechanisms of Carcinogenesis, International Agency for Research on Cancer, Lyons, France

Dr J. Parizek, International Programme on Chemical Safety, World Health Organization, Geneva, Switzerland

Dr Z. Gregorievskaya, Centre of International Projects, Moscow, USSR

NOTE TO READERS OF THE CRITERIA DOCUMENTS

Every effort has been made to present information in the criteria documents as accurately as possible without unduly delaying their publication. In the interest of all users of the environmental health criteria documents, readers are kindly requested to communicate any errors that may have occurred to the Manager of the International Programme on Chemical Safety, World Health Organization, Geneva, Switzerland, in order that they may be included in corrigenda, which will appear in subsequent volumes.

* * *

ENVIRONMENTAL HEALTH CRITERIA FOR PYRROLIZIDINE ALKALOIDS

A WHO Task Group on Environmental Health Criteria for Pyrrolizidine Alkaloids met in Tashkent, USSR, on 1-5 December 1986. Dr M. Gounar opened the meeting on behalf of the three co-sponsoring organizations of the IPCS (UNEP/ILO/WHO). The Task Group reviewed and revised the draft criteria document and made an evaluation of the health risks of exposure to pyrrolizidine alkaloids.

Access to the original papers on the subject published in the USSR was made possible by PROFESSOR M. ABDULLAHODJAEVA. DR A.R. MATTOCKS wrote the first drafts of the sections on Properties and Analytical Methods, Metabolism, and Mechanisms of Toxicity and Other Biological Actions. DR C.C.J. CULVENOR assisted PROFESSOR H.D. TANDON in the finalization of the document after the Task Group meeting. Dr J. Parizek, who was originally the IPCS staff member responsible for the preparation of the document, and was to be Secretary of the Task Group, could not attend the meeting because of sudden illness, and the Task Group was assisted in his place by Dr M. Gounar, former IPCS staff member. Dr A. Prost was responsible for the final version of the document.

The Secretariat acknowledge the help of both Professor H.D. Tandon and Dr C.C.J. Culvenor. The Task Group meeting in Tashkent was organized by the Centre of International Projects, USSR State Committee for Science and Technology.

The efforts of all who helped in the preparation and finalization of the document are gratefully acknowledged.

* * *

Partial financial support for the publication of this criteria document was kindly provided by the United States Department of Health and Human Services, through a contract from the National Institute of Environmental Health Sciences, Research Triangle Park, North Carolina, USA - a WHO Collaborating Centre for Environmental Health Effects.

* * *

A comprehensive data base on pyrrolizidine alkaloids has been made available by CSIRO Division of Animal Health, Private Bag No. 1, Parkville, Vic. 3052, Australia. The data base consists of alkaloid occurrence tables and keyworded bibliography readable by SCI-MATE software system (Bibliographic Manager, Institute for Scientific Information), but adaptable to other systems. It is available from CSIRO on IBM - PC diskettes; price on application to L.W. Smith.

PREFACE

A disease caused by the consumption of plants containing pyrrolizidine alkaloids (PAs) has been recognized independently as an endemic disease in certain parts of the West Indies and in Uzbekistan in the USSR. Outbreaks of the disease have affected significant segments of populations or large numbers of people in geographically confined areas in Afghanistan, India, and Uzbekistan. The outbreaks have been caused through contamination of the staple food crops with the seeds of plants containing PAs, growing among the crops; such plants are likely to thrive following periods of drought.

It is notable that the same family of plants that caused endemic disease and large-scale outbreaks in Uzbekistan also caused another outbreak of the disease in adjacent Afghanistan, long after the chemical etiology of the disease (through consumption of toxic seeds in the food) had been identified in the USSR. This happened because there was a lack of general awareness of the causal relationship between the chemical present in the plant and the disease. Sporadic cases continue to occur in different parts of the world through the consumption of seeds or plant parts containing toxic PAs, as home remedies, beverages, or food.

The IPCS recognized that this was a health problem that might be lethal, and that it was entirely preventable, provided that it was recognized in time. It was also recognized that the dissemination of knowledge, about both the disease and the sources of the chemicals involved, would be a critical step in its prevention.

Accordingly, the IPCS invited Professor H.D. Tandon, who was responsible for establishing such a causal relationship in the outbreaks in Afghanistan and India, to prepare a draft criteria document and to assist in its further development and finalization after the Task Group meeting, which was held in Tashkent, USSR, on 1-5 December, 1986.

In most episodes of toxic human disease caused by PAs, the liver has been the principal target organ, except for an outbreak in the USSR caused by Trichodesma alkaloids, in which the symptoms were mostly extra-hepatic. The Environmental Health Criteria document provides comprehensive coverage of the hepatotoxic PAS, but lack of relevant documentation prevented the Task Group from analysing the role of Trichodesma alkaloids in detail.

INTRODUCTION - PYRROLIZIDINE ALKALOIDS AND HUMAN HEALTH

Pyrrolizidine alkaloids (PAs) are found in plants growing in most environments and all parts of the world. The main sources are the families Boraginaceae (all genera), Compositae (tribes Senecionae and Eupatoriae), and Leguminosae (genus Crotalaria), and the potential number of alkaloid-containing species is as high as 6000, or 3% of the world's flowering plants (Culvenor, 1980). They have long been known to be a health hazard for livestock, at least since 1902 (Schoental, 1963), and loss of livestock in various parts of the world has been traced to their grazing on certain plants growing in pastures, especially following periods of drought or in arid climates. They have been found to be toxic for all species of animals tested (Schoental, 1963), though some species, notably the guinea-pig, are resistant (Chesney & Allen, 1973a; White et al., 1973). Human disease caused by PA toxicity has been known to be endemic in the central Asian republics of the USSR, at least since the early thirties (Ismailov, 1948a,b; Mnushkin, 1949) when several outbreaks occurred, and the cause was discovered to be the seeds of plants of Heliotropium species (Dubrovinskii, 1947, 1952; Khanin, 1948), which contaminated the staple food crops. A spate of reports followed, mostly from the West Indies, of acute and chronic liver disease (Bras et al., 1954, 1961; Bras & Hill, 1956; Stirling et al., 1962), associated with the ingestion by people of herbal infusions for the treatment of certain ailments. Schoental (1961) and Davidson (1963) suggested that, in view of the evidence of the hepatotoxicity of PAs, consumption of plants containing them could be of etiological significance in human liver disease, especially in developing countries where they are consumed as food or herbal medicines. In spite of this, and the fact that such an ubiquitous source of toxic material is capable of producing animal and human disease and that there have been more recent reports, the PAs have not attracted much attention in the world as a health hazard. In fact, a recent handbook on naturally occurring toxic agents in food (Rechicigl, Jr, 1983) refers to them only in passing and makes no mention of human disease caused by them. Veno-occlusive disease (VOD) (Bras & Hill, 1956), which is characterized by the dominant occlusive lesion of the centrilobular veins of the liver lobule and is caused by these alkaloids, has since been reported from all parts of the world, in both man and animals (Hill, 1960; Bras, 1973). It has been attributed to the accidental contamination of food by toxic plant products or the ingestion of herbal infusions. There have been reports of stray cases and of small outbreaks from both developing and developed countries. However, in the most recent studies from Afghanistan (Tandon & Tandon, 1975; Mohabbat et al., 1976; Tandon, B.N. et al., 1978; Tandon, H.D. et al., 1978) and India (Tandon, B.N. et al., 1976; Tandon, R.K. et al., 1976; Krishnamachari et al., 1977; Tandon, H.D. et al., 1977; Tandon, B.N. et al., 1978), the disease has been reported to affect large masses of the population, resulting in high mortality, and has been attributed to the accidental contamination of their staple food crops by PA-containing seeds of plants, following periods of drought.

There is conclusive evidence from studies on experimental animals that the effects of a single exposure to PAs may progress relentlessly to advanced chronic liver disease and cirrhosis (Schoental & Magee, 1957, 1959; Nolan et al., 1966), following a long interval of apparent well-being, and without any other latent or provocative factor (Schoental & Magee, 1959). The lowest levels of such alkaloids administered thus far to experimental animals, e.g., 1 - 4 mg/kg diet, have produced chronic liver disease and tumours (Hooper & Scanlan, 1977; Culvenor & Jago, 1979). Pyrrolizidine alkaloids have also been shown to act synergistically with aflatoxin, another environmental toxin present in agricultural products, in causing cirrhosis and hepatoma in primates (Lin et al., 1974). Though there is no conclusive evidence yet of a carcinogenic role of PAs in man, such a possibility has been suspected on the basis of experimental data (Hill, 1960; Williams et al., 1967; IARC, 1976, 1983; Huxtable, 1980; Culvenor, 1983), and experimental studies have demonstrated carcinogenicity in rats given dosages equivalent to those reported to have been ingested in human cases (Cook et al., 1950; Culvenor, 1983).

Alkaloids/toxic metabolites have been shown to be secreted in the milk of lactating dairy cattle (Dickinson et al., 1976) and rats, and the young of both sexes have been shown to suffer toxic damage, even when suckled by mothers treated with retrosine, who apparently are not affected themselves (Schoental, 1959). Such suckling animals may also be in apparent good health while the livers show toxic effects. Protein-deficient and young suckling animals are particularly vulnerable (Schoental, 1959).

Chromosomal aberrations have been demonstrated in rats and humans with veno-occlusive disease (Martin et al., 1972).

Alkaloids have been found in the honey secreted by bees feeding on the toxic plants (Deinzer et al., 1977). According to Culvenor and his co-workers, populations in some countries are exposed to low levels of alkaloids in commonly used foodstuffs, e.g., honey in Australia (Culvenor et al., 1981; Culvenor, 1983, 1985) and comfrey in many countries (Culvenor et al., 1980a; Culvenor, 1985).

Human cases of acute disease following the brief ingestion of the alkaloids have been known to progress to cirrhosis (Stuart & Bras, 1957; Braginskii & Bobokhadzaev, 1965; Stillman et al., 1977; Tandon, B.N. et al., 1977; Tandon, H.D. et al., 1977) in as short a period as 3 months from the acute phase (Stuart & Bras, 1957). The initial disease may be cryptic (Braginskii & Bobokhadzaev, 1965) and may not be ascribed to herbal consumption, and yet may progress to cirrhosis (Huxtable, 1980). Veno-occlusive disease was stated to be the most common cause of cirrhosis in infants in Jamaica (Bras et al., 1961) and has been believed to be a significant etiological factor for adult cirrhosis, especially in developing countries (Gupta et al., 1963).

Plants known or suspected to contain toxic alkaloids are widely used for medicinal purposes as home remedies all over the world, without systematic testing for safety (Schoental, 1963; Smith & Culvenor, 1981) and some are even used as food (Schoental & Coady, 1968; Culvenor, 1980). There are several reports of the continued use of such herbs for medicinal purposes in technically advanced countries (Culvenor, 1980). Senecio jacobaea continues to be sold at herbalists shops in the United Kingdom (Schoental, 1963; Burns, 1972), and Symphytum spp. (comfrey) are still used as a vegetables, beverages, or remedies (Mattocks, 1980). Both these herbs are known to be carcinogenic (IARC, 1976; Hirono et al., 1978). Young flower stalks of Petasites japonicus Maxim, the pre-bloom flower of coltsfoot, Tussilago farfara, the leaf and root of comfrey, Symphytum officinale, and the young leaves and stalks of Farfugium japonicum and Senecio cannabifolius, which are all used in Japan as human food or herbal remedies, are known to be carcinogenic for rats (Hirono et al., 1983). Symphytum x uplandicum Nyman (Russian comfrey), which contains several toxic PAs (Culvenor et al., 1980b) echimidine and 7 acetylycopsamine being the main constituents, is used as a salad plant, green drink, and medicinal herb. It has been estimated that the rate of ingestion of alkaloids from this herb may, over a period of time, exceed the levels reported to have been taken during the Afghan outbreak. There is a report of at least one patient who developed toxic effects as a result of consuming a comfrey preparation (Culvenor et al., 1980a; Ridker et al., 1985). Arseculeratne et al. (1981) found that 3 of the 50 medicinal herbs commonly used in Sri Lanka contained PAs that had been proved to be hepatotoxic for animals. They suggested that consumption of such herbs might contribute to the high incidence of chronic liver disease, including primary liver cancer, in Asian and African countries, especially as they may act synergistically with aflatoxin and hepatitis B virus. The risk of toxic effects due to these alkaloids may be particularly high in children

(Schoental, 1959; Jago, 1970) and protein malnutrition, which exists in some countries, may potentiate them (Schoental & Magee, 1957). Recent studies from Hong Kong (Kumana et al., 1985; Culvenor et al., 1986), the United Kingdom (McGee et. al, 1976; Ridker et al., 1985), and the USA (Stillman et al., 1977; Fox et al., 1978; Ridker et al., 1985) report instances of human disease that have been caused by the use of such herbs, resulting in fatality or the development of cirrhosis, even in countries with well-developed health services and among the higher economic and educated strata of society. Indeed, Stillman et al. (1977), from the USA, called PA toxicosis the "iceberg disease", implying that cases of this disease might be more frequent than reported in the USA, especially among populations of Mexican-American origin. In general, the use of herbal remedies is not elicited in the clinical history and patients do not volunteer this information themselves. Furthermore, the alkaloids are eliminated within 24 h (Huxtable, 1980) and, even though methods are available for their detection in biological tissues and fluids, the suspicion cannot be confirmed, as the symptoms may take several days or months to appear.

Contamination of food crops is particularly likely to occur in parts of the world with arid climates, poor or uncertain rainfall, poor irrigation facilities, and following periods of drought, all of which promote the growth of the PA-containing plants that grow as weeds among cultivated crops, as has been found in studies on the outbreaks in Afghanistan, India, and the USSR (Terekhov, 1939; Dubrovinskii, 1947; Ismailov, 1948a,b; Tandon & Tandon, 1975; Mohabbat et al., 1976; Tandon, B.N. et al., 1976; Tandon, R.K. et al., 1976; Tandon, H.D. et al., 1978) and in grazing pastures. The use of traditional medicines is common in these countries and there is insufficient awareness of this hazard, the disease condition, and its diagnostic pathological picture. Furthermore, health services are poorly developed. Thus, many of the cases or even outbreaks may go unnoticed or unrecorded and may even be ascribed to malnutrition (Lancet, 1984). Also, many of the reported cases of so-called "Budd-Chiari syndrome", a condition associated with obstruction of major hepatic veins and/or inferior vena cava, may actually be cases of veno-occlusive disease (Sherlock, 1968), in which only the central veins of the liver lobule or sublobular veins are occluded.

Another type of PAs, <u>Trichodesma</u> alkaloids, has been known to cause a human outbreak of disease in the USSR, through contamination of the staple cereal with the seeds containing these PAs; in this outbreak, the symptoms were principally extra-hepatic (Ismailov et al., 1970).

This document is aimed at focusing on a health menace that is insufficiently recognized, in order to evaluate the health risks on the basis of published data, and to draft a set of recommendations that would help in its recognition, prevention, and control.

1. SUMMARY AND RECOMMENDATIONS

1.1 Summary

The ingestion of pyrrolizidine alkaloids (PAs) in foods and medicinal herbs results in acute and chronic effects in man, affecting mainly the liver. Data from experimental animal studies indicate that PAs represent a potential cause of cancer in man.

The alkaloids are produced by numerous plant species and occur throughout the world. In the present document, the alkaloids and their properties are described together with the sources of human exposure and the diseases that they produce in man and animals. The risks for human health are evaluated and recommendations are made for reducing such risks.

1.2 Sources and Chemical Structure

The known pyrrolizidine alkaloids, most of which are hepatotoxic, are produced by plant species within the following families: Boraginaceae (<u>Heliotropium</u>, <u>Trichodesma</u>, <u>Symphytum</u>, and most other genera), Compositae (<u>Senecio</u>, <u>Eupatorium</u>, and other genera of the tribes Senecioneae and Eupatoriae), Leguminosae (genus <u>Crotalaria</u>), and Scrophulariaceae (genus <u>Castilleja</u>). These genera are mainly herbaceous and very widely distributed, some species being found in most regions of the world. The majority of the species within these genera have not yet been investigated, but are expected to contain pyrrolizidine alkaloids.

The hepatotoxic alkaloids have a 1,2-double bond in the pyrrolizidine ring and branched chain acids, esterifying a 9-hydroxyl and preferably also the 7-hydroxyl substituent. Modified seco-pyrrolizidine alkaloids, in which the central bond between the N and C8 atoms is broken, are also hepatotoxic. Some <u>Senecio</u> species contain non-basic derivatives that are 5-oxopyrroles. The toxicity of these derivatives may be similar to that of the alkaloids, but this aspect has not been investigated. The alkaloids occur as free bases and <u>N</u>-oxides. The latter are reduced to the free bases in the gastrointestinal tract of animals and have a similar toxicity when ingested orally.

Suitable analytical procedures are available for screening plant species, including a simple field test for toxic alkaloids. Thin-layer chromatography (TLC), high-performance liquid (HPLC), gas chromatography (GC), and gas chromatography-mass spectrometry (GC-MS) have been applied for separating, characterizing, and quantifying the alkaloids present. Effective use of these procedures requires authentic alkaloids for standards, few of which are available. Improved analytical methods are required for the determination of very low levels of alkaloids in some foodstuffs.

1.3 Mechanisms and Features of Toxicity

The toxic effects of pyrrolizidine alkaloids are due to activation in the liver. Metabolism of the alkaloids by mixed-function oxidases leads to pyrrolic dehydro-alkaloids, which are reactive alkylating agents. Reaction of initial metabolites with constituents of the liver cell in which they are formed are probably the main cause of liver cell necrosis. Metabolites are released into the circulation and are believed to pass beyond the liver to the lung causing vascular lesions characteristic of primary pulmonary hypertension, especially when alkaloids, such as monocrotaline, are administered to animals.

In experimental animals, PAs are quickly metabolized and are almost completely excreted in 24 h, so that no residual products are detectable in the biological fluids or body tissues after this period.

The rate of formation of pyrrolic metabolites is influenced by the induction or inhibition of the mixed-function oxidases in the liver, but the relationship between the rate of metabolism and expression of toxicity is uncertain.

Several pyrrolizidine alkaloid-derivatives and related compounds are known to cause chromosome aberrations in plants, leukocyte cell cultures of the marsupial (Potorus tridactylus), and in hamster cell lines. Some pyrrolizidine alkaloids induce micronuclei formation in erythrocytes in the bone marrow and fetal liver in mice, sister chromatid exchanges in a Chinese hamster cell line and human lymphocytes in vitro, and repair DNA synthesis in rodent hepatocyte cell cultures. Chromosome aberrations have been reported in the blood cells of children suffering from veno-occlusive disease VOD, presumably caused by fulvine.

A number of pyrrolizidine alkaloids have been shown to be mutagenic in the Salmonella typhimurium assay, after metabolic activation. The carcinogenic activity of pyrrolizidine alkaloids appears to parallel their mutagenic behaviour, but not their hepatotoxicity.

Heliotrine at doses of 50 mg/kg body weight or more, administered to rats during the second week of gestation, has been shown to induce several abnormalities in the fetus. Doses of 200 mg/kg body weight resulted in intrauterine deaths or resorption of fetuses. Dehydroheliotridine, the metabolic pyrrole derivative of heliotrine, was 2.5 times more effective on a molar basis than its parent PA in inducing teratogenic effects.

The ability of PAs to cross the placental barrier in the rat and to induce premature delivery or death of litters has been demonstrated. The embryo in utero appears to be more resistant to the toxic effects of pyrrolizidine alkaloids than the neonate. PAs are known to have passed through the mother's milk to the sucklings.

Megalocytosis, the presence of enlarged hepatocytes containing large, hyper-chromatic nuclei, is a characteristic feature of pyrrolizidine alkaloid-induced chronic hepatotoxicity in experimental animals. The enlarged hepatocytes arise through the powerful antimitotic action of the pyrrole metabolites of pyrrolizidine alkaloids. This change has not been observed in the human liver, though human fetal liver cells in vitro culture become enlarged when exposed to PAs, indicating susceptibility to the antimitotic effect of the alkaloids.

In experimental animals, protein-rich and sucrose-only diets have given some measure of protection against the effects of the alkaloids, as has pre-treatment of animals with thiols, anti-oxidants, or zinc chloride.

PAs are noted mainly for the poisoning of livestock due to the animals grazing on PA-containing toxic weeds, and large-scale outbreaks have been recorded. Such episodes have been reported from most parts of the world, including those with temperate or cold climates. Studies carried out on a wide variety of farm and laboratory animals have revealed generally common features of toxicity with some species variations. The liver is the principal target organ. In small laboratory animals, doses approaching a lethal dose produce a confluent, strictly zonal haemorrhagic necrosis in the liver lobule, within 12 - 48 h of administration of PAs. Simultaneously in non-human primates, or after a short time in the rat, chicken, and swine, changes begin to occur, and later become organized, in the subintima of the central or sublobular veins in the liver resulting in their occlusion. The reticulin framework in the central zone of the lobule collapses following necrosis leading to scarring. Repeated administration of suitable doses leads to chronic liver lesion characterized by megalocytosis, and increasing fibrosis, which may result in cirrhosis. Chronic liver disease including cirrhosis has been shown to develop in the rat following administration of a single dose of a PA. In a number of animal species, the lungs develop vascular lesions characteristic of primary pulmonary hypertension with secondary hypertrophy of the right ventricle of the heart. In rats, appropriately low repeated doses of several alkaloids have been shown to induce tumours, mainly in the liver. In some studies, a single dose has been carcinogenic.

The central nervous system is the target organ of the toxic PAs contained in Trichodesma, which produce spongy degeneration of the brain.

1.4 Effects on Man

In man, PA poisoning is usually manifested as acute veno-occlusive disease characterized by a dull dragging ache in the right upper abdomen, rapidly filling ascites resulting in marked distension of the abdomen, and sometimes associated with oliguria, and massive pleural effusion. It can also manifest as subacute disease with vague symptoms and persistent hepatomegaly. Children are particularly vulnerable. Many cases progress to cirrhosis and, in some cases, a single episode of acute disease has been demonstrated to progress to cirrhosis, in spite of the fact that the patient has been removed from the source of toxic exposure and has been given symptomatic treatment. Mortality can be high with death due to hepatic failure in the acute phase or due to hematemesis resulting from ruptured oesophageal varices caused by cirrhosis. Less severely affected cases may show clinical, or even apparently complete, recovery. The Task Group was not aware of any substantiated report of primary pulmonary hypertension resulting from PA toxicity. However, in view of the evidence in experimental animals and circumstantial evidence in one case report, the possibility of the development of toxic pulmonary disease in man cannot be ruled out. There is a report of an outbreak of Trichodesma poisoning in the USSR in which the symptoms were mainly neurological.

1.4.1 Nature and extent of health risks

The two main sources of pyrrolizidine alkaloid poisoning reported in human beings are the consumption of cereal grain contaminated by weeds containing the alkaloids and the use of alkaloid-containing herbs for medicinal and dietary purposes. A third form of exposure, with the potential to affect large populations is the possible low-level contamination of some foodstuffs, such as honey and milk, but the Task Group was not aware of any cases of human toxicity having been caused through the contamination of these foods.

Liver disease caused by the contamination of cereal grains has been reported in rural populations in Afghanistan, India, South Africa, and the USSR. A contributing factor appears to be abnormally dry weather, resulting in the growth of an exceptionally high proportion of the alkaloid-containing weeds in the crops, the seeds of which contaminate the cereal grain on harvesting. The weeds responsible for known outbreaks have been Heliotropium, Trichodesma, Senecio, and Crotalaria species. Mortality in such outbreaks has been reported to be high. In the largest reported outbreak in northwestern Afghanistan, an estimated 8000 people were affected in a total population of 35 000 with 1600 - 2000 deaths.

Human poisoning through the medicinal use of herbs containing pyrrolizidine alkaloids has been reported from all parts of the world. PAs were responsible for a common liver disease in children in Jamaica, and individual cases in Ecuador, Hong Kong, India, the United Kingdom, and the USA. The plants involved were species of Crotalaria, Heliotropium, Senecio, Symphytum, and Gynura. Symphytum-containing preparations present a particular hazard because of their widespread use and the generally high levels of individual exposures. The use of herbs is almost universal in traditional folk medicine and is increasing in developed countries. Some of the herbs used contain pyrrolizidine alkaloids and have a long-term toxicity that is unsuspected by the people taking them. Knowledge of the species used in herbal medicine and the frequency of such use is very limited in the scientific literature. About 40 such species are listed in this report, about one-third of which are in use in developed countries. They are often prescribed by herbalists, naturopaths, and other non-orthodox practitioners. The extent of the contribution to acute and chronic liver disease cannot be accurately assessed. It may also constitute an etiological factor in cirrhosis of the liver and, once this stage is reached, it may not be possible to identify the cause as a PA.

PAs are known to be transmitted from the feed of dairy animals into milk and to cause toxic damage in the suckling young. One instance of large-scale contamination of honey is known to have been caused by a common weed rich in PAs, which was the source of nectar and pollen for the honey-secreting bees. No reports of cases of acute toxicity caused by consumption of contaminated dairy products or honey were available to the Task Group. Furthermore, no information is available on the possible presence of PAs or their metabolites in the meat of animals fed toxic weeds before slaughter; however, the possibility of toxic disease being caused through this medium is considered to be low.

There are no substantial, long-term follow-up data to assess whether exposure to PAs results in increased incidence of chronic liver disease or cancer in man. Available clinical and experimental data suggest that a single episode of PA toxicity and possibly also a long-term low level exposure may lead to cirrhosis of the liver. PAs could also be possible carcinogens in man, since a number of them have been demonstrated to induce cancer in experimental animals, the main target organ being the liver. These include some which have caused episodes of human toxicity, and some others which are found in herbs traditionally used as items of food. Also, in several instances of human toxicity, the reported daily rates of intake of such PAs were in close range of those known to induce tumours in rats. However, these risks cannot be adequately assessed on a quantitative basis. There are indications that PA intoxications leading to liver disease are more prevalent than the reported frequency of cases would seem to indicate.

Because of their known involvement in human poisoning and their possible carcinogenicity, exposure to pyrrolizidine alkaloids should be kept as low as practically achievable. The setting of regulatory tolerance levels for certain food products may be required in some situations.

1.5 Methods for Prevention

The only known method of prevention is to avoid consumption of the alkaloids. In the USSR, a set of agricultural (or agrotechnical) legislative, phyto-sanitary and educational measures has prevented new outbreaks of poisoning due to <u>Heliotropium</u> and <u>Trichodesma</u>, since 1947.

1.6 Recommendations

1.6.1 General recommendations

1. Cereal crops should be assessed throughout the world for possible contamination by weeds likely to contain pyrrolizidine alkaloids. Appropriate grain inspection systems are desirable in order to achieve near-zero levels of contamination by such weeds.

2 There is a need to create awareness, among the general population and those responsible for the delivery of health services, with regard to the hazards of consuming such plants as contaminants in food or as food, or for medicinal purposes. Advice on hazards should include mention of possible increased risks, if the alkaloid intake is associated with drug treatment, (e.g. phenobarbitone) or foods which increase the level of liver metabolizing enzymes.

3. Ethnobotanical and taxonomic studies are required in many countries to provide specific information on the use of plant species containing pyrrolizidine alkaloids for medicinal and dietary purposes. There may be a need to control the sale of some species, and their prescription by herbalists and other practitioners of traditional systems of medicine.

4. Honey and dairy products, both local and bulk supplies, should be assayed for pyrrolizidine alkaloids in all regions where a risk of contamination of these foodstuffs has been identified.

1.6.2   Recommendations for research

1. Long-term follow-up studies of the survivors of both alkaloid poisoning in human beings and animal outbreaks are required, in order to determine the possible development of chronic liver disease or cancer. Similar studies are also desirable on individuals who regularly consume comfrey or other PA-containing herbs over a substantial period of time.

2. Epidemiological studies should be carried out in countries with a high incidence of primary liver cancer, in order to determine whether there is an association with the intake of herbs containing pyrrolizidine alkaloids.

3. A network of reference laboratories is needed to assist member states in identifying plants and their seeds suspected of producing toxic effects and for the assay and identification of PAs. Provisions may be made for the easy availability of pure alkaloids for use as reference standards for assays.

4. It is necessary to develop improved assay procedures, suitable for the purposes of recommendation (4) in section 1.6.1, particularly using fluorescence and immunochemical methods.

5. There is a need for further toxicological studies, such as studies on the carcinogenicity of echimidine and the toxicity of the 5-oxopyrrole constituents of Senecio species, and for studies that would provide more quantitative information on the various adverse biological effects of PAs. A study of the carcinogenicity of the alkaloids in the pig is also indicated, since the pig exhibits a high sensitivity to acute and subacute toxicity similar to that seen in man.

6. Study is required of the possible alkaloid content of the meat, organs, and fat of animals that have recently consumed plants containing pyrrolizidine alkaloids.

7. Experimental studies are needed on the influence of nutritional status on the metabolism, and acute and chronic effects of PAs.

8. Further metabolic studies are required to define more specifically the enzymes involved in the microsomal activation

and detoxification of PAs, to determine whether organelles other than microsomes are involved, and to explore further, quantitative relationships between different routes of metabolism.

9. The maximum no-observed-adverse-effect dose levels for repeated long-term administration in the rat and the pig need to be determined.

10. Experimental studies should be conducted to determine:

   (a) whether pyrrolizidine alkaloid $\underline{N}$-oxides may be metabolized directly into the pyrrolic dehydroalkaloid in mitochondria, especially in tumour cells; and

   (b) which P450 enzymes are involved in the activation and $\underline{N}$-oxidation of PAs and thence in the selective induction of $\underline{N}$-oxidation enzymes.

11. A study might be conducted of human variability and its genetic aspects in relation to factors that influence susceptibility to PAs; for example, the study of mixed-function oxidase levels in the liver by metabolism of appropriate test substances recognized as harmless.

2. PROPERTIES AND ANALYTICAL METHODS

2.1 Chemical Structure and Properties

The chemical structure of PAs in relation to their toxic effects has been reviewed recently by Mattocks (1986). The pyrrolizidine alkaloids with which this document is concerned are those that have previously been called "hepatotoxic" or "nucleotoxic". Here it is proposed to refer to them as "toxic" PAs, because of the weight of evidence now available that they produce damage in other organs as well as the liver, and the need to avoid a restrictive term. There are other types of pyrrolizidine alkaloids, such as those that occur in the plant family Orchidaceae, which are not toxic and are not discussed here.

The toxic PAs are esters of the amino-alcohols derived from the heterocyclic nucleus. The pyrrolizidine molecule is made up of two 5-membered rings inclined to each other as shown in Fig. 1 so that geometric isomerism is possible, and which share a common nitrogen at position 4.

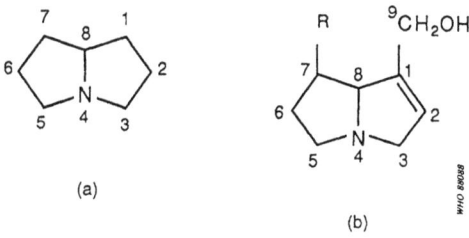

Fig. 1. Molecular structure of pyrrolizidine molecule.

Most hepatotoxic alkaloids are esters of molecules similar to that shown in Fig. 1(b) (1-hydroxymethyl-1:2-dehydropyrrolizidine). However, a few hepatotoxic alkaloids are esters of the amino-alcohol otonecine, e.g., petasitenine (Fig. 2, No.7). The unsaturated pyrrolizidine nucleus itself is not toxic, but esters of branched-chain acids are. Ester linkages may be at positions 9, 7, or (rarely) 6. Some esters have an "open" molecule, e.g., heliotrine, whereas others are macrocyclic diesters, e.g., monocrotaline and retrosine. Examples of some pyrrolizidine alkaloid structures are shown in Fig. 2.

The ring nucleus contains a double bond at the 1:2 position, which is essential for the toxic effects of the alkaloid, but not for unrelated effects.

1. **Echimidine**
   Chemical structure:

   Chemical formula: $C_{20}H_{31}NO_7$
   Relative molecular mass: 397
   CAS registry number: 520-68-3

2. **Heliotrine**
   Chemical structure:

   Chemical formula: $C_{16}H_{27}NO_5$
   Relative molecular mass: 313
   CAS registry number: 303-33-3

3. **Indicine-N-oxide**
   Chemical structure:

   Chemical formula: $C_{15}H_{25}NO_6$
   Relative molecular mass: 315
   CAS registry number: 41708-76-3

4. **Jacobine**
   Chemical structure:

   Chemical formula: $C_{18}H_{25}NO_6$
   Relative molecular mass: 351
   CAS registry number: 6870-67-3

5. Lasiocarpine
Chemical structure:

Chemical formula: $C_{21}H_{33}NO_7$
Relative molecular mass: 411
CAS registry number: 303-34-4

6. Monocrotaline
Chemical structure:

Chemical formula: $C_{16}H_{23}NO_6$
Relative molecular mass: 325
CAS registry number: 315-22-0

7. Petasitenine
Chemical structure:

Chemical formula: $C_{19}H_{27}NO_7$
Relative molecular mass: 381
CAS registry number: 60132-19-6

8. Retrorsine (retrosine N-oxide = isatidine)
Chemical structure:

Chemical formula: $C_{18}H_{25}NO_6$
Relative molecular mass: 351
CAS registry number: 480-54-6

9. <u>Senecionine</u>
Chemical structure:

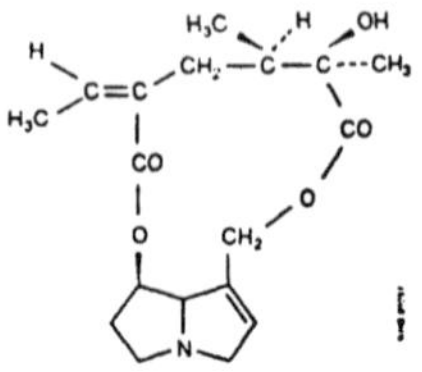

Chemical formula: $C_{18}H_{25}NO_5$
Relative molecular mass: 335
CAS registry number: 130-01-8

10. <u>Symphytine</u>
Chemical structure:

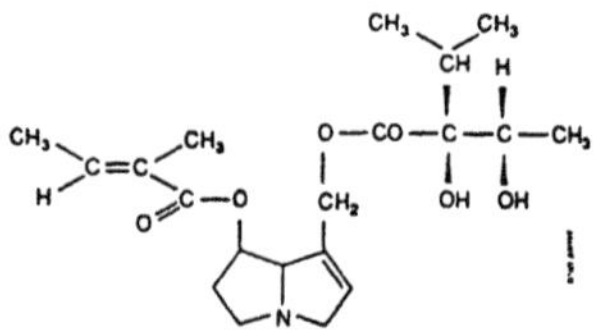

Chemical formula: $C_{20}H_{31}NO_6$
Relative molecular mass: 381
CAS registry number: 22571-95-5

11. <u>Trichodesmine</u>
Chemical structure:

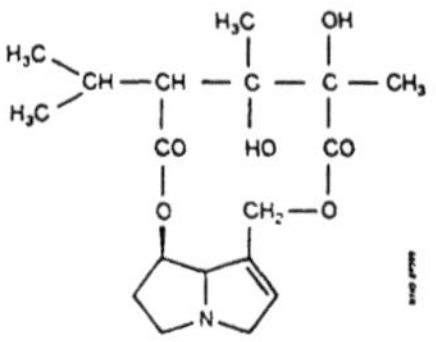

Chemical formula: $C_{18}H_{27}NO_6$
Relative molecular mass: 353
CAS registry number: 548-90-3

12. <u>Incanine</u>
Chemical structure:

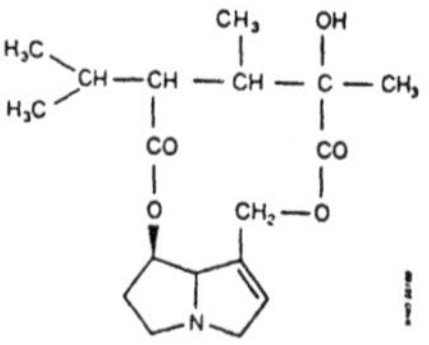

Chemical formula: $C_{18}H_{27}NO_5$
Relative molecular mass: 337
CAS registry number: 480-77-3

Fig. 2. Molecular structure and chemical data for some pyrrolizidine alkaloids.

As the Task Group met in Tashkent, it is of historical interest to recall that the structures of heliotrine and lasiocarpine, the main alkaloids of Heliotropium lasiocarpum, were worked out by Dr G.P. Men'shikov and associates in Moscow in the 1930s. This work included determining the structure of heliotridine, the parent compound of the amino-alcohol, heliotridane. Dr Men'shikov's studies were carried out at essentially the same time, but independently of studies by English and American authors on retronecine-based alkaloids.

The alkaloids in plants are often found together with their N-oxides, which are also toxic, when ingested orally. The pyrrolizidine alkaloids acquire their toxic properties only through the toxic pyrrolic intermediates (the general structure of which is shown in Fig. 3) formed by the mixed-function oxidases of the hepatocytes. To form these pyrrolic derivatives, the alkaloid molecule should have:

(a) a double bond at the 1:2 position of the ring nucleus;

(b) esterified hydroxyl groups in the nucleus at the C 9 and/or C 7 positions; and

(c) a branched carbon chain in at least one of the ester side-chains (McLean, 1974).

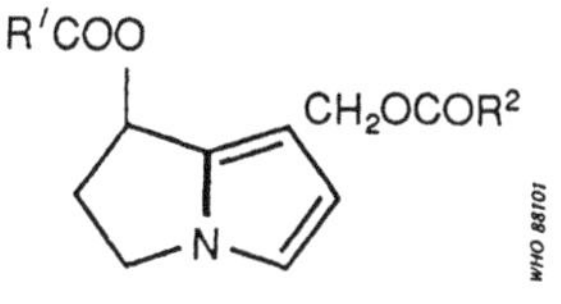

Fig. 3. Molecular structure of toxic pyrrolic intermediates.

Substitution at the α position of the acid and esterification of the C-7 hydroxy group both enhance the toxicity of the alkaloid (Robins, 1982).

A group of related alkaloids, isolated from Senecio species by Bohlmann et al. (1979), have non-basic pyrrolic structures similar to those of toxic pyrrolizidine alkaloid metabolites, but they are chemically deactivated by the presence of a carbonyl group at position 3 of the pyrrolizidine nucleus, e.g., senaetnine (Fig. 4). Senaetnine does not possess the acute hepatotoxic characteristics of basic pyrrolizidine alkaloids. However, it had a direct irritant action on tissues near the site of intra-peritoneal administration and caused damage to pulmonary vascular tissue when given intra-veinous to rats (Mattocks & Driver, 1987).

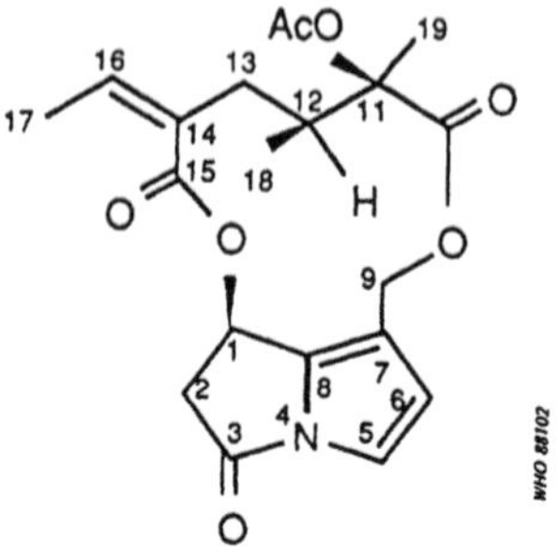

Fig. 4. Molecular structure of senaetnine, a non-alkaloidal pyrrolic constituent of some Senecio species.

The alkaloids are fairly stable chemically, but the ester groups may undergo hydrolysis under alkaline conditions. Some alkaloids in plant material may decompose during drying (Bull et al., 1968), but others appear to be stable under similar conditions (Pedersen, 1975; Birecka et al., 1980). The N-oxides of unsaturated pyrrolizidines are more readily decomposed by heat than the basic alkaloids, especially when dry. However, the stability of the alkaloids and N-oxides in hot water as, for example, in cooking, is not known.
Some pyrrolizidine alkaloids have a limited water solubility, unless neutralized with acid; but others (e.g., indicine), and all the N-oxides, are readily soluble.

2.2 Analytical Methods

When analysing for PAs, it is important to recognize that this group consists of many different compounds (section 2.1) and that these often occur as mixtures in plants or in materials of plant origin. They may vary in structure, relative molecular mass, response to analytical procedures, and toxicity. Both basic alkaloids and corresponding N-oxides may be present at the same time. Thus, where such mixtures are present, analyses will inevitably be approximate, unless the individual components are separated and identified.

Nevertheless, such estimates can be useful. In particular, all hepatotoxic PAs are unsaturated in the sense that they possess a 1:2-double bond in the pyrrolizidine nucleus, and analytical methods that are specific for this structure can be of value in screening for potential toxicity. A simple qualitative field test for screening plant materials for the presence of such alkaloids and their N-oxides, without the need of high technology equipment, is described in section 2.2.2.5.

2.2.1 Extraction

2.2.1.1 Plant tissue

Pyrrolizidine alkaloids are usually extracted from dried, milled plant material with hot or cold alcohol. The alcohol is evaporated, the bases taken up in dilute acid, and fats extracted with ether or petroleum. It is usual, at this stage, to reduce any N-oxides present to the corresponding basic alkaloids with zinc, before making the solution alkaline and extracting the alkaloids with chloroform (Koekemoer & Warren, 1951). Alternatively, alcohol can be continuously circulated through the plant material and then cation exchange resin, and the alkaloids subsequently eluted from the resin (Mattocks, 1961; Deagen & Deinzer, 1977). PAs can also be extracted by soaking plant material in dilute aqueous acid (Briggs et al., 1965; Craig et al., 1984).

2.2.1.2 Biological fluids and tissues

Pyrrolizidine alkaloids have been extracted for analytical purposes from honey (Deinzer et al., 1977), milk (Dickinson et al., 1976), blood-plasma (Ames & Powis, 1978; McComish et al., 1980), urine (Mattocks, 1967a; Jago et al., 1969; Evans et al., 1979), and bile (Jago et al., 1969; Lafranconi et al., 1985).

When attempting to isolate PAs from animal tissues, it must be appreciated that the toxic alkaloids are often metabolized very rapidly in animals, so that the amounts that are recoverable (except from urine), only a few hours after alkaloid ingestion, may be extremely small. Various methods have been used to separate PAs, but some mixtures are extremely difficult to separate. On the analytical scale, the most useful methods are thin-layer chromatography (TLC), high-performance liquid chromatography (HPLC), and gas chromatography (GC) (section 2.2.2).

2.2.2 Analysis for pyrrolizidine alkaloids

2.2.2.1 Thin-layer chromatography (TLC)

For TLC, silica plates are usually used, eluted with chloroform:methanol:aqueous ammonia mixtures (Sharma et al., 1965; Chalmers et al., 1965); solvents suitable for the N-oxides, which are more water-soluble, have been described by Mattocks (1967b) and Wagner et al. (1981). The most sensitive methods for detecting PAs on TLC are those using Ehrlich reagent (4-dimethylaminobenzaldehyde) (Mattocks, 1967b). The unsaturated alkaloids are best visualized by spraying the plates first with a solution of orthochloranil, then with Ehrlich reagent, heating after each spray (Molyneux & Roitman, 1980). The N-oxides of unsaturated pyrrolizidines are detected by spraying a solution of acetic anhydride, heating the plate, and then spraying Ehrlich reagent (Mattocks, 1967b).

Pyrrolizidine alkaloids with a saturated base moiety must be detected in other ways (which are not specific for pyrrolizidines), e.g., by exposing the dried plates to iodine vapour, or by spraying with an iodobismuth (Dragendorff) reagent (Munier, 1953).

2.2.2.2 High-performance liquid chromatography (HPLC)

Analytical or preparative scale HPLC separation of pyrrolizidine alkaloids has been described by Segall (1979a,b) and Dimenna et al. (1980), and an improved method has been reported by Ramsdell & Buhler (1981). Alkaloids from Symphytum officinale (comfrey) have been separated on an analytical scale by Tittel et al. (1979), and partially separated on a preparative scale by Huizing et al. (1981). UV detectors are usually used for the HPLC of pyrrolizidine compounds (Mattocks, 1986).

2.2.2.3 Gas chromatography (GC) and mass spectrometry (MS)

The GC characterization of PAs using packed columns has been described by Chalmers et al. (1965) and Wiedenfeld et al. (1981). Mixtures of alkaloids from comfrey (Symphytum sp.), normally hard to separate, were resolved by Culvenor et al. (1980a) and Frahn et al. (1980) by GC of the methylboronate derivatives.

Gas chromatography combined with mass spectrometry (GC-MS) has become a valuable and highly sensitive means for both the identification and the quantitative determination of pyrrolizidine alkaloids. Thus, alkaloids extracted from honey were separated and identified by Deinzer et al. (1977) and (as butylboronate derivatives) by Culvenor et al. (1981). Deinzer et al. (1978) described a method for the recognition (but not the individual identification) of retronecine-based pyrrolizidine alkaloids, by hydrolysing them to retronecine (the amino alcohol moiety) followed by GC-MS of its bis-trifluoroacetate. The use of capillary GC has greatly improved the sensitivity of pyrrolizidine alkaloid analysis, especially when used with MS (Luthy et al., 1981). The MS of pyrrolizidine compounds has been reviewed (Bull et al., 1968; Mattocks, 1986).

Pyrrolizidine N-oxides generally undergo thermal decomposition, when subjected to GC, but they can first be reduced to the corresponding basic alkaloids (Koekemoer & Warren, 1951). Alternatively they may be derivatised. Thus, trimethylsilylation of indicine N-oxide or heliotrine N-oxide can lead either to the trimethylsilyl (TMS) derivative of the parent alkaloid or to the TMS derivative of the dehydro-alkaloid (pyrrolic derivative), depending on the reagents used, and these products will run successfully on GC-MS (Evans et al., 1979, 1980).

2.2.2.4 Nuclear magnetic resonance (NMR) spectrometry

A convenient, but relatively insensitive, method, specifically for the determination of unsaturated PAs, has been described by Molyneux et al. (1979). The basic alkaloids are extracted, then subjected to NMR spectrometry along with an internal standard (p-dinitrobenzene). This enables quantitative measurements to be made of the signal(s) representing the H2 proton(s) in unsaturated pyrrolizidines, and thus the alkaloid(s) can be determined. Quantitative NMR analysis of pyrrolizidine alkaloid mixtures from Senecio vulgaris has been described by Pieters & Vlietinck (1985) and compared with an HPLC method by the same authors (1986). Qualitative aspects of the NMR spectrometry of pyrrolizidine alkaloids have been reviewed by Bull et al. (1968) and Mattocks (1986).

2.2.2.5 The Ehrlich reaction

This method (Mattocks, 1967a, 1968b) is specific for unsaturated pyrrolizidine alkaloids and is not suitable for other alkaloids. Thus, it is the most useful colorimetric method for potentially hepatotoxic pyrrolizidine compounds. The procedure converts the alkaloid into its N-oxide, using hydrogen peroxide. The product reacts with acetic anhydride to form a pyrrolic derivative (dehydro-alkaloid) that gives a magenta colour with a specially modified Ehrlich reagent. The latter contains boron trifluoride to give maximum sensitivity. As little as 5 µg of most unsaturated pyrrolizidines can be measured by this method. If the oxidation stage is omitted, only the unsaturated pyrrolizidine N-oxides can be determined. The determination of pyrrolizidine N-oxides has also been discussed by Mattocks (1971b).

A simplification of the above colorimetric procedure was described by Mattocks (1971d) to provide a qualitative test that could be used to screen large numbers of plant samples for the presence of unsaturated pyrrolizidine alkaloid N-oxides. An improved version of this field test is now available (Mattocks & Jukes, 1987). It is suitable for any plant parts, such as leaves, stems, flowers, seeds, or roots, or materials of plant origin, such as cereals or herbal teas, but has not yet been applied to cooked food.

The plant material (0.2 - 1 g) is extracted by grinding it with aqueous ascorbic acid (5%) and a small amount of sand. The solution is filtered and divided into two equal portions ("test" and "blank"). An aqueous solution (0.2 ml) of sodium nitroprusside (5%) containing sodium hydroxide (10^{-3} mol) is added to the "test" sample. Both portions are heated for approximately 1 min at 70 - 80 °C; then Ehrlich reagent is added and heating is continued for 1 min. The Ehrlich reagent contains 4-dimethylaminobenzaldehyde (5 g) dissolved in a mixture of acetic acid (60 ml), water (30 ml), and 60% perchloric acid (10 ml). A magenta colour in the "test" compared with the "blank" indicates the presence of an unsaturated PA N-oxide. The "blank" may show a colour if the plant contains compounds, such as indoles or pyrroles, which can themselves give a colour with Ehrlich reagent. The intensity of colour in the "sample" compared with the "blank" can give a rough idea of the amount of alkaloids present, and indicate whether further chemical or toxicological testing of the plant material is adviseable.

In practice, the majority of PA-containing plants contain enough alkaloid in the N-oxide form (often a large proportion) to react positively in this test. The main exceptions are some seeds (Crotalaria), which may contain much alkaloid base, but little or no N-oxide. These (and any other sample not containing chlorophyll) can be tested for basic PAs by grinding them with chloroform, heating the filtered extract with a solution (0.1 ml) of orthochloranil (0.5%) in acetonitrile, and then heating it with Ehrlich reagent. A magenta colour indicates the presence of an unsaturated PA. Non-toxic pyrrolizidine alkaloids having a saturated pyrrolizidine nucleus, and pyrrolizidine alkaloids that are otonecine esters, such as petasitenine, will not respond to this test.

2.2.2.6 Indicator dyes

A method generally applicable to tertiary bases has been adapted for pyrrolizidine alkaloids by Birecka et al. (1981). It is sensitive, but is not specific for this group of alkaloids, and it does not distinguish between the saturated and unsaturated alkaloids. A chloroform solution of the alkaloid is shaken with acidified aqueous methyl orange. The yellow alkaloid:dye complex is subsequently released from the chloroform phase, using ethanolic sulfuric acid, and measured spectrophotometrically.

2.2.2.7 Direct weighing

An insensitive way to determine the alkaloids in, for example, a plant sample, providing enough is available, is to extract the alkaloids (section 2.2.1) and weigh them. This will provide a rough measure of the total bases present in the sample; however, these may not necessarily be PAs. Nevertheless, the sample can then be subjected to further tests, e.g., GC-MC, nuclear magnetic resonance (NMR), or colorimetric analysis. Furthermore, pyrrolizidine N-oxides are generally too water soluble to be appreciably extractable from aqueous solution by chloroform. Thus, if two portions of the sample are extracted, and one of them is reduced to convert N-oxides to bases, the weight difference between the two products will represent the alkaloid existing in the form of N-oxide in the original sample.

2.3 Determination of Metabolites in Animal Tissues

Important metabolites of toxic pyrrolizidine alkaloids in animals include "pyrrolic" derivatives (dehydro-alkaloids) and N-oxides. A procedure for measuring pyrrolic metabolites in tissue samples (such as liver or lung) has been described by Mattocks & White (1970). The sample (usually 0.5 g) is homogenized in an ethanolic solution of mercuric chloride; the solids are separated by centrifugation and heated with Ehrlich reagent to give a soluble colour that can be measured spectrophotometrically.

The measurement of pyrrolic and N-oxide metabolites, formed by the action of hepatic microsomal preparations on PAs in vitro, is an improvement described by Mattocks & Bird (1983).

3. SOURCES AND PATHWAYS OF EXPOSURE

3.1 Hepatotoxic Pyrrolizidine Alkaloids and Their Sources

Plants constitute the only natural source of pyrrolizidine alkaloids (PAs) that cause toxic reactions in man and animals. PAs occur in a number of species in the families Boraginaceae, Compositae, Leguminosae (genus Crotalaria), Ranunculaceae (genus Caltha), and Scrophulariaceae (genus Castilleja) (Table 1). The most important genera of PA-containing toxic plants are Crotalaria (Leguminosae), Senecio (Compositae), Heliotropium, Trichodesma, Amsinckia, Echium, and Symphytum (Boraginaceae) (Hooper, 1978). The recorded cases of human toxicity have mainly been caused by at least 12 different pyrrolizidine alkaloids, mostly derived from Heliotropium, Senecio, and Crotalaria genera. The Senecio spp. grow throughout the world; the Crotalaria spp. are mainly found in the tropics and subtropics (Culvenor, 1980).

Table 1. List of plant genera containing toxic pyrrolizidine alkaloids (with number of species investigated)

Family	Genera
Apocynaceae	Fernaldia (1), Parsonsia (4),
Boraginaceae	Alkanna (1), Amsinckia (4), Anchusa (2), Asperugo (1), Borago (1), Caccinia (1), Cynoglossum (9), Echium (3), Hackelia (1), Heliotropium (25), Lappula (2), Lindelofia (7), Lithosperum (1), Macrotomia (1), Messerschmidtia (1), Myosotis (2), Paracaryum (1), Paracynoglossum (1), Rindera (5), Solenanthus (4), Symphytum (7), Tournefortia (2), Trachelanthus (2), Trichodesma (2), Ulugbekia (1)
Compositae	Adenostyles (3), Brachyglottis (1), Cacalia (4), Conoclinium (1), Crassocephalum (1), Doronicum (2), Echinacea (2), Emilia (2), Erechtites (1), Eupatorium (8), Farfugium (1), Gynura (2), Ligularia (5), Petasites (4), Senecio (142), Syneilesis (1), Tussilago (1)
Leguminosae	Crotalaria (60)
Ranunculaceae	Caltha (2)
Scrophulariaceae	Castilleja (1)

An alphabetical list of pyrrolizidine alkaloids with their plant sources has been published by Smith & Culvenor (1981) and Mattocks (1986). An updated version is attached as Appendix I. The plant genera containing toxic PAs are listed in Table 1 indicating the number of species investigated for PAs. A comprehensive list of species of plants belonging to each of these genera, the alkaloids isolated from each, and the part of the plant containing the alkaloid are presented in Appendix II. Table 1 in Appendix II includes species known to contain alkaloids of proved hepatotoxicity, or of a molecular structure that would make them very probably hepatotoxic. Table 2 in Appendix II includes species containing pyrrolizidine amino-alcohols or esters, which, while not having all the features of hepatotoxicity, would need only minor structural modifications to render them hepatotoxic. Plants of the same taxonomic groups as the plants of proven hepatotoxicity are listed in part (a) of the table. There is a possibility that, on further examination, hepatotoxic alkaloids may be found, as minor constituents, in strains or parts of these plants not yet investigated or under specific conditions of growth. It should be noted that the species that have been investigated and are listed are only few compared with the total number of species in each genera. It has been recommended by Smith & Culvenor (1981) that it would be prudent to regard all species in the family Boraginaceae and the genera Crotalaria, Senecio, and Eupatorium as potentially hepatotoxic.

It is pertinent to note that the alkaloid content in different parts of the plant (e.g., roots, leaves, stalks, flowers, and buds) varies and is subject to fluctuations according to the climate, soil conditions, and time of harvesting (Danninger et al., 1983; Hartmann & Zimmer, 1986). Mattocks (1980) demonstrated that the alkaloid content of the leaves of Symphytum spp. (Russian comfrey), which are used as an item of food, varies with their maturity. The toxic PA content is highest at the beginning of the vegetative period and declines as the leaves mature. The PA content of the roots is much higher than that of the leaves, and dried leaves contain a higher concentration than fresh leaves (Mattocks, 1986). According to Danninger et al. (1983), in some species (Symphytum asperum), relatively long storage may lead to a reduction in the alkaloid content, presumably because enzymes are released during drying. Candrian et al. (1984b) studied the stability of PAs in hay and silage containing various amounts of Senecio alpinus. The PA content of hay remained constant for several months, but the PAs in silage were mainly degraded. However, the degradation of PAs was much less complete in the lower concentration range. A quantitatively significant PA-degradation product in silage was identified as retronecine. Silage with an S. alpinus percentage of 3.5 - 23 still contained macrocyclic PAs at a concentration of about

20 mg/kg wet weight. Such silage was not considered safe for cattle bearing in mind that a 600-kg calf eats about 30 kg silage/day, amounting approximately to a daily intake of about 1 mg PAs/kg body weight. In feeding trials with Senecio jacobaea, Johnson (1979) found that the minimum lethal dose for cattle was between 1 and 2 mg PAs/kg body weight per day.

PAs known to have been associated with instances of human toxic liver disease in different parts of the world are listed in Table 2. Two groups of alkaloids that, according to Culvenor (1983), are consumed in significant amounts by people in different parts of the world include:

(a) Echimidine, acetyllycopsamine, and related alkaloids (many countries)

Leaves of plants of the Symphytum sp. (Symphytum officinale (comfrey) and Symphytum x uplandicum) are used traditionally as a salad and as a medicinal herb in Australia, many countries of Europe, and the USA. S. officinale has been shown to be carcinogenic for rats (Hirono et al., 1978). Leaves of Russian comfrey contain a concentration of alkaloids (mainly echimidine) of 0.1 - 1.5 g/kg. The highest level of daily consumption of the alkaloids has been estimated to be 5 - 6 mg (Culvenor, 1983).

(b) Echimidine and related alkaloids (Australia)

PAs derived from Echium plantagineum, with echimidine as the major component, have been found in honey secreted by bees feeding on the plant (Culvenor et al., 1981). The plant is a major source of honey (section 3.3.4).

3.2 Pneumotoxic and Other Toxic Pyrrolizidine Alkaloids

Not all hepatotoxic alkaloids are pneumotoxic. The commonest ones used to produce experimental lung injury are fulvine (Barnes et al., 1964; Kay et al., 1971a; Wagenvoort et al., 1974a,b) and monocrotaline (Lalich & Ehrhart, 1962; Chesney & Allen, 1973b; Huxtable et al., 1977). These are also the most active (Mattocks, 1986). The seeds of Crotalaria spectabilis, which contain monocrotaline, have also been used to study pneumotoxic effects on experimental animals (Turner & Lalich, 1965; Kay & Heath, 1966; Kay et al., 1967a) and C. spectabilis has been called the pulmonary hypertension plant (Kay & Heath, 1969), because of the pulmonary hypertensionogenic properties of the PAs it contains. Culvenor et al. (1976a) screened 62 PAs for hepatotoxicity and pneumotoxicity. Chronic lung lesions were produced by most compounds that induced chronic liver lesions, though high

Table 2. Instances of human toxicity caused by pyrrolizidine alkaloids[a]

Principal alkaloid	Plant	Country/Region	Cause of intake	Reference
Heliotrine and other alkaloids similar to lasiocarpine	Heliotropium popovii	Afghanistan	contamination	Tandon & Tandon (1975); Tandon, B.N. et al. (1978); Tandon, H.D. et al. (1978); Mohabbat et al. (1976)
Senecionine	Senecio illiciformis; Senecio-burchelli	South Africa	contamination	Wilmot & Robertson (1920)
	Senecio spp.	South Africa	contamination	Selzer & Parker (1951)
Alkaloids of trichodesmine and senecionine type	Crotalaria juncea	Ecuador	medicine	Lyford et al. (1976)
Heliotrine and lasiocarpine	Heliotropium lasiocarpum	Hong Kong	medicine	Kumana et al. (1985); Culvenor et al. (1986)
Crotananine and cronaburmine	Crotalaria nana	India	contamination	Tandon, R.K. et al. (1976); Krishnamachari et al. (1977); Siddiqui et al. (1978a,b)
Heliotrine N-oxide	Heliotropium eichwaldii	India	medicine	Datta et al. (1978a,b)
Monocrotaline fulvine	Crotalaria retusa; Crotalaria fulva	West Indies	medicine	Bras et al. (1954, 1957) Stuart & Bras (1957)
	Ilex sp.	United Kingdom	medicine	McGee et al. (1976)
Riddelline retrorsine N-oxide (with others)	Senecio longilobus	USA	medicine	Stillman et al. (1977); Fox et al. (1978); Huxtable (1980)

Table 2 (contd).

Indicine N-oxide	purified chemical	USA	medicine	Letendre et al. (1984)
Symphytine, symglandine, and other symphytum alkaloids	Symphytum sp.	USA	medicine	Ridker et al. (1985); Huxtable et al (1986)
Lasiocarpine and heliotrine	Heliotropium lasiocarpum	USSR	contamination	Dubrovinskii (1952); Mnushkin (1952)
Trichodesmine and incanine	Trichodesma incanum	USSR	contamination	Shtenberg & Orlova (1955); Yunosov & Plekhanova (1959)

<u>a</u> Adapted from: Culvenor (1983) and Mattocks (1986). Refer also to Table 15 for details and section 7.

doses were required in some instances. It is possible that chronic lung lesions may not occur in experimental animals because of early death due to acute toxicity. However, the authors identified a number of PAs that were particularly prone to produce chronic lung damage in rats including crispatine, senecionine, seneciphylline, and usaramine (12-membered macrocyclic, retronecine diesters), anacrotine and madurensine (crotonecine esters), and the heliotridine esters, heliosupine, lasiocarpine, and rinderine.

The molecular structure-activity requirements for pneumotoxicity are the same as those for hepatotoxicity. This is consistent with their both being caused by the same toxic metabolites and by the metabolic activation of the alkaloids in the liver cells to form a reactive pyrrolic dehydro-alkaloid (Culvenor et al., 1976a).

Trichodesmine and incanine, found in the seeds of Trichodesma incanum (Yunusov & Plekhanova, 1959), are believed to have been the causative factors of the "Ozhalangar encephalitis" that was endemic in Uzbekistan, USSR (1942-51), in which the symptoms and signs were related primarily to the central nervous system (Shtenberg & Orlova, 1955) (section 7.7).

3.3 Pathways of Exposure

Naturally-occurring animal disease is caused by the alkaloid-containing plants growing in fields and pastures or being fed accidentally as fodder. They are mostly herbaceous or small shrubs and many thrive in dry and arid climates. One such plant containing toxic PA alkaloids has been reported to grow in the western desert of Egypt (Hammouda et al., 1984). The growth of this group of plants is particularly prolific during, and following, periods of drought, as has been reported in association with the outbreaks of human disease in Afghanistan (Tandon & Tandon, 1975; Mohabbat et al., 1976) and India (Tandon, B.N. et al., 1976). Alkaloid-containing plants are widespread in the tropics, especially Crotalaria, of which there are over 300 species in Africa. Ordinarily, the alkaloid-containing plants have a bitter taste and grazing animals will reject them, unless their normal fodder is scarce. However, PAs often occur largely as $\underline{N}$-oxides, which are said not to be bitter, and plants containing PAs are readily eaten by some animal species.

Human intoxication may result from the ingestion of the toxic substance in either contaminated food or herbal infusion.

3.3.1 Contamination of staple food crops

The products of pyrrolizidine alkaloid-containing plants, generally seeds, may contaminate the staple food and may be eaten over long periods of time. The fact that these plants may cause disease is generally not recognized by the people and such contamination is known to have resulted in large-scale outbreaks of poisoning (Dubrovinskii, 1952; Mnushkin, 1952; Shtenberg & Orlova, 1955; Tandon & Tandon, 1975; Mohabbat et al., 1976; Tandon, B.N. et al., 1976, 1977; Tandon, R.K. et al., 1976; Krishnamachari et al., 1977; Tandon, H.D. et al., 1977) (Table 2, section 3.1).

3.3.2 Herbal infusions

Plants have been used traditionally for medicinal purposes all over the world. Herbs have been the mainstay of the indigenous systems of medicine, especially in China, Greece, and India, since ancient times. Table 3 includes a list of some plants that are suspected, or known, to contain PAs and have been used as herbal medicines in different countries (Mattocks, 1986).

Several PA-containing plants are included among the list of plants used in indigenous systems of medicine in India (Chopra, 1933). As a part of a research study on the etiological factors of chronic liver disease in Sri Lanka, Arseculeratne et al. (1981) chemically screened the first 50 plants used in Sri Lanka's traditional medicine pharmacopoaea, and found that 3 of them contained PAs. All 3 were hepatotoxic in rats. Of the 3, the presence of alkaloids in Cassia

Table 3. Some plants containing (or suspected of containing) PAs, which have been used by people either as herbal medicines (M) or foods (F)

Plant	Mode of use	Country or region	Reference[a]	
BORAGINACEAE				
Anchusa officinalis	M	Europe	Broch-Due & Aasen (1980)	B
Borago officinalis	M	USA	Delorme et al. (1977)	A
Cynoglossum geometricum	M	East Africa	Schoental & Coady (1968)	A
Cynoglossum officinale	M	Iran	Coady (1973)	B
Heliotropium eichwaldii	M	India	Gandhi et al. (1966a); Datta et al. (1978a,b)	B A
H. europaeum	M	India, Greece	IARC (1976)	A
H. lasiocarpum	M	Hong Kong	Kumana et al. (1985); Culvenor et al. (1986)	A A
H. indicum	M	India, Africa, South America, and elsewhere	Schoental (1968a); Hoque et al. (1976)	B B
H. ramossissimum (ramram)	M	Arabia	Macksad et al. (1970); Coady (1973)	B B
H. supinum	M	Tanzania	Schoental & Coady (1968)	A
Pulmonaria spp.	M	USA	Delorme et al. (1977)	A
Symphytum officinale (comfrey)	F, M	Japan and elsewhere	Hirono et al. (1978, 1979b)	A
	M	USA	Furuya & Hikichi (1971); Delorme et al. (1977)	A A
S. x uplandicum	F, M	General	Hills (1976)	B
		USA	Culvenor et al. (1980a,b)	A
S. asperum	M	USA	Pedersen (1975)	A
COMPOSITAE				
Cacalia decomposita (matarique)	M	USA	Sullivan (1981)	B
C. yatabei	F	Japan	Hikichi & Furuya (1978)	B

Table 3 (contd).

Plant	Mode of use	Country or region	Reference[a]	
Farfugium japonicum	M	Japan	Furuya et al. (1971)	B
Ligularia dentata	F	Japan	Asada & Furuya (1984)	B
Petasites japonicus	F, M	Japan	Hirono et al. (1973, 1979b)	A
Senecio abyssinicus	M	Nigeria	Williams & Schoental (1970)	B
S. aureus	M	USA	Wade (1977)	B
S. bupleuroides	M	Africa	Watt & Breyer-Brandwijk (1962)	A
S. burchelli	F, M	South Africa	Rose (1972)	A
S. coronatus	M	South Africa	Rose (1972)	A
S. discolor	M	Jamaica	Asprey & Thornton (1955)	B
S. doronicum	M	Germany	Roeder et al. (1980a)	B
S. inaequidens	F	South Africa	Rose (1972)	B
S. jacobaea (ragwort)	M	Europe	Schoental & Pullinger (1972); Wade (1977)	B B
S. longilobus (S. douglassi)	M	USA	Stillman et al. (1977); Huxtable (1979a)	A B
S. monoensis	M	USA	Huxtable (1980)	A
S. nemorensis spp. fuchsii	M	Germany	Habs et al. (1982)	A
S. pierotti	F	Japan	Asada & Furuya (1982)	B
S. retrorsus (S. latifolius)	M	South Africa	Rose (1972)	A
S. vulgaris (common groundsel)	M	Europe	Watt & Breyer-Brandwijk (1962)	A
		Netherlands	Wade (1977)	B
	M	Iran	Coady (1973)	B
Syneilesis palmata	F	Japan	Hikichi & Furuya (1976)	B
Trichodesma africana	M	Asia	Omar et al. (1983)	B

Table 3 (contd).

Plant	Mode of use	Country or region	Reference[a]	
Tussilago farfara (coltsfoot)	M	Japan	Culvenor et al. (1976a)	A
	M	China	Hirono et al. (1976b)	A
	M	Norway	Borka & Onshuus (1979)	B
	M	USA	Borka & Onshuus (1979); Culvenor et al. (1976b);	B B
LEGUMINOSAE				
Crotalaria brevidens	F	East Africa	Coady (1973)	B
C. fulva	M	Jamaica	Barnes et al. (1964); McLean (1970, 1974)	A A
C. incana	M	East Africa	Schoental & Coady (1968)	A
C. juncea	M, F	India	Chopra (1933); Watt & Breyer-Brandwijk (1962)	A A
C. laburnifolia	M	Tanzania	Schoental & Coady (1968)	A
	F	Asia	Coady (1973)	B
C. mucronata	M	Tanzania	Coady (1973)	B
C. recta	M, F	Tanzania	Schoental & Coady (1968); Coady (1973)	A B
C. retusa	M, F	Africa	IARC (1976)	A
		India	Watt & Breyer-Brandwijk (1962)	A
C. verrucosa	M	Sri Lanka	Arseculeratne et al. (1981)	A

[a] A = Reference in the reference list of this document.
B = Reference in Mattocks (1986).

all, of the plants reported to be etiological agents in human cases of veno-occlusive disease can be found in an inventory of medicinal plants used in different countries (WHO, 1980), which also indicates the countries that they are used in. The above lists may not be complete as many such plants may be used in folk medicine but have not been mentioned in the scientific literature. However, the lists do indicate the wide and varied use of such toxic herbs in all parts of the world.

Lately, there has been a growing interest in the developed countries in organically grown products for food, as well as home remedies (Table 3), and some of the PA-containing herbs have been freely available in herbal shops (Schoental, 1968; Burns, 1972). Danninger et al. (1983) listed plants containing PAs that are commonly used in the Federal Republic of Germany as medicaments (Table 4). He also listed 9 plants in which alkaloids have only been identified qualitatively, the toxicity of which has not been, or has been insufficiently, investigated (Table 5). Similarly, Roitman (1983) listed 10 plants, in which the presence of PAs is suspected or has been proved and which are used as herbal teas in the USA. The lists include 10 plants containing PAs, most of which have been proved hepatotoxic experimentally, some having highly carcinogenic promoter activity. Some of these alkaloids have been associated with human case reports of PA toxicity. The more recent reports (Table 2) of instances of PA poisoning through the use of herbal medicines are from developed countries (Lyford et al., 1976; Stillman et al., 1977; Fox et al., 1978; Kumana et al., 1985; Ridker et al., 1985). Such use of the herbs is the reason that veno-occlusive disease is endemic in Jamaica (Bras et al., 1954; Jellife et al., 1954a,b; Bras & Watler, 1955; Stuart & Bras, 1955, 1957). There are obvious difficulties in exercising any kind of control to restrict this use only to plants that have been tested and certified as safe for human use. It is impossible to identify many such herbs, as they are sold as plants or their amorphous products in the herbal shops.

Manufactured preparations may also contain PA-containing herbs, e.g., comfrey-pepsin capsules sold as a digestive aid (Huxtable et al., 1986).

3.3.3 Use of PA-containing plants as food

Several PA-containing plants are used as food as can be seen in Table 3 (Mattocks, 1986). **Petasites japonicus** Maxim, **Tussilago farfara** L. (coltsfoot), and **Symphytum officinale** L. (comfrey or Russian comfrey) are known as edible plants in Japan, and have been proved to contain carcinogenic

Table 4. Medicinal plants containing PAs of known hepatotoxicity, reported as commonly used in the Federal Republic of Germany, and the PAs contained in them[a]

Family	Genus	Species	Pyrrolizidine alkaloids
Compositae	Eupatorium	E. cannabinum (hemp agrimony)	amabiline± supinine[b]
	Petasites	P. hybirdus	senecionine[b,c] integerrimine[b] senkirkine[b]
	Senecio (groundsel)	S. nemorensis sp. fuchsii (Fuch's groundsel)	fuchsisenecionine senecionine[b,c]
		S. vulgaris (groundsel)	senecionine[b,c] seneciophylline[b] retrorsine[b] riddelline[b,c]
		S. Jacobaea (ragwort)	jacobine[b] senecionine[b,c] seneciphylline[b] jacoline, jaconine chlorinated PAs[d]
		S. aureus (American golden ragwort)	senecionine[b,c]
	Tussilago (coltsfoot)	T. farfara (coltsfoot)	senkirkine[b] senecionine[b,c] tussilagine
Boraginaceae	Alkanna	A. tinctoria	7-angelylretronecine triangularine dihydroxytriangularine
	Anchusa	A. officinalis	lycopsamine
	Borago	B. officinalis (borage)	lycopsamine/intermedine± acetyllycopsamine/ acetylintermedine amabiline supinine
	Symphytum (comfrey)	S. officinale (comfrey)	symphytine[b] echimidine(?) lycopsamine acetyllycopsamine[b] lasiocarpine[b,c] heliosupine N-oxide

Table 4 (contd).

Family	Genus	Species	Pyrrolizidine alkaloids
		S. peregrinum S. x uplandicum	lycopsamine[b] intermedine[b] symphytine[b] echimidine[b] 7-acetyllycopsamine 7-acetylintermedine symlandine uplandicine
		S. asperum (prickly comfrey)	asperumine heliosupine N-oxide echimidine[b] echinatine
	Cynoglossum (hound's tongue)	C. officinale (hound's tongue)	heliosupine N-oxide echinatine acetyl heliosupine[b] O-7-angelylheliotridine[b]
	Heliotropium (Heliotrope)	H. europaeum (common heliotrope)	heliotrine[b,c,e] lasiocarpine[b,c,e] supinine heleurine europine acetyllasiocarpine[b]

[a] Modified from: Danninger et al. (1983).
[b] Toxic alkaloids.
[c] Alkaloids known to have caused human toxicity.
[d] Alkaloids with highly carcinogenic promoter activity.
[e] Used only in homeopathy.

pyrrolizidine alkaloids (Hirono et al., 1973, 1979a,b). The young flower-stalks of P. japonicus and the buds of coltsfoot have been used in Japan as human food or herbal remedies. The leaf and root of comfrey are also used as an edible vegetable or tonic (Hirono et al., 1978) in Japan and other countries (Culvenor, 1985). The carcinogenic PAs in these plants are petasitenine (P. japonicus), senkirkine (coltsfoot), and the group including symphytine (comfrey). They were also mutagenic in the Ames system of Salmonella typhimurium and V79 hamster cell line and induced transformation in cryo-preserved hamster embryonic cells (Hirono et al., 1979b). Other such PA-containing plants, used as food in Japan, include young leaves of Syneilesis palmata, various Cacalia species, and young Senecio pierotti (Mattocks, 1986). According to

Table 5. Medicinal plants containing PAs, reported as commonly used in the Federal Republic of Germany, the toxicity of which has not been, or has been insufficiently, investigated[a]

Family	Genus	Species
Compositae	Eupatorium	E. perforatum
	Brachyglottis	B. repens
	Arnica	A. montana (mountain arnica)
Boraginaceae	Lappula	L. intermedia (stickseed)
	Pulmonaria	P. officinalis (lungwort)

[a] Modified from: Danninger et al. (1983).

Culvenor (1985), consumers of comfrey could be ingesting up to 5 mg PAs per day. Rose (1972) listed a number of plants of the genus Senecio that are used as spinach in South Africa. These include S. burchelli, which is known to have caused an episode of PA poisoning through the ingestion of contaminated bread (Wilmot & Robertson, 1920).

3.3.4 Contaminated honey

In the USA, Deinzer et al. (1977) reported the presence of all PAs contained in Senecio jacobaea (ragwort) and proved to be hepatotoxic, in the honey secreted by bees feeding on the plant. The total alkaloid content ranged from 0.3 to 3.9 mg/kg. It has been estimated that an average annual human intake of honey (600 g) at the highest alkaloid level quoted would contain less than 3 mg of PAs (Mattocks, 1986). Culvenor et al. (1981) and Culvenor (1983, 1985) drew attention to the same potential hazard in honey from Echium plantagineum, a weed that grows widely in Southern Australia and is a major source of honey, yielding an estimated 2000 - 3000 tonnes per annum for human consumption. Echimidine is the major component of the alkaloids of Echium, which are present in concentrations of up to 1 mg/kg. Culvenor (1983) estimated that individuals may consume up to 80 g honey/day with a corresponding alkaloid intake of 80 µg/day, if only the Echium honey were used. No reports of acute human toxicity through this source are available.

3.3.5 Milk

PAs have been shown to produce toxic effects via transference into the milk of dams (Schoental, 1959). Retrorsine was administered orally to 17, and intraperitoneally to 6, lactating rats weighing 185 - 350 g in 5 - 10 mg doses daily, the first dose being given during the first 24 h after parturition. The rats received from 1 to 14 doses, the total intake amounting to 21 - 335 mg/kg body weight. The litters were separated from the mothers for 1/2 h following the administration of PA to avoid direct contamination of the former by licking. Apparently the milk production was not affected as the stomachs of many of the young, examined postmortem, were distended with milk. All animals whose mothers had received a total dose of 138 mg PA or more died within 30 days. Many of the young whose mothers had received smaller doses survived until they were killed at 6 months. Biopsy of the liver of the young at various intervals or at autopsy showed marked changes, even in cases where the mothers did not appear to be affected. Animals dying at 18 - 30 days showed hydropic or fatty vacuolation of liver cells. In the liver of animals dying or killed later, various degrees of haemorrhagic necrosis and increase in the centrilobular reticulin of the liver, and some thickening of centrilobular veins were seen. In animals that survived 6 months, the appearance was less abnormal, but some hyperplastic nodules and bile-duct proliferation were seen. The lactating rats dosed with the PAs generally survived longer than the suckling animals and usually did not show any ill effects, suggesting that the susceptibility of the suckling rats was greater than that of the mothers.

Dickinson et al. (1976) demonstrated the presence of PAs in the milk of dairy cattle fed or dosed with ragwort (Senecio jacobaea). When 4 cows were administered the dried plant material at levels of up to 10 g/kg body weight per day through rumen cannula, PA levels of up to 0.84 mg/kg were observed in the milk. However, only one (jacoline) of the several PAs contained in the plant was secreted. Calves, bucket fed on the milk did not show any signs of PA toxicity.

Dickinson (1980) repeated the study on goats. Four milk goats were freshly prepared with rumen cannulae. The kids were separated from their dams and were fed milk twice a day. Dried tansy ragwort plant material with a PA content of 0.16% (dry weight) was administered through the cannulae to each goat at a dosage rate of 10 g/kg body weight per day over 125 days. During this period, each of the 4 kids received milk from their dams at approximately 125 ml/kg per day in addition to ad lib feeding on alfalfa grass hay. Six PAs were isolated from the plant material: jacobine, jaconine, jaconline, jacozine, senecionine, and seneciphylline. Milk samples collected twice daily showed PA contents of 225 - 530 µg/litre with a mean of 381 µg/litre. No apparent health effects were noted in the kids, and only mild hepatic damage was suspected in the dams, on the basis of liver function tests. Fifty percent of the kids were killed after 10 weeks. No lesions of PA toxicity were seen. The dams were rebred and appeared normal throughout the gestation period. However, three dams aborted at almost full term, and the fetuses were born dead. One of the dams died shortly after parturition and showed evidence of severe liver damage characteristic of PA toxicity. Another, which delivered normally, also showed a lesser degree of liver damage at biopsy.

Data relating to PA secretion were compared with similar earlier data on cows. Mean secretion of PAs in cows appeared much higher, e.g., 684 µg/litre. The authors concluded that the amount of PAs secreted in the goat's milk did not cause any serious deleterious effects in the kids.

Johnson (1976) fed long-term lethal doses of Senecio jacobaea, by stomach tube, to 6 cows. Feeding started at term or within 30 days post-partum, and continued until what was considered to be a lethal dose had been fed. The daily dose of the plant ranged from 1 to 4.4 g/kg body weight, the total amount fed representing 5 - 15% of body weight over a period of 54 - 126 days. Five cows died within 98 days; one, in a moribund state, was killed on day 126. The calves suckled for 30 - 126 days. Suckling started immediately after birth in the case of 4 calves and 10 and 30 days later, respectively, in the 2 remaining calves. Three calves were killed with their dams or soon after, and 3 were retained for 1 year for observation. Milk samples from 2 cows were collected and pooled in 14- to 16-day lots during 64 days of feeding of the Senecio plant. Each pooled sample was administered intragastrically to a group of rats in daily doses of 12 ml for 15 - 30 days. A control group of rats were fed raw milk from cows not fed Senecio. Blood samples of the dams and the calves were analysed for glutamic oxaloacetic transaminase (GOT), lactic dehydrogenase (LDH), and gamma-glutamyl transpeptidase (GGTP). Serum-enzyme levels in all cows indicated statistically significant deviations suggesting liver dysfunction, and the livers at autopsy had characteristic features of PA toxicosis. The LDH and GOT levels in calves were generally abnormal after 20 - 45 days of suckling. The abnormalities ranged from mild to a 15- to 170-fold increase. One calf was autopsied at the peak increase of serum-enzymes and was found to have mild focal hepatitis. No significant pathological features were seen in the livers of other animals, nor of the rats, some of which were retained for up to 150 days.

Goeger et al. (1982) fed dried Senecio jacobaea (tansy ragwort) to lactating goats in a proportion of 25% of the feed. The milk contained 7.5 µg PA/kg dry weight. The milk produced by the goats was pooled and then bottle fed to appetite to 2 Jersey bull calves (1 day old) that also had access to tansy ragwort-free hay for 109 and 124 days, respectively. They were then weaned and given normal feed and observed for 6 months, after which they were killed and autopsied. In another study, rats were fed a diet containing the freeze-dried milk at 80% level for 180 days with a calculated total PA intake of 0.96 mg/rat. Other groups of rats were fed tansy ragwort at dietary levels of 0.01 - 10 g/kg (corresponding to PA intakes of 39.77, 5.04, 0.52, and 0.05 mg/rat). The calf livers only showed very mild non-specific changes, but the livers of rats fed tansy ragwort or the milk from tansy ragwort-fed goats showed definite, but mild, changes including swollen hepatocytes, megalocytosis, biliary hyperplasia, and fibrosis. Histopathological changes in milk-fed rats were similar to those in the group fed tansy ragwort in the diet at 0.01 g/kg. The authors concluded that there was evidence of PA transfer into milk, which proved hepatotoxic for rats. It was also noted that the goats had been fed high levels of tansy ragwort at the upper limit of their acceptance, and that the hepatic changes observed in rats fed high levels of milk, for extensive periods, were slight.

Luthy et al. (1983) produced direct evidence of excretion of macrocyclic esters of retronecine of the senecionine and seneciphylline-type into rat milk. ^{3}H-retronecine, an ^{3}H-necic acid-labelled senecionine, and seneciphylline were prepared biosynthetically with seedlings of Senecio vulgaris L. Two lactating rats (Ivanovas, Sprague Dawley), weighing 300 - 400 g, were fed the first of the second compound by stomach tube, in doses of 2.7 mg/kg and 5.5 mg/kg body weight, respectively. Samples of blood were examined 1, 3, and 6 h after treatment, and those of milk 1 and 3 h after. Animals were killed after 6 h. They were found to have excreted approximately 0.08% of the applied radioactivity in the milk within 3 h, mainly as unidentified retronecine-derived metabolites, and approximately 0.02% as unchanged PAs. The highest levels of PAs and metabolites in tissues were found in the liver and lungs, 6 h after administration.

Candrian et al. (1984a) also demonstrated that Drosophila melanogaster flies fed on milk from lactating rats that had been administered an oral dose of seneciphylline showed 1.2% sex-linked recessive lethals, compared with 0.3% in controls, indicating the transfer of the mutagenic properties of the PA via milk (section 6.4.7).

The implications of the above studies on the possibility of carry-over of PAs into foodstuffs of animal origin are obvious. However, no reports of human cases of acute PA toxicity, ascribed to the consumption of contaminated milk, are available.

3.3.6 Meat

There have not been any reports of the detection of PAs in meat products from livestock exposed to them.

3.3.7 Use of PAs as chemotherapeutic agents for cancer

An alkaloid of Heliotropium indicum L. (indicine N-oxide) has been found to have antitumour activity and has been used in experimental clinical chemotherapy for cancer (section 7.9).

4. METABOLISM

4.1 Absorption, Excretion, and Tissue Distribution

4.1.1 Absorption

There have been few studies on the absorption of PAs in man, but absorption has been inferred from studies on tissue distribution and the amounts of alkaloids and their metabolites excreted in the urine, faeces, and bile of animals (section 4.1.2).

Swick et al. (1982c) measured the transfer of a mixture of pyrrolizidine alkaloids extracted from Senecio jacobaea, across isolated intestine and stomach segments from rabbits. The alkaloid mixture contained seneciphylline, jacobine, jacozine, jacoline, and senecionine. The alkaloids were transferred across the ileum and jejunum, but not the stomach. Brauchli et al. (1982) compared the oral and percutaneous absorption in rats of a crude alkaloid mixture obtained from comfrey (Symphytum officinale L.). The mixture consisted of N-oxides of 7 alkaloids, principally 7-acetylintermedine and 7-acetyl-lycopsamine. A dose of 194 mg/kg was either given by gavage, or was applied to the shaved skin and left for 44 h. After the dermal application, the excreted N-oxides in urine (up to 48 h) amounted to 0.1 - 0.4% of the dose. After oral dosage the excreted level of N-oxides and alkaloid bases was quoted as being 20 - 50 times greater.

4.1.2 Excretion and distribution

The excretion and distribution of heliotrine in rats has been reported in Bull et al., 1968. Young rats (150 g), given the LD_{50} of heliotrine by ip injection, were killed at intervals, bled quickly, and the organs and tissues analysed. Heliotrine was present in the liver after 2 min (3.7% of total dose), the level peaking at 5 min (6.3%), and dropping to 2.2% at 1 h and 0.5% at 2.5 h. In adult rats, the level in the liver at 5 h was 0.07% of the total dose. Five min after dosing, 30 - 40% of the initial dose remained in the peritoneal cavity, and the blood level of heliotrine was 60 mg/litre, dropping to 3 mg/litre at 1 h. The urinary excretion of base and metabolites other than pyrrolic metabolites, collected and measured 16 h after administration of several alkaloids by ip injection, is shown in Table 6. The proportion of base excreted unchanged increased with the hydrophilicity of the alkaloid, being 62% for heliotrine N-oxide, 30% for heliotrine, and only 1 - 1.5% for lasiocarpine. Heliotridine, the hydrolysis product from heliotrine

Table 6. Urinary metabolites of pyrrolizidine bases in the rat (16-h urine)[a]

Urine constituent (amount in percentage of dose injected)

Base administered (ip injection)	Unchanged base	Base N-oxide	Heliotridine trachelanthate	Heliotridine trachelanthate N-oxide	Heliotridine	Heliotridine N-oxide
Heliotrine	30	Trace	10	5	3	15
Heliotrine N-oxide	62	(62)	2.7	ca. 6	ca. 1	ca. 10
Lasiocarpine	1-1.5			1.5-3		6
Heliotridine trachelanthate	35	ca. 1	(35)	(ca. 1)	5	20
Heliotridine	40	20			(40)	(20)

[a] From: Bull et al. (1968).

and lasiocarpine, was excreted in the form of the N-oxide in larger quantities after the administration of each of these alkaloids.

The distribution and excretion of monocrotaline was studied in rats by Hayashi (1966) who found that 50 - 70% was excreted in the urine within the first day. However, the analysis was by a non-specific chemical method that did not distinguish between the unchanged alkaloid and its metabolites. Mattocks (1968a) gave toxic pyrrolizidine alkaloids intraperitoneally to male rats and measured the urinary excretion of the unchanged alkaloid, and of N-oxide and pyrrolic metabolites. The excretion of N-oxide and unchanged alkaloid was rapid and almost complete in the first 24 h. Excretion of pyrroles was also rapid but continued for a little longer. For example, in rats given retrosine (60 mg/kg body weight), the urine in the first 24 h contained 10.6% unchanged alkaloid, 13.3% N-oxide, and 13.4% pyrrolic metabolites. During the second day, only 0.1% alkaloid, 0.2% N-oxide, and 1.8% pyrroles were excreted. Biliary excretion also occurred. About one-quarter of an iv dose of retrosine in rats was excreted in the bile as pyrrolic metabolites, and 4% as unchanged alkaloid; most of this excretion occurred during the first hour after the injection (White, 1977).

Jago et al. (1969) gave heliotrine iv to sheep; urinary excretion of the unchanged alkaloid together with metabolites (N-oxide, and demethylation and hydrolysis products) occurred rapidly and continued for up to 8 h. Excretion in the bile was only 2% of that in the urine.

The tissue distribution of radioactivity from a tritiated toxic pyrrolizidine alkaloid analogue was studied by Mattocks & White (1976) using synthanecine A bis-N-ethylcarbamate (40 mg/kg body weight). The highest concentrations of radioactivity were seen in the liver (where metabolism occurs), lungs, kidneys, and spleen (respectively, 3.9%, 0.19%, 0.18%, and 0.27% of the dose given), and about 69% of the dose was eliminated in the urine during the first day. Radioactivity in the expired air was negligible. The binding of radioactivity in the liver, and especially the lungs, was more persistent than in other organs. Similar results were given by the semisynthetic pyrrolizidine alkaloid analogue, retronecine bis-N-ethylcarbamate (Mattocks, 1977).

The distribution of the uniformly ^{14}C-labelled natural pyrrolizidine alkaloid senecionine in lactating mice was studied by Eastman et al. (1982). After 16 h, 75% of the radioactivity had been recovered in the urine, 14% in the faeces, but only 0.04% was in the milk; the liver contained 1.92%. The mice were milked using teat cups. Candrian et al. (1985) studied the distribution of radioactivity in rats given small doses of senecionine or seneciphylline (0.3 - 3.3 mg/kg), tritiated in the pyrrolizidine (retronecine) moiety. Most radioactivity was eliminated in the urine and faeces within 4 days. Using mass spectrometry, Dickinson et al. (1976) found a concentration of up to 0.84 mg PAs/litre in the milk of cows fed Senecio jacobaea. Blood levels of senecionine in rats given 0.1 LD_{50} ip were determined by Culvenor (1978). The levels were 0.38, 0.32, and 0.14 mg/litre at 0.5, 1, and 2 h after injection, respectively.

To summarize, the available evidence suggests that ingested toxic pyrrolizidine alkaloids are rapidly metabolized and that the excretion of unchanged alkaloid and of most metabolites is also rapid. Thus, within a few hours, only a relatively small proportion of the dose remains in the body, much of this in the form of metabolites bound to tissue constituents. It appears improbable that a significant amount of unchanged alkaloid will remain in the body after the first day.

Pyrrolizidine N-oxides are much more water soluble than their parent alkaloids. Indicine N-oxide (which is exceptionally water soluble) is very rapidly excreted, either unchanged or conjugated. Thus, indicine N-oxide given iv to mice, monkeys, or rabbits disappeared from the serum with initial half-lives ranging from 3 to 20 min (Powis et al., 1979; El Dareer et al., 1982). Over 80% of tritium-labelled indicine N-oxide given iv was excreted in the urine of mice or monkeys within 24 h (El Dareer et al., 1982); at 2 h, the highest concentrations of radioactivity were in the kidneys, liver, and intestines. Urinary excretion of indicine N-oxide was also rapid in rabbits, but somewhat slower in human beings (Powis et al., 1979).

4.2 Metabolic Routes

The major metabolic routes of unsaturated pyrrolizidine alkaloids in animals are: (a) hydrolysis (of the ester groups); (b) N-oxidation; and (c) dehydrogenation (of the pyrrolizidine nucleus) to dehydro-alkaloids (pyrrolic derivatives). Other minor routes of metabolism are known, but the three pathways account for the major known toxic effects of these alkaloids (Fig. 5). Routes (a) and (b) are believed to be detoxification mechanisms. Route (c) leads to toxic metabolites and appears to be the major activation mechanism. Route (a) may occur in various tissues, including the liver and blood. Routes (b) and (c) are brought about in the liver by the microsomal mixed-function oxidase system.

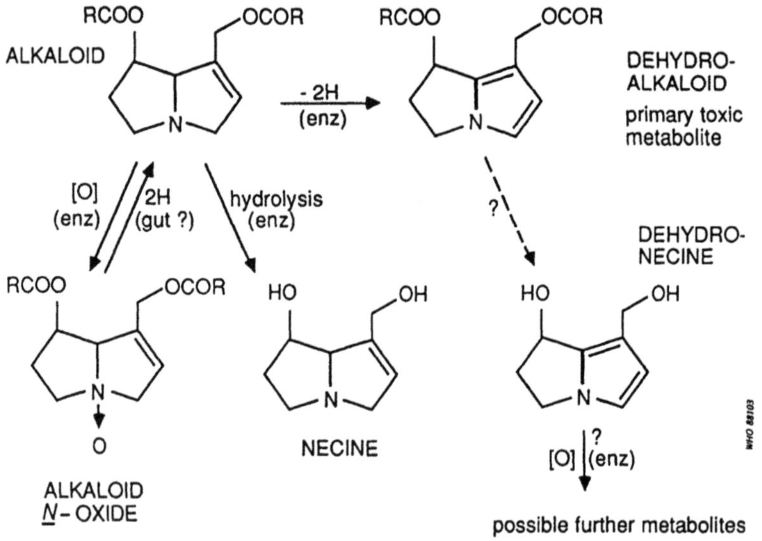

Fig. 5. Major metabolic routes of unsaturated pyrrolizidine alkaloids.

4.2.1 Hydrolysis

The hydrolysis of a PA leads to the formation of the amino-alcohol moiety (necine base) and the acid moiety. Neither of these is hepatotoxic (Schoental & Mattocks, 1960; Culvenor et al., 1976a). The highly water-soluble necine base is readily excreted, is not accessible to the microsomal system, and is not activated to a toxic metabolite. Thus, pyrrolizidine alkaloids that are very susceptible to (enzymic) hydrolysis have low toxicity (Mattocks, 1982). A major factor contributing to resistance to esterase is the steric hindrance in the acid moiety. Thus, the chain branching near the carbonyl groups slows hydrolysis allowing the formation of relatively high levels of pyrrolic metabolites; a conformation of the basic moiety, which brings the two ester groups close together, thus leading to mutual steric hindrance, can also prevent hydrolysis (Mattocks, 1981a).

The influence of hydrolysis in vivo on alternative metabolic pathways is demonstrated by the fact that treatment of rats with an esterase inhibitor, before giving pyrrolizidine alkaloids (or synthetic analogues), can lead to greatly increased production of pyrrolic metabolites from alkaloids that are normally susceptible to hydrolysis, but little increase in those from alkaloids normally resistant to hydrolysis (Mattocks, 1981a).

4.2.2 N-oxidation

The $\underline{N}$-oxidation of pyrrolizidine alkaloids is induced by the hepatic microsomal enzymes. The $\underline{N}$-oxide metabolites are highly water soluble and are rapidly excreted in the urine (Mattocks, 1968a). Pyrrolizidine $\underline{N}$-oxides are not converted to any significant extent to pyrrolic metabolites by microsomal enzymes (Jago et al., 1970; Mattocks & White, 1971a), and there is no evidence that they are toxic, unless first reduced to the corresponding basic alkaloids, which can then be activated by the microsomal system (Mattocks, 1971c). Thus, it appears that the formation of $\underline{N}$-oxides represents a detoxification pathway.

4.2.3 Conversion to pyrrolic metabolites

In laboratory animals, toxic pyrrolizidine alkaloids are metabolized to pyrrolic derivatives, so-called because the unsaturated ring of the pyrrolizidine system loses 2 hydrogen atoms to form what is in effect a pyrrole ring (though the structure is more correctly a dihydropyrrolizidine). Pyrrolic metabolites are easily detectable in the tissues shortly after giving a toxic pyrrolizidine alkaloid to an animal, by treating the tissue with an Ehrlich reagent containing boron trifluoride, when a red colour is produced; this reaction also occurs with the urine (Mattocks, 1968a; Mattocks & White, 1970). In rats given retrosine, pyrrolic metabolites were found principally in the liver, with highest levels associated with the microsomal and solid debris fractions and less in the mitochondrial fraction; low levels were found in the lungs, heart, spleen, and kidneys, within 4 h of giving retrosine. Rats given 60 mg retrosine/kg body weight excreted 14% of the dose in the urine, within 48 h.

Pyrrolic metabolites are formed by the hepatic mixed-function oxidase system, with a requirement for cytochrome P450, oxygen, and NADPH, as has been demonstrated in vitro (Jago et al., 1970; Mattocks & White, 1971a). Conversion of pyrrolizidine alkaloids to pyrrolic metabolites by the lung tissue of the human embryo (Armstrong & Zuckerman, 1970), rat (Mattocks & White, 1971a; Hilliker et al., 1983), or rabbit (Guengerich, 1977) was negligible. The formation of pyrrolic metabolites does not proceed via N-oxide intermediates, but appears to result from an initial hydroxylation of the unsaturated pyrrolizidine ring adjacent to the nitrogen atom (Mattocks & White, 1971a; Mattocks & Bird, 1983). This would lead to a chemically unstable intermediate that would be expected to decompose spontaneously to the pyrrolic product. The primary pyrrolic metabolites (or dehydro-alkaloid) formed by dehydrogenation of pyrrolizidine alkaloids are chemically dehydropyrrolizidine esters (Fig. 5). These are highly reactive compounds that can rapidly react with tissue constituents or hydrolyse to the corresponding pyrrolic alcohols, or dehydro-necines, which can thus be regarded as secondary metabolites. The latter can also react with tissue constituents, but more slowly. Because of their high chemical reactivity, the primary metabolites would be expected to have a short life in the liver cell (minutes or seconds) before they are hydrolysed or react with nucleophilic tissue constituents. Some might escape into the blood stream and reach other organs, especially the lungs. Dehydro-necines are more stable and also more water soluble, and can become more widely distributed throughout the body. However, they are also capable of reacting with tissue constituents. Thus, measurements of pyrroles formed from pyrrolizidine alkaloids in tissue samples, using a colour reaction (Mattocks & White, 1970), will not represent a single metabolite, but mixtures of the metabolites together with various reaction products of these with tissue constituents. It will be seen (section 5) that pyrrolic metabolites are believed to be responsible for major toxic actions of pyrrolizidine alkaloids (Mattocks, 1972a).

A pyrrolic metabolite with reactivity midway between that of dehydromonocrotaline and dehydroretronecine has been reported to be formed from monocrotaline in isolated, perfused rat liver (Lafranconi et al., 1985). Studies on this metabolite (isolated from bile) indicated that it is a monoester, and that it is toxic in perfused rat lung. This suggests that monoester pyrrolic metabolites may play a part in the toxic actions of PAs in extra-hepatic tissues.

When rats or other laboratory animals are given a toxic pyrrolizidine alkaloid, pyrrolic metabolites accumulate rapidly in the liver (Mattocks, 1973; White et al., 1973), reaching a peak within 1 - 2 h, then falling slowly during the next 24 h; the metabolites may still be detectable after 2 days. Accumulation is especially rapid after intraperitoneal injection, a very high level of pyrrolic metabolites being attained within 20 min; this indicates how rapid the metabolism of pyrrolizidine alkaloids can be.

The level of pyrrolic metabolites in rat liver is generally directly related to the amount of alkaloid given 2 h previously, at least up to an acute LD_{50} dose. The pyrrole level depends on the alkaloid used, and is related to the acute hepatotoxicity of the alkaloid (Mattocks, 1972a).

To be converted to the type of chemically reactive, toxic pyrrolic metabolites described above, an alkaloid must possess a 1-hydroxymethyl pyrrolizidine system, unsaturated in the 1,2-position (this makes the ring susceptible to dehydrogenation), and at least one hydroxyl group must be esterified, usually by a branched-chain acid. Otonecine esters are converted to similar pyrrol metabolites by a different media involving N-demethylation. Pyrrolizidine amino-alcohols (e.g., retronecine) are not metabolized to more than small amounts of pyrroles (Jago et al., 1970; Mattocks, 1981a), possibly because they are too water soluble to reach the microsomal enzymes.

The metabolic formation of pyrroles is catalysed by cytochrome P450 and specificity exists in the various isozymes (Guengerich, 1977; Juneja et al., 1984).

A few non-toxic pyrrolizidine alkaloids (e.g., rosmarinine and hygrophylline) are converted to pyrrolic metabolites in vivo (Mattocks, 1973). Such metabolites are chemically different from the pyrroles of toxic alkaloids, and they are neither reactive nor toxic (Mattocks & White, 1971b). The balance of structural features necessary for a pyrrolizidine alkaloid to be converted to give high concentrations of toxic pyrrolic metabolite has been discussed by Mattocks (1981a); the optimum conditions appear to be met in some alkaloids that are macrocyclic diesters, such as retrosine.

4.3 Effects of Treatments Affecting Metabolism

The formation of pyrrolic metabolites (and of N-oxides) is altered by treatments that affect the hepatic microsomal enzymes. Such effects have been studied by measuring rates of metabolism of pyrrolizidine alkaloids in vitro using microsomal preparations from animals pre-treated in various ways (Table 7). For example, microsomes from rats given the microsomal enzyme inducers phenobarbitone or DDT (but not

Table 7. Effect of pre-treatment of male rats on the conversion of PAs to pyrrolic derivatives and to N-oxides by liver microsomes in vitro

Alkaloid	Pre-treatment, and time before enzyme measurements	Enzyme activity as % of control values for formation of:		Reference
		Pyrroles	N-oxides	
retrorsine	phenobarbitone, ip, 3 x 100 mg/kg, 1 - 3 days	311	232	Mattocks & White (1971a)
retrorsine	DDT, ip, 75 mg/kg, 3 days	407	203	Mattocks & White (1971a)
retrorsine	3-methylcholanthrene, ip, 3 x 20 mg/kg, 1 - 3 days	95 (ns)	116 (ns)	Mattocks & White (1971a)
retrorsine	retrorsine, ip, 35 mg/kg, 20 h	63	-	Mattocks & White (1971a)
retrorsine	protein-free diet, 3 days	39	-	Mattocks & White (1971a)
monocrotaline	phenobarbitone, sc, 4 x 75 mg/kg, 1 - 4 days	448	165	Chesney et al. (1974)
monocrotaline	chloramphenicol, sc, 200 mg/kg, 1 h	10	109 (ns)	Chesney et al. (1974)
monocrotaline	SKF 525A, ip, 75 mg/kg, 1 h	10	87 (ns)	Chesney et al. (1974)

ns = not significantly different from controls.

those from rats given 3-methylcholanthrene) induce greatly increased pyrrole formation and smaller increases in N-oxide formation, from the alkaloid retrorsine (Mattocks & White, 1971a). Enzyme preparations from rats treated with inhibitors of microsomal enzymes, including SKF 525A and chloramphenicol, are much less active in converting monocrotaline to pyrroles (Chesney et al., 1974). The ability to metabolize retrorsine is diminished in microsomes from rats fed a protein-free diet, or from rats acutely poisoned with retrorsine (Mattocks & White, 1971a).

The effects of _in vivo_ treatment with several types of enzyme inducers on the toxicity of lasiocarpine and senecionine for primary rat hepatocyte cultures was investigated by Hayes et al. (1985). Pre-treatment with phenobarbitone potentiated the cytotoxicity of senecionine towards the cultured cells, whereas pre-treatment with 3-methylcholanthrene diminished the toxic action of senecionine, but had little effect on lasiocarpine cytotoxicity. The cytocidal effects of both alkaloids were substantially inhibited in the presence of SKF 525A.

4.4 Other Factors Affecting Metabolism

Variations between animal species have been investigated by White et al. (1973) and Shull et al. (1976). For instance, metabolism to form pyrroles is high in rats and very low in guinea-pigs, which, however, have higher rates of N-oxidation. For example, 2 h after an ip dose of retrosine (100 mg/kg body weight), the liver-pyrrole level in male rats was 13 times higher than that in male guinea-pigs (White et al., 1973). Liver microsome preparations from male rats were 28 times more active than microsomes from male guinea-pigs in the dehydrogenation of monocrotaline (Chesney & Allen, 1973a).

The development with age of the ability of Wistar rats to metabolize retrosine was studied by Mattocks & White (1973). The ability to form pyrroles is very low in new-born rats, but, by 5 days of age, it is nearly as high as in adult males. This activity continues at a similar level in male rats, but, in females, it falls after the age of about 20 days until, by 60 days, it is about one-eighth that in males. Such a sex difference was not observed in mice (White et al., 1973).

4.5 Other Metabolic Routes

The actions of hepatic microsomal enzymes on pyrrolizidine alkaloids can produce other metabolites as well as pyrroles and N-oxides, but there are few reports of these. Eastman & Segall (1982) demonstrated hydroxylation of the acid moiety of senecionine by liver microsomes from female mice. Such metabolism should not prevent the subsequent conversion of the product to pyrrolic or N-oxide metabolites. The formation of other microsomal metabolites of senecionine has been reported by Segall et al. (1984).

The O-demethylation of the acid moiety of heliotrine has been demonstrated by Jago et al. (1969) and represents a partial detoxification mechanism, since the product is about half as toxic as heliotrine. Other detoxification mechanisms exist in the rumen of sheep (Dick et al., 1963; Lanigan &

Smith, 1970a,b), which are, thus, particularly resistant to the effects of pyrrolizidine alkaloids.

4.6 Metabolism of Pyrrolizidine N-Oxides

As mentioned in section 4.2, the N-oxides of pyrrolizidine alkaloids are not converted to pyrrolic metabolites by liver microsomes. It appears that their main route of metabolism in animals is reduction to the corresponding basic alkaloids, which may then be further metabolized as already described. This reduction has been shown to occur in the rat or rabbit gut (Mattocks, 1971c; Powis et al., 1979), and may be brought about by intestinal bacteria or possibly by gut enzymes. Such reduction can also be brought about by hepatic microsomal fractions (Powis et al., 1979) in the presence of NADH or of NADPH, and by sheep rumen fluid (Lanigan et al., 1970a, b).

The reduction of pyrrolizidine N-oxides in vivo is of great importance as a step in the bioactivation of these compounds (Mattocks, 1971c), as shown in section 4.2.2.

4.7 Metabolism in Man

Powis et al. (1979) found that indicine N-oxide given iv to 3 human patients as an antitumour drug was partially reduced to indicine base, detectable in the urine and plasma. Armstrong & Zuckerman (1970) showed that human embryo liver slices, but not lung slices, converted the pyrrolizidine alkaloids lasiocarpine, retrorsine, and fulvine to pyrrolic metabolites in vitro.

5. MECHANISMS OF TOXICITY AND OTHER BIOLOGICAL ACTIONS

5.1 Metabolites Responsible for Toxicity

5.1.1 Metabolic basis of toxicity

The toxic effects of pyrrolizidine alkaloids are mediated through their toxic metabolites and not by the alkaloids themselves. The following observations are evidence for the above statement (Mattocks, 1972a):

(a) The alkaloids are chemically rather unreactive and it is hard to envisage reactions with cell constituents that they could undergo readily under physiological conditions. On the other hand, chemically prepared derivatives, similar or identical to known metabolites of these alkaloids, are highly reactive and are capable of causing toxic effects similar to those of PAs, often at dose levels much lower than those required by the alkaloids themselves.

(b) The liver is usually the main organ affected, whatever the route of administration of the alkaloid. The alkaloids are known to be metabolized in the liver.

(c) Direct application of these alkaloids to the skin does not cause local toxic effects (Schoental et al., 1954), nor do cytotoxic effects occur at sites of injection.

(d) The susceptibility of animals to the toxic actions of PAs is related to the ability of the animal to metabolize the alkaloids. For example, the hepatic microsomal enzymes of rats less than 1 h old have very low activity towards retrorsine and these rats are relatively resistant to it, whereas rats aged several days have a high enzyme activity and are highly susceptible to the alkaloid (Mattocks & White, 1973). Guinea-pigs are very resistant to retrorsine, unless they have been given phenobarbitone, which potentiates the enzymes that metabolize it (White et al., 1973). Rats pre-treated with microsomal enzyme inhibitors, such as SKF 525A or chloramphenicol, have increased resistance to retrorsine or monocrotaline (Allen et al., 1972; Mattocks, 1973). In general, there is a good relationship between the rate of hepatic metabolism of PAs to pyrrole in vitro (Shull et al., 1976) and chronic toxicity. Highly resistant species, e.g., guinea-pigs, Japanese quail, and sheep, have a low rate of pyrrole formation, while susceptible species, such as the horse, cattle, and rat, have a high rate. Notable exceptions are the rabbit and hamster, which have high rates of pyrrole formation, but are resistant. It is possible that this may be due to changes in the balance between activation and the involvement of other factors, such as activity of detoxification. For example, sheep have a high epoxide hydrolase activity in the liver (Swick et al., 1983), which may affect PA detoxification (Cheeke & Pierson-Goeger, 1983).

5.1.2 Isolation of pyrrolic metabolites

There is plenty of evidence that many unsaturated PAs are converted into pyrrolic esters (dehydro-alkaloids) in the mammalian liver (section 4.2.3). These primary pyrrolic metabolites cannot be isolated, because of their high reactivity and rapid rate of hydrolysis. However, their more stable hydrolysis products (pyrrolic alcohols; dehydronecines) have been isolated and identified. Thus dehydroheliotridine has been obtained from the in vitro incubation of both the heliotridine-based alkaloids, lasiocarpine and heliotrine, with rat liver microsomes (Jago et al., 1970) and dehydroretronecine was found to be the main detectable pyrrolic metabolite in the liver, blood, and urine of rats injected with the retronecine-based alkaloid, monocrotaline (Hsu et al., 1973). There is evidence that these materials are identical, i.e., the (±)-form resulting from racemization during hydrolysis of the parent pyrrolic esters (Kedzierski & Buhler, 1985).

The results of these studies confirm that rat liver enzymes convert PAs into metabolites with known cytotoxic activity (section 5.2), and imply that these metabolites are formed via the yet more toxic and short-lived dehydroalkaloids (Jago et al., 1970).

5.1.3 Chemical aspects of pyrrolic metabolites

5.1.3.1 Preparation

Chemical methods are available for converting unsaturated PAs into pyrrolic esters (dehydro-alkaloids), the putative primary toxic metabolites, enabling the physical, chemical, and toxicological properties of the latter to be studied.

Small amounts of dehydro-pyrrolizidine alkaloids are usually prepared by the reaction of the corresponding alkaloid N-oxides with either acetic anhydride (Mattocks, 1969; Culvenor et al., 1970a) or methanolic ferrous sulfate (Mattocks, 1969). The products must be protected from moisture and from acids, which can cause their immediate decomposition.

A variety of reagents can dehydrogenate the alkaloid bases to pyrrolic derivatives, these include manganese dioxide (Culvenor et al., 1970a,b; Mattocks, 1969), potassium permanganate (Culvenor et al., 1970a), chloranil (Culvenor et al., 1970a), 2,3-dichloro-5,6-dicyanobenzoquinone (Mattocks, 1969), iodine (Culvenor et al., 1970b), and aryl thiols (Juneja et al., 1984). Some PAs are slowly oxidized to pyrroles by molecular oxygen (Bick et al., 1975).

The more stable pyrrolic alcohol, dehydroretronecine (Fig. 6), is prepared from retronecine using chloranil (Culvenor et al., 1970a) or aqueous potassium nitrosodisulfonate (Mattocks, 1981c) or from retronecine N-oxide (isatinecine) using ferrous sulfate (Mattocks, 1969). The enantiomeric dehydroheliotridine can be prepared from heliotridine in similar ways. Racemic dehydro-heliotridine has been synthesized (Viscontini & Gilhof-Schaufelberger, 1971; Bohlmann et al., 1979).

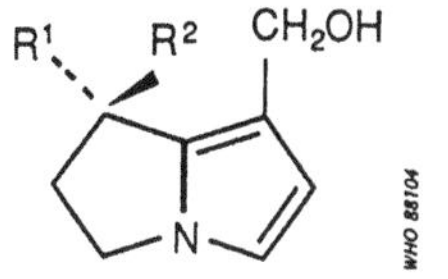

Fig. 6. Structures of dehydroheliotridine (R^1 = OH, R^2 = H) and dehydroretronecine (R^1 = H, R^2 = OH).

5.1.3.2 Chemistry associated with toxic actions

Dehydro-pyrrolizidine alkaloids and dehydronecines (pyrrolic esters and alcohols) act chemically as alkylating (electrophilic) agents, i.e., they can react with compounds possessing electron-rich (nucleophilic) groups, such as amines, thiols, and some hydroxyl compounds. The products of alkylation consist of the "pyrrole" moiety covalently bonded to the substrate molecule. The mechanism of alkylation is illustrated in Fig. 7 (Mattocks, 1972a). An ester (R = COR) or hydroxyl group (R = H) attached to the pyrrole ring via one carbon atom (i.e., at C7 or C9) is highly reactive, being easily cleaved, leaving a positively charged pyrrole moiety with a high affinity for electron-rich substrates. Pyrrolic esters are the most reactive, RCOO being a better "leaving group" than HO. When 2 oxygen functions are present (as illustrated), either (in turn) can act as an alkylating centre. Such bifunctional alkylation could lead to cross linking of macromolecules (Mattocks, 1969; White & Mattocks,

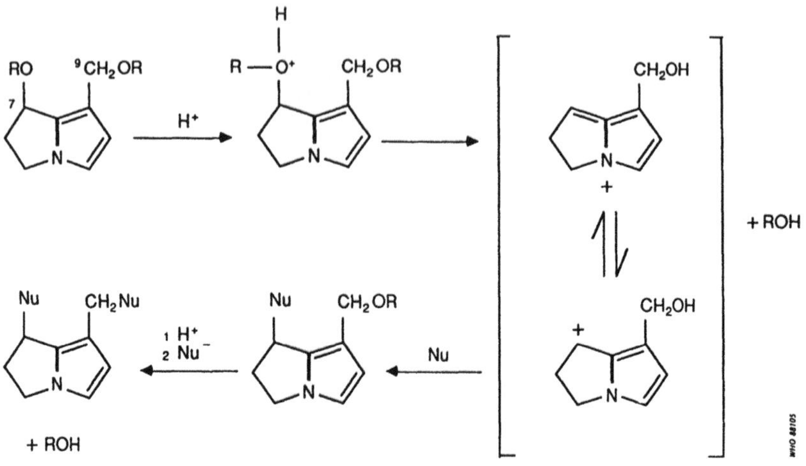

Fig. 7. Mechanism of reaction of a pyrrolic alcohol or ester with a nucleophil (Nu).

1972; Petry et al., 1984, 1986). When the groups (R) are the same or similar, C7 is the more reactive site. Examples of such alkylations using pure chemicals (amines or alcohol) have been given by Mattocks (1969) and Culvenor et al. (1970a). Mattocks & Bird (1983) showed that a variety of nucleophiles of biological interest could be alkylated by dehydroretronecine. Black & Jago (1970) demonstrated the in vitro alkylation of DNA by dehydroheliotridine, and Robertson (1982) and Wickramanayake et al. (1985) the alkylation of deoxyguanosine by dehydroretronecine. The alkylation of mouse or rat liver DNA by pyrrolizidine alkaloids has been shown in vivo by Eastman et al. (1982) and Candrian et al. (1985).

5.1.4 Possible further metabolites

The possibility that pyrrolic metabolites of PAs might themselves be metabolized by microsomal enzymes to further cytotoxic derivatives was suggested by Guengerich & Mitchell (1980). These authors showed that the tritium-labelled model compounds 1,2,3-trimethylpyrrole and 1-methyl-3-4 bishydroxymethylpyrrole could be metabolized in rats or by rat liver microsomes to unidentified derivatives able to bind covalently to proteins and nucleic acids. It is possible that liver damage, seen in some rats given iv injections of pneumotoxic pyrrolic esters, might have been due to metabolites of the latter formed in the liver (Mattocks & Driver, 1983). Segall et al. (1985) have identified trans-4-hydroxy-2-hexenal in an in vitro mouse liver microsomal system metabolizing the PA senecionine and suggested that it might have been formed from the alkaloid via a pyrrolic intermediate. The compound is capable of causing liver damage and might contribute to the acute hepatotoxicity of senecionine and other alkaloids. However, this has not been proved, and the highly reactive and toxic primary pyrrolic metabolites from PAs are themselves capable of causing the known hepatotoxic effects of these alkaloids.

5.2 Toxic Actions of Pyrrolic Metabolites

Pyrrolic derivatives prepared chemically from PAs, as well as some analogous compounds, have been tested in experimental animals and in vitro systems, and shown to have a variety of toxic actions.

5.2.1 Animals

5.2.1.1 Pyrrolic esters (dehydro-alkaloids)

Dehydro-pyrrolizidine alkaloids are very reactive and their effects in vivo are largely confined to the first tissues they encounter. When given orally to rats, they are destroyed almost immediately in the aqueous acid of the stomach and show no toxic action. When given ip, they cause severe local irritation and peritonitis (Mattocks, 1968a; Butler et al., 1970); subcutaneous injection leads to skin lesions (Hooson & Grasso, 1976). After iv injection of pyrroles, such as dehydromonocrotaline (monocrotaline pyrrole), into the tail veins of rats, the toxic injuries appear principally in the lungs. Depending on the dose, these include vascular lesions and pulmonary oedema (Plestina & Stoner, 1972); a progressive alveolar proliferation similar to that produced by very much larger doses of the parent alkaloid (Butler et al., 1970) and pulmonary hypertension (Hilliker et al., 1983). Dehydromonocrotaline does not require further metabolism to express its pneumotoxicity, and it is rapidly rendered inactive after exposure to aqueous media (Bruner et al., 1986). Similar pneumotoxicity is produced by totally synthetic pyrrolic esters having a simpler structure but the same type of chemical reactivity as the alkaloid derivatives (Mattocks & Driver, 1983), thus confirming the chemical mechanism of this action.

Injections of dehydro-pyrrolizidine alkaloids or synthetic analogues into mesenteric veins of rats lead to liver damage after smaller doses than the alkaloids themselves (Butler et al., 1970; Shumaker et al., 1976). The liver damage differs somewhat from the alkaloid damage, consistent with the toxin being introduced via the hepatic vascular system rather than being produced within the hepatocytes, as is the case with the alkaloids. Nevertheless, the progressive liver lesions are very similar to those produced by PAs (Butler et al., 1970). The lung damage after tail vein injections bears a closer resemblance to pyrrolizidine damage, since the latter is also believed to be caused by metabolites entering the lungs via the bloodstream (Barnes et al., 1964).

5.2.1.2 Pyrrolic alcohols (dehydro-necines)

Dehydroheliotridine (Fig. 6), a secondary pyrrolic metabolite from heliotridine-based PAs, such as heliotrine and lasiocarpine, is less acutely toxic than its parent alkaloids; it has an LD_{50} (7 days) of about 250 mg/kg body weight in mice (Percy & Pierce, 1971). Its effects on 14-day-old rats were studied by Peterson et al. (1972). All rats given ip doses of 0.4 mmol/kg body weight survived, but a dose of 0.6 mmol/kg killed most animals within 10 days. Toxic effects were mainly found in rapidly developing tissues. In young rats, it caused fur loss, tooth defects, and atrophy of hair follicles, gut mucosa, spleen, thymus, testis, and bone marrow. The lungs were not affected. Pathological effects in the liver were confined to necrosis of isolated cells and antimitotic action, which was manifested as a mild megalocytosis (development of giant hepatocytes) in rats surviving 4 weeks or more. The persistent antimitotic action of dehydroheliotridine and of its parent alkaloid lasiocarpine in the liver of rats was investigated by Samuel & Jago (1975), who located the mitotic block as being either late in the DNA synthetic (S) phase or early in the post synthetic (G2) phase of the cell cycle.

Dehydroheliotridine is also carcinogenic. Peterson et al. (1983) showed that rats given 9 ip injections of this compound (60 - 76.5 mg/kg body weight) over 23 weeks had a shorter life span and suffered a significantly higher incidence of tumours than control rats. The authors concluded that dehydroheliotridine is responsible for some, or possibly all, of the carcinogenicity of its parent alkaloids.

Dehydroheliotridine was found to be teratogenic when given ip to female hooded rats on the 14th day of pregnancy. A dose of 40 mg/kg body weight produced effects similar to those produced by the alkaloid heliotrine at a dose of 200 mg/kg (Peterson & Jago, 1980). For the immunosuppressant activity of this compound, see section 6.4.10.

The toxic actions of dehydroretronecine (DHR) (Fig. 7) when given sc to rats are similar to those of dehydroheliotridine (Hsu et al., 1973; Shumaker et al., 1976). Repeated large doses also caused ulceration of the glandular stomach. Daily sc doses (4 mg/kg body weight), administered to rats for 1 week, caused lung damage leading to right ventricular hypertrophy (Huxtable et al., 1978). DHR was carcinogenic when applied repeatedly to mouse skin (Johnson et al., 1978; Mattocks & Cabral, 1982).

5.2.2 Cell cultures

Dehydroheliotridine and dehydrosupinidine both have an inhibitory action in cultures of KB cells (human epidermoid carcinoma of the nasopharynx) with ED_{50} concentrations of 10^{-4} mol and 10^{-5} mol, respectively (Culvernor et al., 1969).

Bick & Culvenor (1971) found dehydroheliotridine (DHR) to be considerably more effective than the alkaloid heliotrine in suppressing cell division and causing chromosome breaks, in cultures of leukocytes from the marsupial Potorus tridactylus; at a concentration of 6×10^{-5} mol, the mitotic index was zero, and more than half the cells had disintegrated. In a study by Mattocks & Legg (1980), dehydroretronecine and several synthetic analogues completely inhibited cell division in a cultured rat liver cell line at a concentration of 10^{-4} mol. Ord et al., (1985) found that DHR induced sister chromatid exchange in human lymphocytes without the need for metabolic activation. Analogous pyrroles with only one functional (reactive) group were much less effective. DHR was also weakly active in inducing mutations in the Salmonella typhimurium base substitution strain, TA92, and gave positive results in an in vitro cell transformation test using a culture derived from hamster kidney cells (Styles et al., 1980).

The toxicity of the pyrrolic ester, dehydromonocrotaline, for cultures of mouse fibroblasts was studied in vitro by Johnson (1981). The level of exposure was approximately 1 ng per cell. Cell death was preceded, first by the swelling and disruption of organelles, including mitochondria, and then by the rupture of plasma membranes with the release of cell components.

Bick et al., (1975) investigated whether the effects of PAs on leukocyte cultures of Potorus tridactylus were due to pyrrolic metabolites. Levels of dihydropyrrolizines, which could be demonstrated in the culture media, were insufficient to account for the observed effects of heliotrine, lasiocarpine, and monocrotaline on the cells, but the actual amounts formed within the cells may have been higher than those observed.

5.2.3 Possible participation of membrane lipid peroxidation

Distinct increases in NADPH- and ascorbate-dependent peroxidation of microsomal membrane lipids were found in rats given heliotrine subcutaneously (300 mg/kg body weight) (Savin 1983). The primary biochemical interactions and cellular macromolecular targets for the pathogenesis of PA-induced toxicity remain unidentified.

5.3 Chemical and Metabolic Factors Affecting Toxicity

The toxicity of an alkaloid depends on the extent to which it is converted into active metabolites and on the disposition and reactivity of these metabolites, once formed.

5.3.1 Structural features of a toxic alkaloid

The essential structural features of a hepatotoxic PA (Fig. 8) are:

(a) a 1-hydroxymethylpyrrolizidine ring system unsaturated in the 1:2-position, with preferably a second hydroxyl group in the 7-position;

(b) esterification of at least one of the hydroxyls, though toxicity is much greater when both hydroxyls are esterified; and

(c) ester groups that are resistant to enzymic hydrolysis, which usually means that there is a high degree of chain branching in the acid moiety.

The above requirements apply to natural PAs but, strictly speaking, only the right hand (pyrroline) ring is essential, being the ring that is metabolized to a pyrrole derivative. Thus, esters of 2,3-bis-hydroxymethyl-1-methyl-3-pyrroline (synthanecine A) (Fig. 9) have pyrrolizidine-like hepatotoxicity (Mattocks, 1971a; Driver & Mattocks, 1984).

Structural requirements for N-oxides are the same as those for the hepatotoxic alkaloids. However, it is important to note that a PA N-oxide is not hepatotoxic itself; toxicity depends on it being reduced to the corresponding basic alkaloid, chiefly in the gut (Mattocks 1971c), but possibly in other organs, such as the liver (Powis et al., 1979).

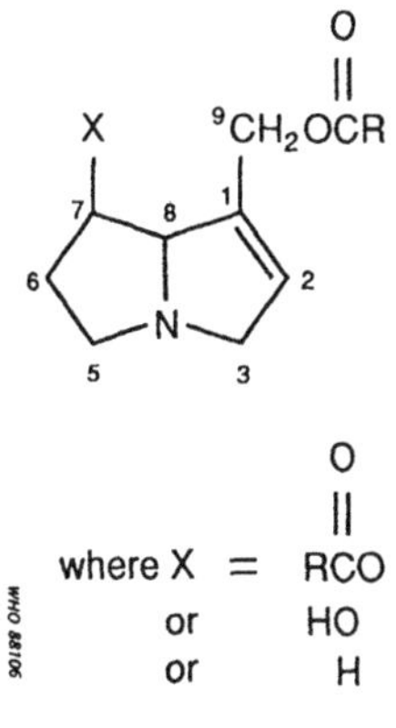

Fig. 8. Essential structural features of a hepatotoxic pyrrolizidine alkaloid.

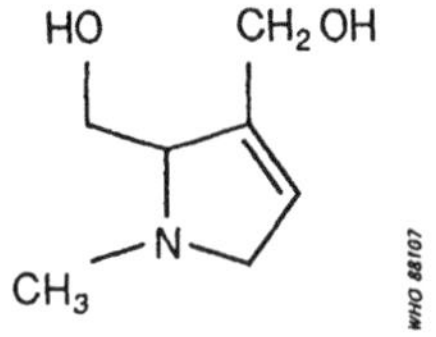

Fig. 9. Structure of synthanecine A.

5.3.2 Activation and detoxication

Factors affecting the proportion of an ingested alkaloid that is converted into toxic metabolites in an animal include the following:

(a) **Lipid solubility**

Highly water-soluble alkaloids (such as indicine) are easily excreted and have low toxicity. Alkaloids that are more lipophilic are more open to activation by liver microsomes (Mattocks, 1981a).

(b) Subceptibility to hydrolysis

This is determined by the molecular structure and conformation of the alkaloid (Mattocks, 1981a,b). If the alkaloid is open to esterase attack, it may be largely detoxified by hydrolysis.

(c) Susceptibility to N-oxidation

The relative amounts of an alkaloid converted by hepatic microsomal enzymes to N-oxide and to pyrrolic metabolites depends on its molecular structure and conformation (Mattocks & Bird, 1983). N-oxidation represents a detoxication pathway (Mattocks, 1972b).

5.3.3 Factors affecting the toxicity of active metabolites

5.3.3.1 Reactivity of the metabolite

Toxic metabolites are formed in liver cells. Primary pyrrolic metabolites (dehydro-alkaloids) are very reactive and, thus, are quickly hydrolysed or deactivated by reaction with cell constituents. To damage tissues other than the cells in which they are formed, active metabolites must cross the cell membrane and survive while being transported in the bloodstream. The more stable pyrrolic metabolites, such as dehydromonocrotaline from the alkaloids monocrotaline, are able to reach, and become bound to, lung tissue (Mattocks, 1973). Thus, monocrotaline frequently damages the lungs, whereas retrorsine, which yields a more reactive pyrrolic metabolite, normally does not.

Secondary metabolites (pyrrolic alcohols, e.g., dehydroretronecine), formed by the hydrolysis of primary pyrrolic metabolites, are water soluble, relatively stable compounds that can become more widely distributed throughout the body or excreted; these are not acutely toxic.

5.3.3.2 The number of reactive groups

The toxicity of a pyrrolic alkylating agent is affected by the number of reactive ester or hydroxyl groups (1 or 2) present as the following examples show:

(a) Many pyrrolic esters can cause acute lung damage when given iv to rats, but only bifunctional ones also cause delayed effects on the lungs (Mattocks & Driver, 1983).

(b) Bifunctional pyrrolic alcohols are more effective inhibitors of mitosis in cultured cells than monofunctional pyrroles (Mattocks & Legg, 1980).

(c) Bifunctional pyrrolic alcohols are much better inducers of sister chromatid exchange (SCE) in lymphocytes than monoalcohols (Ord et al., 1985).

Reasons for these differences might be that the bifunctional pyrroles are able to crosslink macromolecules or simply that they can bind more strongly to target molecules.

5.4 Metabolites Associated with the Biological Actions of Pyrrolizidine Alkaloids

5.4.1 Acute hepatotoxicity

The following is good evidence that acute liver necrosis is caused by primary pyrrolic ester metabolites (dehydro-alkaloids):

(a) The liver, in which these metabolites are formed, is the only organ exposed to them in relatively high concentrations.

(b) There are good correlations between amounts of pyrroles bound to liver tissue and acute hepatotoxicity (Mattocks, 1973).

(c) Pyrrolic alcohols are not acutely hepatotoxic, even when given to animals in very large amounts.

(d) Pyrrolic esters injected iv into the liver are much more acutely hepatotoxic than the parent alkaloids (Butler et al., 1970).

It is possible that other metabolites, such as 4-hydroxy 2,3-unsaturated aldehydes, might also contribute to the acute hepatotoxicity of some PAs (Segall et al., 1985). However, this has still to be confirmed.

5.4.2 Chronic hepatotoxicity

The persistent antimitotic action on the liver that leads to the formation of giant hepatocytes can be produced both by pyrrolic ester metabolites, such as dehydromonocrotaline (Hsu et al., 1973), and by pyrrolic alcohols, such as dehydroheliotridine (Peterson et al., 1972). Both kinds of metabolites can lead to similar alkylation products and both are likely to be present in the liver when the alkaloids are metabolized. Thus, either could be responsible for chronic hepatotoxic effects. However, the antimitotic action alone is not sufficient. It must be accompanied or followed by a stimulus of cell division. This may be provided by the acute necrotic effect of primary pyrrolic metabolites or by any other cause of acute liver injury that leads to tissue regeneration. In very young animals, the stimulus can be the enhanced rate of replication that already exists in them.

5.4.3   Pneumotoxicity

Characteristic pyrrolizidine lung damage is produced by iv injections of pyrrolic ester metabolites, which are effective at much lower doses than the parent alkaloids. The latter are not metabolized in lung tissue; thus, lung damage from PAs is believed to be due to pyrrolic esters reaching the lungs from the liver (Butler et al., 1970). Chronic lung damage appears to be caused by bifunctional rather than by monofunctional pyrrolic alkylating agents (Mattocks & Driver, 1983) (section 5.3.3.2).

There is some evidence that pyrrolic alcohol metabolites might also be able to contribute to chronic (but not acute) pneumotoxicity (Huxtable et al., 1978).

5.4.4   Toxicity in other tissues

Chronic heart damage including right ventricular hypertrophy is a consequence of pyrrolizidine lung damage (pulmonary hypertension) (Hayashi et al., 1967). Brain damage is attributed to ammonia intoxication secondary to severe pyrrolizidine liver injury (Hooper, 1972). This view has been contested and some PAs are known to have direct effects on the central nervous system (section 6.4.3). There is no evidence that PAs are appreciably metabolized in tissues other than the liver. Thus, damage to other organs is probably due to metabolites transported from the liver. For example, in the relatively uncommon cases of chronic kidney damage after pyrrolizidine intoxication (Hooper, 1974; Hooper & Scanlan, 1977) megalocytosis in this organ suggests that pyrrolic metabolites (either ester or alcohol) are involved. Overall, patterns of disease, as observed in extra-hepatic sites, are consistent with a "spillover" effect of the pyrroles produced in the liver (Hooper, 1978). Toxicity of an alkaloid reflects its rate of metabolism to a pyrrole (Tuchweber et al., 1974) and so the spillover effect is likely to be more evident at higher doses. Studies of Culvenor et al. (1976a) suggest that the PAs that are hepatotoxic for rats should also be pneumotoxic when administered at higher doses. In acute poisoning, the hepatotoxic effects could outweigh the pneumotoxic effects or those on other organs, to such a degree that the latter are not manifested. Variation in expression of disease (primarily hepatic or extra-hepatic) also depends on the reactions of host tissues in different species of animals, in addition to the quantities of the pyrroles (Hooper, 1978). The sensitivity of the blood vessels might explain severe interstitial pneumonias in some animals, or severe nephroses in pigs (McGrath et al., 1975).

5.4.5 Carcinogenicity

The pyrrolic alcohols dehydroretronecine and dehydroheliotridine are known carcinogens (Johnson et al., 1978; Peterson et al., 1983), whereas the pyrrolic esters dehydromonocrotaline and dehydroretrorsine are only carcinogenic in conjunction with a tumour promotor (Mattocks & Cabral, 1979, 1982). This suggests that the more persistent secondary metabolites (pyrrolic alcohols) might account for the rather weak carcinogenicity of some PAs.

5.4.6 Antitumour activity

Some PAs and their N-oxides are active as tumour inhibitors in test systems (Culvenor, 1968; Suffness & Cordell, 1985). Indicine N-oxide, in particular, showed high activity against B16 melanoma, mammary xenograft, M5076 sarcoma, P388 leukaemia, and Walker 256 carcinoma. In clinical studies, indicine N-oxide has shown significant activity against some forms of leukaemia, with dosage limited mainly by myelosuppression and sometimes by hepatotoxicity. It is tempting to suppose that this action is related to the powerful antimitotic action of their pyrrolic metabolites, even though some of these alkaloids and derived pyrroles are themselves carcinogenic. On the other hand, there is evidence suggesting that indicine N-oxide owes it activity to a property of the compound itself rather than to the pyrrolic metabolites, which could be formed through reduction to indicine (Powis et al., 1979). The evidence, that indicine is less effective than indicine N-oxide, is not conclusive and other structure-activity data (Milkowsky, 1985) point to a need for a structural capability to form a pyrrolic metabolite. It is also possible that indicine N-oxide is converted directly to dehydroindicine by mitochondrial enzymes in liver or tumour cells, since the type of reaction required has been observed in the mitochondrial metabolism of the N-oxides of tryptamine alkaloids and certain methylated amino acids (Fish et al., 1956; Smith et al., 1962).

5.5 Prevention and Treatment of Pyrrolizidine Poisoning

There is no known way to prevent pyrrolizidine liver damage, once a hepatotoxic dose of the alkaloid has been ingested. A number of dietary regimes have been found to partially protect animals (chiefly rodents) from the acute effects of subsequent alkaloids ingestion. None of these are of any practical use for preventing pyrrolizidine intoxication in livestock. Furthermore, chronic toxic effects in the liver or in other organs are sometimes more severe in animals receiving higher doses of alkaloids after being protected against acute hepatotoxicty.

5.5.1 Modified diets

The mechanism of action of modified diets is not clear, but they may be associated with the decreased metabolic activation of the alkaloids. Some examples follow:

(a) A protein-rich diet can give some protection to rats against Senecio jacobaea alkaloids (Cheeke & Gorman, 1974). Rats fed a high casein diet survived longer than rats given a normal diet, when poisoned with retrorsine or riddelline, but the survivors were more liable to develop liver tumours (Schoental & Head, 1957). However, whether this was simply due to a prolongation of life of the animals by the diet is open to question.

(b) Male rats previously fed a sucrose-only diet for 4 days were considerably protected against the acute hepatotoxicity of retrorsine (LD_{50} 120 mg/kg body weight compared with 34 mg/kg in normal rats). However, lung damage, rare in control rats, was frequently seen in "protected" rats given high doses of retrorsine (Mattocks, 1973).

(c) Restriction of feed intake to 40% of normal attenuated the increase in lung weight and lavage protein concentration in cell-free bronchopulmonary lavage fluid and abolished the right ventricular hypertrophy in monocrotaline-treated rats. Furthermore, the percentage of diet-restricted animals that survived was significantly higher than that in animals that had eaten ad libitum up to day 28, but, from this time onwards, there was no difference. Alterations of dietary sodium intake alone did not affect the results of monocrotaline-induced toxicity (Ganey et al., 1985).

5.5.2 Pretreatment to enhance the detoxication of active metabolites

Treatments that have afforded some protection against pyrrolizidine hepatotoxicity (probably by increasing the liver level of sulfydryl compounds, which are known to react with pyrrolic metabolites) (White, 1976) include the following:

(a) Pre-treatment of rats with mercaptoethylamine (150 mg/kg body weight ip) partially protected rats against the acute hepatotoxicity of monocrotaline given 15 min later (Hayashi & Lalich, 1968); it gave no protection when administered 2 h after the alkaloid. Mercaptoethylamine, when given orally (300 mg/kg body weight) at the same time as the lasiocarpine, also increased the resistance of rats to the alkaloid (Rogers & Newberne, 1971).

(b) Cysteine (1% in the diet) partially protected rats against Senecio jacobaea alkaloids (Buckmaster et al., 1977) and mice against monocrotaline (Miranda et al., 1981c).

(c) The antioxidant ethoxyquin fed at a level of 2.5 g/kg diet to female mice for 38 days, increased the liver thiol concentration and raised the acute LD_{50} of monocrotaline, given ip on the 10th day, to 364 mg/kg compared with 243 mg/kg in control mice (Miranda et al., 1981a).

(d) Rats or mice also had increased resistance to acute pyrrolizidine hepatotoxicity when fed the antioxidant butylated hydroxyanisole (BHA) (up to 7.5 g/kg diet) (Miranda et al., 1981c, 1982a,b; Kim & Jones, 1982).

(e) Heliotrine-induced toxicity can be modified by the co-administration of cupir (a copper-containing complex) at a level of 1 mg/kg per day for 20 days. It prevented the exit of hepatic cytosolic enzymes into the blood and improved all the energy reactions studied in the mitochondria of heliotrine-intoxicated rats (Yuldasheva & Sultanova, 1983). Inhibition of lipid peroxidation by cytoplasmic copper was shown later (Wittig & Stephen, 1964). Savin (1983) found that lethality to rats of heliotrine (300 mg/kg sc) was completely prevented by co-administration of α-tocopherol (6 ml/kg ip).

(f) Rats pre-treated with ip doses of zinc chloride (72 μmol/kg body weight) had increased resistance to the hepatotoxicity of Senecio jacobaea alkaloids, as assessed by histology and enzyme measurements (Miranda et al., 1982c). The zinc treatment increased the liver level of metallothionein, a sulfhydryl-rich protein that might react with pyrrolic metabolites.

Metabolic inhibitors of the microsomal P450 mixed-function oxidase system, SKF 525A, metyrapone, and allylisopropyl acetamide, which inhibit the formation of toxic pyrroles in the liver, have been tried successfully in the prevention of the toxic effects of monocrotaline in rats (Eisenstein & Huxtable, 1979). The use of P450 inhibitors was stated to show "potential therapeutic promise". However, this would seem impracticable considering that, at least in the rat, PAs undergo a high rate of metabolism commencing a few minutes after ingestion (Mattocks, 1972b). In some instances, they have been known to lead to an increase in toxicity, e.g., with lasiocarpine as reported by Tuchweber et al. (1974).

5.5.3 Other treatments

Lanigan & Whittem (1970) attempted, unsuccessfully, to protect sheep against Heliotropium europaeum poisoning by treating them with cobalt, in the hope that this would enhance the vitamin B_{12}-mediated detoxication of the alkaloids in the rumen (Dick et al., 1963).

Lanigan et al. (1978) found that the resistance of sheep to dietary Heliotropium europaeum was increased by giving them large daily doses of the antimethanogenic drug, iodoform. However, Swick et al., (1983) found that Senecio jacobaea alkaloids were not detoxified by incubation for 48 h with sheep rumen fluid in vitro.

6. EFFECTS ON ANIMALS

6.1 Patterns of Disease Caused by Different Plant Genera and of Organ Involvement in Different Species

The most important genera of PA-containing plants listed in section 3.1 are all hepatotoxic. Among these, Crotalaria spp. cause damage in the broadest range of tissues in most domestic species. In pigs, they are known to be severely nephrotoxic (Peckham et al., 1974; McGrath et al., 1975; Hooper & Scanlan, 1977). Some species are known to be pneumotoxic for horses (Watt & Breyer-Brandwijk, 1962; Gardiner et al., 1965), cattle (Sanders et al., 1936; Berry & Bras, 1957), sheep (Laws, 1968), and pigs (Peckham et al., 1974; Hooper & Scanlan, 1977), as well as hepatotoxic.

Although several Crotalaria spp. are known to be pneumotoxic for horses (Gardiner et al., 1965), C. retusa is an exception. It is an important cause of disease in horses in northern Australia (Hooper, 1978) and has been shown to be pneumotoxic for pigs in the same area (Hooper & Scanlan, 1977); yet it produces only hepatic disease in horses (Rose et al., 1957a,b).

Similarly, Senecio spp. are primarily hepatotoxic, but S. jacobaea has been demonstrated to be pneumotoxic for pigs (Harding et al., 1964), though it could probably be an inconsistent change (Bull et al., 1968). This plant is also known to cause pulmonary disease in rats and mice (Hooper, 1974). However, there are no reports of its affecting the lungs in cattle, sheep, horses, or chicken. Renal megalocytosis and mild nephrosis are reported in most species poisoned with S. jacobaea (Harding et al., 1964; Bull et al., 1968). Heliotropium spp., Amsinckia spp., and Echium spp. are all mainly hepatotoxic.

Roitman (1983) summarized the pattern of organ involvement observed in man and different species of farm and experimental animals affected by pyrrolizidine alkaloids (Table 8). Even within a single species, the nature of a toxic effect, as well as the organ affected, can be altered by changing the dose rate and duration.

6.2 Field Observations - Outbreaks in Farm Animals

The veterinary problem of PA toxicity has been reviewed by Bull et al. (1968) and McLean (1970). Mattocks (1986) listed the cases of livestock poisoning and feeding trials since 1968, and cited relevant literature. Peterson & Culvenor (1983) produced a useful and comprehensive table of the plant species known or suspected of causing natural outbreaks of

Table 8. Animal species and organs affected by pyrrolizidine alkaloids[a]

Species	Liver	Lung	Kidney	Heart	Pancreas	Gastric mucosa	Muscle
Man	+						
Monkey	+	+	+	+			
Horse	+	+	+				
Pig	+	+	+		+	+	
Sheep	+	+	+				
Goat	+	+					
Cattle	+	+					
Dog	+						
Mouse	+		+				
Rat	+	+		+		+	
Chicken	+	+	+			+	+
Turkey	+	+					+

[a] From: Roitman (1983).

poisoning in each animal species. The influence of factors such as species, age, sex, and diet, on toxicity is also reviewed in the same paper.

The first cases of pyrrolizidine poisoning were described in cattle as early as 1903 (Gilruth, 1903). Since then, there have been numerous reports from most parts of the world, of poisoning among farm animals caused by grazing or feeding on PA-containing weeds (Bull et al., 1968; Mattocks, 1986). One of the first clues to the etiology of the human disease in Jamaica, came from a study in which calves fed with Crotalaria fulva (Bras et al., 1957) developed characteristic venoocclusive disease in the liver.

Laws (1968) described a field outbreak in sheep in a herd of 100 adult merino ewes, which developed within 2 weeks of moving into a coastal farm in Australia, where they grazed Crotalaria mucronata. The etiology was confirmed by feeding the plant to 6 sheep, 4 of which died within 24 h of feeding, with severe pulmonary oedema. However, the rapidity of poisoning and the atypical lung lesions suggest that possibly a toxin other than pyrrolizidine alklaoids was also present.

An outbreak of Crotalaria retusa poisoning was observed in a piggery near Darwin, Northern territory, Australia (Hooper & Scanlan, 1977) containing approximately 350 sows. It was caused by feeding sorghum contaminated by Crotalaria seeds at the rate of about 0.1% by weight for about 3 weeks, and at a rate of about 0.05% for a further week. The disease was indicated by reduced body weight gain and inappetence. The dominant pathological features at autopsy were severe nephrosis with chronic uraemia, and to a lesser degree, severe diffuse interstitial pneumonia. Both were accompanied by microscopic disease in the liver, and both the liver and kidney showed megalocytosis.

Walker & Kirkland (1981) reported outbreaks in the Hunter river valley of New South Wales in Australia, in cattle that had been grazing a pasture in which Senecio lautus was growing. There were sporadic deaths among the cattle as well as two protracted outbreaks affecting calves 3.5 months of age and older animals, in which groups of 3 - 16 head of cattle died in addition to sporadic deaths of animals over periods of 1 and 6 months. Clinical signs were weakness, emaciation with recumbency, aimless wandering and ataxia, which suggested neurological involvement. At autopsy, the liver showed characteristic megalocytosis, periportal fibrosis, and focal necroses. An aged animal had cirrhosis. The etiology of S. lautus was proved in feeding studies on 3 calves.

Knight et al. (1984) reported the deaths of 10 horses during a 3-year period after being fed hay from the same pasture. The animals became sick in the spring after being fed only the suspect hay throughout the winter. The hay was found to be contaminated with Cynoglossum officinale (hound's tongue), which contained two PAs (heliosupine and echinatine) in much higher quantities than is generally reported in Senecio species. The animals developed weight loss, icterus, ataxia, and symptoms of hepatic failure. At autopsy, there was diffuse severe megalocytosis, biliary hyperplasia, and fibrosis. The C. officinale was proved as the etiological factor in a feeding trial on a 14-year-old pony, that developed clinical features and pathological changes in the liver suggesting PA poisoning. The signs of PA toxicity in horses are mostly neurological, though non-specific gastric and oesophageal lesions have also been reported (McLean, 1970) (section 6.4.3). Rose et al. (1957a,b) described a disease in which the dominant symptom was "compulsive" walking in a straight line. It occurred in areas where Crotalaria retusa was growing, and was ascribed to a steep rise in blood-ammonia levels, which accompanied chronic liver failure.

Farm animals differ widely in their sensitivity to PAs, sheep and goats being fairly resistant, cattle and horses, less so, and poultry and pigs, rather sensitive (section 6.4.1.2). Sheep are not immediately affected and generally survive one season, after feeding on heliotrope and Senecio (Bull & Dick, 1959). During the second season of feeding, they die of neurological symptoms caused presumably by rising blood-ammonia levels associated with chronic liver disease, or with haemoglobinuria and very high copper levels in the blood (Bull et al., 1956, 1958) (section 6.4.1.2). With Crotalaria, the lung seems to be the target organ (Hooper, 1978). Similar acute responses to a single feed of the plant Crotalaria spectabilis were described in cattle by Emmel (1948) and in the chicken by Piercy & Rusoff (1946).

Poisoning of cattle in northwestern USA has reached such proportions that it has become a considerable economic problem (Johnson, 1982). Culvenor (1985) has reviewed the problem of livestock losses due to PA toxicosis in Australia, where it has been estimated that about 10 million sheep are exposed to heliotrope and Echium plantagineum (Paterson's curse) to a greater or lesser extent and may suffer a shortening of their productive life by as much as two years. Most of the PA-containing plants are reported to grow in fallow fields and pastures and thrive particularly in a dry climate or following periods of drought. However, instances of cattle poisoning have been reported from most parts of the world, including countries with temperate or cold climates, which do not ordinarily suffer drought. Three herds of cattle have been reported to have been affected in the alpine/subalpine region of Switzerland, after they grazed pastures that had Senecio alpinus growing on them. Nine different hepatotoxic PAs were found in the weed, the main one being seneciphylline. Analyses of urine samples from one cow confirmed the presence of PA metabolites. Several cows had to be slaughtered, because of cirrhosis of the liver (Luthy et al., 1981).

6.3 Studies on Farm Animals

There are several reports of the production of disease characteristic of PA toxicity in farm animals, by feeding them PA-containing plants.

Senecio jacobaea (tansy ragwort) is a weed that commonly grows in pastures and has been the cause of extensive livestock losses in the United Kingdom (Forsyth, 1968) and the USA (Johnson, 1982). Extensive studies on this plant have been carried out using a variety of farm animals. Dickinson et al. (1976) fed tansy ragwort to cows through a rumen canula at the rate of 10 mg/kg body weight per day, for 2 weeks. Liver biopsies showed characteristic megalocytosis and fibroplasia, and autopsy also showed centrilobular necrosis. A PA, jacoline, was found in the milk, but when the calves were bucket fed the milk, there were no detectable effects on them. Thorpe & Ford (1968) made similar observations in 5 calves fed ragwort in their diet. Animals eventually dying of toxicity showed characteristic necrosis, megalocytosis, and veno-occlusive lesions. In a study by Goeger et al. (1982), goats were fed dried ragwort mixed in the diet at 250 g/kg. Four of the 11 goat kids and lactating dairy goats died. Characteristic megalocytosis was seen in the liver. Goats are more resistant to tansy ragwort toxicosis than cattle and horses, the chronic lethal dose for cattle or horses being 0.05 - 0.2 kg ragwort/kg body weight and, for goats, 1.25 - 4.04 kg/kg body weight. The alkaloid levels in the plant, and thus its toxicity, varies with season and locality.

Hooper & Scanlan (1977) studied the long-term effects of feeding very low levels of ground C. retusa seeds, mixed with the feed, to pigs and chickens. Seven groups of 4 pigs each (sex not mentioned) bred from Saddleback-large white cross sows and Large White or Landrace boars, were maintained on diets containing 0 (control), 0.004%, 0.01%, 0.02%, 0.05%, 0.1%, or 0.5% body weight ground seeds. Another 8 pigs were fed a diet containing 0.1% C. retusa for 21 days followed by 0.05% for 7 days and then kept on a C. retusa-free diet. Pigs either died or were killed when moribund, or at the end of 136 days of feeding.

In a second study, groups each of 4 2-week-old chickens were fed diets containing 0 (control), 0.005%, 0.01%, 0.05%, 0.1%, and 0.5% ground seeds of C. retusa. Chickens fed 0.5% started dying 12 days after the commencement of feeding and were all dead by day 45. Five out of 8 birds fed 0.1% or more died between days 22 and 56. No deaths occurred in animals fed 0.01% - 0.005%.

In the study on pigs, all animals died between days 63 and 107 except for 2 that survived 136 days. In these animals, pulmonary disease was the main cause of death. Hepatic and renal megalocytosis was seen in almost all animals in both the field outbreak and study group. The lungs showed extensive consolidation and oedema. Besides megalocytosis in the glomeruli and tubules, the kidneys showed glomerular atrophy and tubular necrosis. In the study on poultry, the major disease was hepatic necrosis of irregular distribution. The kidneys showed mild megalocytosis.

In the above study, the low levels of contamination that produced serious disease are worthy of note.

Johnson & Molyneux (1984) fed 55 cattle, by gastric lavage, with hay mixed with threadleaf groundsel (Senecio douglasii var. longilobus), which grows commonly in the pastures of southwestern USA. The PA dosage in different groups ranged from 5 to 40 mg/kg, daily, and the total intake ranged from 80 to 284 mg/kg body weight. The groups were fed for periods ranging from 2 to 20 days. One hundred percent mortality occurred in 3 out of 9 groups, each consisting of 2 - 8 calves, receiving doses of 13 mg/kg or more. Mean survival time was generally inversely proportional to the dosage received. All sick calves had typical clinical signs of seneciosis. At autopsy, the principal lesion was seen in the liver and consisted of swelling of the hepatocytes, necrosis, biliary hyperplasia, and marked fibrosis. The estimated minimum lethal dose of the PA was 13 mg/kg body weight for 15 days, or a total intake of approximately 200 mg PA/kg, over a 15-day period. Cattle that consumed up to 600 mg PA/kg, in hay, in a 20- to 100-day period, were unaffected or only slightly affected. The authors concluded that the time-dose relationship for PA toxicosis in cattle is important and that there is a threshold level that must be exceeded for the toxicosis to develop.

A similar study was conducted by Johnson et al. (1985) in which the dry whole or ground leaves of <u>Senecio riddelli</u>, mixed with the feed, were fed to calves in gelatin capsules or by gavage. Forty-two female Hereford calves were divided into 3 groups. One group of 12 was fed the leaves mixed with alfalfa hay feed estimated to have 20 - 40 mg PA/kg body weight per day over a 20-day period in different regimes, receiving a total of 400 - 800 mg PAs per animal. The second group of 12 animals received the plant packed into gelatin capsules, receiving an estimated PA content of 10 - 20 mg/kg body weight, daily, over 20 days, with a total of 200 - 400 mg per animal. The third group of 18 animals was administered (by gavage) finely ground leaves in a water slurry at various PA dosages ranging from 10 to 60 mg/kg body weight per day and a total of 200 - 500 mg over 20 days.

Calves that received 10 mg PA/kg body weight per day for 20 days did not develop clinical signs of disease or show any changes in serum-enzyme. However, feed containing the plant that provided 15 - 20 mg PA/kg per day or more, administered by gavage or fed in capsules, resulted in high mortality. Malaise, depression, erratic or unprecedented behaviour, aimless walking, and ataxia, were observed in the affected calves; diarrhoea with tenesmus and rectal collapse were frequently observed. The feed intake decreased progressively and was negligible terminally. The animals that died and those that were moribund or in a state of irreversible wasting, were autopsied. Hepatobiliary lesions were present in all such animals. The most consistent change was portal biliary hyperplasia and periportal fibrosis. Centrilobular or zonal haemorrhage and necrosis were observed in some lobules. Fibrosis of some central veins was common, often encroaching on the lumen, resulting in complete occlusion. Hepatocytes also showed nonspecific changes. Central nervous system changes were present in all animals with clinical signs of seneciosis, consisting mainly of spongy degeneration of the brain.

The plant mixed in the hay ration was eaten slowly and reluctantly and was tolerated at dosages > 20 mg/kg per day, emphasizing that the toxicity depended on the rate at which the dosage was consumed and that mortality was not necessarily dependent on the cumulative dosage.

Burguera et al. (1983) produced the disease in turkey poults by feeding them seeds of C. spectabilis. Simultaneous addition of sodium selenite at doses of 0.1, 5, or 10 mg selenium/kg diet did not provide any protection.

6.4 Experimental Animal Studies

6.4.1 Effects on liver

6.4.1.1 Relative hepatotoxicity of different PAs and their N-oxides

The LD_{50} values for rats, listed in Table 9, are for some of the most commonly used hepatotoxic alkaloids, calculated from data on animals dying from acute haemorrhagic necrosis of the liver, 3 - 7 days after intra-peritoneal administration of a single dose. It is evident that the toxicity varies widely between the alkaloids. The most toxic are certain macrocyclic diesters of retrorsine and the least toxic are the monoesters of heliotridine, retrorsine, and supinine (Mattocks, 1986).

The relative toxicity of N-oxides compared with that of their basic alkaloids depends on the route of administration. The N-oxides of lasiocarpine, monocrotaline, and fulvine were reported to be as toxic as their basic alkaloids (Schoental & Magee, 1959) when administered orally; however, when given by the ip or intravenous (iv) routes to rats, they were much less toxic (Mattocks, 1971c). Similarly, the LD_{50} of retrorsine N-oxide when administered ip to male rats was 250 mg/kg (Table 9) but when given orally, it was 48 mg/kg. This has been explained by the observations on the metabolic pathways of the basic alkaloids and their N-oxides. The PAs or their N-oxides exert toxic effects only after being metabolized to pyrroles by the hepatic microsomal enzymes (section 5.1.1). Hepatic microsomes act directly on the N-oxides (Mattocks & White, 1971b) only after they have been converted to the basic alkaloids (Mattocks, 1986); this mainly occurs in the gut (Mattocks, 1971c; Powis et al., 1979). This matter is of practical importance as the alkaloids are often present as their N-oxides in weeds grazed by farm animals.

6.4.1.2 Factors affecting hepatotoxicity

These factors have been reviewed by Mattocks (1986).

(a) Route of administration

Most studies on LD_{50} values have been carried out using the ip route, and very few experimental data are available on

Table 9. LD_{50}s in male rats after a single intraperitoneal dose of some hepatotoxic alkaloids

Alkaloid	LD_{50} (mg/kg)	Time range (days)	Reference
heliotrine	296	3	Bull et al. (1958)
lasiocarpine	77	3	Bull et al. (1958)
lasiocarpine N-oxide	547	3	Bull et al. (1958)
monocrotaline	175	3	Bull et al. (1968)
retrorsine	34	4 or 7	Mattocks (1972a)
retrorsine N-oxide	250	7	Mattocks (1972a)
senecionine	50	7	Mattocks (1972a)
seneciphylline	77	3	Bull et al. (1968)
senkirkine	220	-[a]	Hirono et al. (1979a)
symphytine	130	-[a]	Hirono et al. (1979a)

[a] Not stated.

toxicity using the oral route. There is a close similarity between the iv data and the ip data. Furthermore, toxicity data on rats administered PAs by the oral route (Schoental & Magee, 1959), including retrorsine, lasiocarpine, heliotrine, and monocrotaline, closely resemble those relating to the LD_{50} values for the same strain administered PAs intraperitoneally (Mattocks, 1972b). Thus, the hepatotoxicity of PAs in rats does not differ very much, irrespective of the route of administration. However, rabbits appear to be less susceptible to PAs in the plant Senecio jacobaea when administered orally than when administered intravenously (Pierson et al., 1977).

(b) Species

Wide differences have been observed in the hepatotoxic effects of PAs and alkaloid-containing plants between different species of both farm animals and laboratory animals, and in the same animal exposed to PAs derived from different plants. Sheep are resistant to PA-containing plants (section 6.2) and when fed Echium plantagineum pellets containing 1.3 g alkaloid/kg as the sole diet for 12 months, over a period of

2 years, showed almost no liver damage (Culvenor et al., 1984). However, adult rats fed the same pellets as only 50% of the diet for 14 days died 4 - 13 weeks later (Peterson & Jago, 1984). Pigs were found to be 5 - 10 times as susceptible to PAs in Crotalaria retusa as chickens (Hooper & Scanlan, 1977). Overall, the approximate ratios of quantities of plant material required to prove toxic in the various species listed are about 200 for the sheep (approximately the same for the goat), 150 for the mouse, 50 for the rat, 14 for cattle (approximately the same for the horse), 5 for the chicken, and 1 for the pig (Hooper, 1978).

Cheeke & Pierson-Goeger (1983) studied the chronic toxicity of Senecio jacobaea for several laboratory animals by feeding the dried plant as a component of a mixed diet. The degree of susceptibility to PA poisoning was compared in terms of the chronic lethal dose of the dried plant as a percentage of the initial body weight among the animals themselves, and with similar data on livestock in other studies. Gerbils, hamsters, and guinea-pigs were resistant to chronic toxicity, gerbils being the most resistant, consuming over 35 times their body weight of the dried plant. Comparison with similar data in other studies indicated that the rabbit (Pierson et al., 1977), Japanese quail (Buckmaster et al., 1977), and goat (Goeger et al., 1982) were resistant, requiring a long-term lethal dose of the plant of 113% or more of the initial body weight, whereas the rat was highly sensitive requiring only 21% (Goeger et al., 1983). Chicks and turkey poults were also susceptible with severe inhibition of growth occurring when there was 5% and 10% contamination of the feed with the plant; survival time was short (Cheeke & Pierson-Goeger, 1983).

In a study by Fushimi et al. (1978), on the carcinogenicity of the flower stalks of Petasites japonicus Maxim in mice and Syrian golden hamsters, species and strain differences were observed, not only with regard to hepatotoxicity, but also with regard to the carcinogenicity of PAs. Mice of ddN, Swiss, and C57BL/6 strains and Syrian golden hamsters were fed on a diet containing young flower stalks of the plant for 480 days. High incidences of lung adenoma and adenocarcinoma were observed in ddN mice, but no significant differences in tumour incidence were observed between the experimental groups of Swiss mice and hamsters and the corresponding control group. No tumours were induced in an experimental group of C57BL/6 strain mice.

These differences have been explained by the differences in the rate of metabolic conversion of PAs to toxic metabolites (pyrroles) by the hepatocyte microsomes in the different animal species (White et al., 1973; Shull et al., 1976; Peterson & Jago, 1984).

The resistance of sheep has been ascribed to destruction of the alkaloids in the rumen by a reductive conversion into non-toxic 1-methylenepyrrolizidine derivatives (Bull et al., 1968; Lanigan, 1971, 1972). It has also been suggested that the resistance of sheep is due to a low level of activation in liver cells (Shull et al., 1976), but this factor was not prominent in some Australian sheep, which were as sensitive as rats to PAs injected intraperitoneally (Hooper, 1974).

Thus, it is possible for ruminants to graze plants containing PAs for a period of months without evident harm, e.g., cattle eating Crotalaria juncea in Africa (Srungboonmee & Maskasame, 1981), but long-term effects may arise in animals exposed over several years.

Considerable differences in LD_{50} values have been reported for the same alkaloids in different species. For example, the LD_{50} for retrorsine varies from 34 mg/kg for male rats to 279 mg/kg for quail and over 800 mg/kg body weight for guinea-pigs (White et al., 1973). Guinea-pigs are also resistant to monocrotaline (Chesney & Allen, 1973a), but not to jacobine or to mixed alkaloids of Senecio jacobaea, which are highly toxic (Swick et al., 1982a).

(c) Sex

Significant differences in the hepatotoxicity of the same alkaloid have been observed between sexes in some species. Male rats are much more susceptible to the acute toxicity of retrorsine or monocrotaline than females (Mattocks, 1972b). Mattocks & White (1973) reported a higher level of metabolic transformation in young male rats to form pyrroles from retrorsine, compared with females (section 4.4). Jago (1971) reported a higher susceptibility in male rats to the chronic hepatotoxic effects of heliotrine, while female rats were more susceptible to lasiocarpine. It is possible that this may be due to the potentiating effect of male sex hormones. Campbell (1957a,b) reported that diethylstilboesterol protects against the effects of seneciphylline and promotes repair of damaged liver in poultry. Protein-deficient rats of both sexes, or female animals pre-treated with testosterone, were more susceptible to monocrotaline (Ratnoff & Mirick, 1949).

(d) Age

Available data on the effects of age are highly conflicting. It has been stated that young rats, particularly suckling animals (Schoental, 1959), are more susceptible than adults to the hepatotoxic effects of some alkaloids (Jago, 1970), such as monocrotaline (Schoental & Head, 1955), and retrorsine and lasiocarpine (Schoental, 1959). Rats, 1 - 4 days old, were far more susceptible to retrorsine and senkirkine than rats aged 25 - 30 days (Schoental, 1970); yet new-born rats (within 1 h of birth) were relatively more resistant to the hepatotoxic effects of retrorsine than 1- to 4-day-old rats (Mattocks & White, 1973). McLean (1970) has critically reviewed the data. In comparing the data on small animals from several studies, newborn and 4-week-old animals appear to have about the same susceptibility as adults. Data for the intervening period obtained by Harris et al. (1957), Schoental (1959), and Hayashi & Lalich (1968) are conflicting, suggesting decreased susceptibility in some studies (Harris et al., 1957 and one series of Schoental's studies, 1959), and increased susceptibility in others (Hayashi & Lalich, 1968 and the second series of Schoental's studies, 1959). Furthermore, Jago (1971) demonstrated that, while rats aged 1 - 2 weeks were more susceptible to the acute effects of heliotrine and lasiocarpine than older rats, sensitivity to the effects of long-term administration of these alkaloids increased with age, after 2 - 3 months.

(e) Diet

Effects of both qualitative and quantitative changes in diet on the hepatotoxicity of PAs have been investigated in several studies. Restriction of protein levels in the diet enhanced the acute hepatotoxic effects of retrorsine in rats (Selzer & Parker, 1951) and the chronic effects of a single dose of orally administered lasiocarpine (Schoental & Magee, 1957) (section 6.4.5.1) as well as the toxicity of PAs in Senecio jacobaea, whereas a high protein diet had a protective effect (Cheeke & Gorman, 1974). Likewise, low lipotrope diet enhanced the toxic effects of orally administered lasiocarpine in pregnant rats and also in the fetal livers (Newberne, 1968). On the other hand, it protected young male rats against the acute toxicity of monocrotaline, because of the reduced metabolic conversion of the alkaloid into pyrrolic metabolites (Newberne et al., 1971, 1974).

Caloric restriction reduced the cardiopulmonary toxicity of a single dose of monocrotaline in rats (Hayashi et al., 1967). This was ascribed to the reduced growth rate in animals on a restricted diet rather than to a reduction in the rate of metabolic conversion of the alkaloid, since dietary restriction started only after administration of the alkaloid. When the animals were put back on the ad libitum feeding regimen, they developed signs of increased toxicity.

A high copper content in the diet has been shown to enhance the toxic effects of PAs (Miranda et al., 1981b). Incorporation of copper sulfate at 50 mg/kg in the diet containing the plant Senecio jacobaea increased the hepatotoxicity in rats, as judged by enzyme measurements. The implications of this observation are obvious if some PA-containing plants being grazed by farm animals also have a high copper content.

Mattocks (1972b) demonstrated the protective effects of sucrose against the acute hepatotoxic effects of retrorsine in rats, if administered for 3 days prior to alkaloid administration (section 5.5.1).

6.4.1.3 Acute effects

Experimental animal studies on the pathological effects of PAs on the liver have been reviewed by Bull et al. (1968) and McLean (1970). Most studies have been carried out on the rat (Schoental & Magee, 1957, 1959; Bull & Dick, 1959; Schoental, 1963; Barnes et al., 1964; McLean et al., 1964; Nolan et al., 1966; Jago, 1969; Butler et al., 1970; Peterson & Jago, 1980), but several other species of animals have been studied including the monkey (Wakim et al., 1946; Rose et al., 1959; Allen & Carstens, 1968, 1971; Allen et al., 1969), turkey (Allen et al., 1963), chicken (Allen et al., 1960), hamster (Harris et al., 1957), mouse, guinea-pig (Chen et al., 1940), quail, cat, rabbit, and pig (Emmel et al., 1935; Hooper & Scanlan, 1977). All animals tested, except the guinea-pig (Chen, 1945), have been found to be susceptible in studies using purified alkaloids and their $\underline{N}$-oxides and crude extracts of PA-containing plants.

Typically, the most common lesion produced in small laboratory animals by doses close to the LD_{50} is a confluent haemorrhagic necrosis in the liver, which appears within about 12 h of exposure and peaks at 24 - 48 h. It is strictly zonal in distribution in different species of animals but may vary within the same animal, depending on the alkaloid used, species, nutritional status, and pretreatment with other chemicals.

Retrorsine produces centrilobular necrosis in the rat, mouse, and guinea-pig, periportal necrosis in the hamster, and focal necrosis in the fowl and in the monkey (White et al., 1973). In the monkey, monocrotaline produces centrilobular necrosis (Allen & Carstens, 1968), but senecionine produces necrosis in the periportal and midzonal areas of the liver lobule (Wakim et al., 1946). Almost simultaneously, or shortly after the development of acute haemorrhagic necrosis of the liver cells, various levels of change appear in the central and sublobular veins of the liver lobules, consisting of subintimal oedema or even necrosis, deposits of fibrin, thrombosis, and occlusion of the lumen, which later becomes organized. While haemorrhagic necrosis is a constant feature, attempts to produce occlusive lesions in the veins of experimental animals have produced variable results (Allen et al. 1967). In man and non-human primates, hepatocellular necrosis and venous occlusion occur simultaneously but, in the rat (McLean et al., 1964), chicken (Allen et al., 1960), and swine (Emmel et al., 1935), the vascular changes follow hepatic necrosis.

Selzer & Parker (1951) produced a lesion comparable to human veno-occlusive disease in albino rats by administrating retrorsine hydrochloride, the active alkaloid of Senecio ilicifolius, as well as the crude plant extract. Four batches of rats were administered alkaloids orally in a single dose of 1 - 1.5 mg/10 g body weight or repeated doses of 5 - 50 mg/kg body weight for 31 days or as single subcutaneous injection of 100 mg/kg body weight. One batch was fed on a diet of Senecio ilicifolius constituting 10% of the ration as crude plant or its extract; the animals lived for 21 - 84 days. Some groups were kept on a normal diet, and others on a diet that was protein-deficient. Animals, administered a single dose orally, developed the earliest degenerative changes in the centrilobular hepatocytes and sinusoidal dilatation, and the vascular lesion appeared after 36 h. Protein deficiency enhanced the toxic effect. Only 5 out of 9 animals administered repeated doses orally showed early centrilobular fibrosis and none showed the vascular lesion, possibly due to scarring.

Bull et al. (1958) studied the effects of a single ip LD_{70} dose of heliotrine (320 mg/kg body weight), lasiocarpine (80 mg/kg), or lasiocarpine $\underline{N}$-oxide (629 mg/kg) in rats of a hooded strain of both sexes. Eighty-one rats were used and 3 rats from each treatment were killed at intervals of 4 - 36 h. Heliotrine produced marked centrilobular necrosis at 24 h, but venous changes were not evident, except for some aggregation of mononuclear macrophages on the endothelium. With lasiocarpine, the hepatic changes were similar, but the necrosis was not clearly centrilobular and, with its $\underline{N}$-oxide, it was midzonal at 34 - 49 h. The earliest toxic effect of the PAs was manifested as a temporary loss of mitochondria at 8 h. The authors concluded that PAs have an early toxic effect on the hepatocytes and that this does not follow vascular injury, as suggested by Davidson's earlier studies (1935).

McLean et al. (1964) administered an aqueous extract of Crotalaria fulva to Wistar rats in a single intragastric dose of 0.8 - 1.5 g/kg body weight. Lesions identical to those of human veno-occlusive disease were produced in the animals by adjusting the dose to permit survival for 8 - 12 days. Loss of cytoplasmic glycogen in the centrilobular cells occurred 3 h following administration. Centrilobular necrosis, which occurred after 24 h, increased with time. The central veins gradually filled up with thickened endothelial cells at about 7 days, later progressing to collagenization. Evidence was presented that the histological occlusion of the central veins was preceded by several days by a functional blocking of the blood flow.

Barnes et al. (1964) observed similar results in rats administered a single oral dose of fulvine N-oxide at 50 mg/kg body weight and studied at intervals of 1 - 4 days after the administration. One hundred and thirty-five rats of both sexes were used. Acute lesions resembling human disease were observed during days 1 - 8. During days 19 - 21, 3 out of 25 animals showed liver damage consisting of some centrilobular haemorrhage and fibrous thickening of the central veins. Of the 78 animals studied at 22 - 44 days, 50% still had centrilobular congestion and some had fibrous thickening of the central veins.

The effects of pyrrolic derivatives of PAs on rats were studied by Butler et al. (1970). Male albino rats of Porton strain were administered solutions of pyrrole derivatives of monocrotaline and retrorsine in dimethyl formamide as a single injection of 0.05 - 0.1 ml solution. When injected in the tail vein at a concentration of 5 mg/kg body weight, it produced progressive proliferation of alveolar epithelium of the lungs and the animals developed signs of respiratory distress in 2 - 3 weeks. When injected in the mesenteric vein at a concentration of 15 mg/kg body weight, as a rule, the animals remained well in the postoperative period and only 1/26 animals died of mesenteric vein thrombosis; the livers developed multiple infarcts in the left lobes that developed into multiple coarse nodules at 6 - 12 weeks. The above studies substantiated the view that PAs act only when converted in hepatocytes to pyrroles. When pyrroles were injected, they affected the smaller vessels at the portal of entry; in animals injected through the mesenteric vein, the main target was the portal vein with only secondary damage to the parenchymal cells, thus sparing the animals from the effects of hepatocellular injury. On the other hand, pyrroles injected through the tail vein went directly to the pulmonary arteries through the heart and damaged the alveolar capillaries.

Acute veno-occlusive disease was produced in monkeys administered monocrotaline (Allen et al., 1967, 1969) and ground Crotalaria spectabilis seed (Allen & Carstens, 1968). In a study published in 1967, these authors used 14 monkeys (Macaca speciosa) of both sexes, each weighing approximately 4 kg. Seven of the animals were administered 1 mg monocrotaline in distilled water by gastric intubation on days 1 and 14. The remaining 7 were used as controls and received distilled water only. Wedge biopsies of liver were examined weekly. The survival time ranged from 14 to 38 days, the mean being 21 days. The livers of treated animals were small and firm and showed changes characteristic of human veno-occlusive disease including centrilobular necrosis, and vascular changes in the central veins of liver lobules ranging from subintimal oedema to progressive collagenization and extension of collagen fibres into the sinusoids. Similar observations were made in studies on Macaca mulatta monkeys (Allen & Carstens, 1968), administered ground Crotalaria spectabilis seed. Sixty-four animals, averaging 6.2 kg in weight, were divided into 3 groups. Group I, comprising 10 experimental animals (4 control animals), received seeds in the diet containing the equivalent of 0.074 mg monocrotaline/kg body weight, daily, up to death. Group II, consisting of 14 treated animals (4 control animals), received a single dose of seeds containing 1.3 g monocrotaline/kg body weight, and Group III, consisting of 26 experimental animals (6 controls), received 3 weekly doses containing the equivalent of 0.116 g monocrotaline/kg body weight, by gastric intubation. Liver biopsies were carried out each month in Group I and each week in Groups II and III. Animals of the last 2 groups were killed when in extremis. The mean survival times for the groups were: Group I, 269 days (176 - 425 days); Group II, 28 days (6 - 43 days); and Group III, 41 days (23 - 91 days). In Group I animals, occlusive lesions of the central and sublobular veins of the liver were seen in 7/10 animals at autopsy. These consisted of oedema, haemorrhages, and fragmentation of the vessel walls, the lumina being filled with fibrin, and degenerating liver cells. The lobular pattern was distorted because of connective tissue bands encircling small groups of liver cells, especially in the central zones of the lobules. In Groups II and III, various levels of focal or centrilobular necrosis were observed and the liver cells were replaced by stromal tissue. Vascular lesions, as described above, were seen in 25 monkeys, but no collagen was demonstrated.

In the studies of Wakim et al. (1946) on the rhesus monkey, senecionine administered iv as a 2% solution at doses of 10 - 30 mg/kg body weight to 4 animals, daily, until they appeared to be sick, produced periportal, or midzonal necrosis in 3 animals accompanied by haemorrhage. No mention was made of any vascular changes.

Electron microscopic studies

Svoboda & Soga (1966) studied the effects of lasiocarpine and Crotalaria fulva on the livers of male Sprague Dawley rats weighing 110 - 150 g each. One group of 22 rats was given an ip injection of lasiocarpine at 80 mg/kg body weight and pairs of animals were killed at various intervals ranging from 15 min to 6 days. A second group of 22 animals was administered a single dose of an aqueous extract of Crotalaria fulva at 0.5 mg/g body weight, by gastric tube, and killed at the same intervals. A third group of 8 rats was administered a total of 3.2 times the LD_{50} dose of lasiocarpine in small doses, 3 times a week, and killed at 9 - 20 weeks. The changes primarily involved the nucleus and interchromatin granules. The first change, seen after 30 min in the nueleoli of the hepatocytes and Kupffer cells of animals receiving lasiocarpine or crotalaria extract, consisted of a separation of the fibrillar and granular components. The hepatocyte nuclei had returned to normal after 72 h and remained so throughout the rest of the study. In animals receiving a single dose of lasiocarpine or crotalaria, round periodic acid schiff (PAS) positive eosinophilic bodies appeared in the cytoplasm after 12 h, consisting of dense masses of cytoplasmic material. Five days after treatment with crotalaria, large cells lined the luminal surface of the central veins; the centrilobular cells had undergone necrosis by this stage. Animals receiving 3.2 times the LD_{50} of lasiocarpine developed megalocytosis after 9 weeks (section 6.4.1.5). The cytoplasm showed vesicles of smooth endoplasmic reticulum with mitochondria of various shapes and sizes. The appearance resembled an exaggerated regenerative response.

Allen et al. (1967, 1969) studied ultrastructural changes in the liver tissue, in general, including the hepatic veins in Macaca speciosa monkeys treated with PAs. In the study on hepatic veins (Allen et al., 1969), 18 treated and 6 control adult animals were used with an average weight of 5.8 kg. Animals were divided into 3 groups of 6 treated and 2 control animals each. The experimental animals received 0.125 g monocrotaline/kg body weight by ip injection. Liver wedge biopsies were examined at various intervals in Group I during hours 1 - 48, in Group II at 4 - 12 days, and in Group III at 16 - 32 days. The earliest changes, observed by light microscopy, were seen at 24 h and consisted of progressive loss of endothelial cells and other associated changes in the lumen and wall leading to occlusion by collagenization by the third week. Under the electron microscope, within 24 h of administration, marked changes were observed in the endothelial cells resulting in their rupture and release of organelles in the lumen. This was followed by penetration of fluid though the vessel walls in the first week and changes in the fibroblasts. By the third to fourth week, hepatic veins showed various levels of occlusion and the vessel was scattered with cell debris, free organelles, etc. The authors concluded that, in this species, hepatocellular necrosis was not the primary factor causing veno-occlusive disease, as the association of cellular necrosis and venous occlusion occurred only in the central area of liver lobules, and the hepatocytes surrounding the sublobular and medium sized hepatic veins were unaffected.

In their study of 1967, Allen et al. also investigated the ultrastructural changes in the liver of M. speciosa monkeys after administering 2 doses of 1 g monocrotaline each, on days 1 and 14. At autopsy, after a mean survival time of 21 days, a wide spectrum of changes was observed in the hepatocytic organelles, many of which were lying, discharged into sinusoids, and also phagocytosed by the Kupffer cells. By the third week, proliferation of connective tissues had started in the sinusoids near the central veins and also in the walls of central veins. The authors concluded that the vascular and parenchymal cell changes were simultaneous and appeared to be equally instrumental in the development of the occlusive lesion.

6.4.1.4 Mechanism of toxic action

The mechanism of toxic action in acute pyrrolizidine hepatotoxicity and the sequence of events, judged from the collective experimental studies, appears to be as follows.

The PA, which is inactive as a cell poison by itself, becomes cytotoxic through its metabolism in the hepatic parenchymal cells to pyrroles, which act preferentially on the hepatocytes and the endothelium of blood vessels in the liver or lungs. In the hepatocytes, the immediate action is a rapid fall in cytoplasmic protein synthesis reaching 30% of control levels at 15 min and 6% at 1 h (Harris et al., 1969). This is manifested as disaggregation of polyribosomes and is followed by failure of pyruvate oxidation, loss of glycogen, structural damage to the mitochondria, lysosomal activity, failure of mitochondrial nicotine-adenine-dinucleotide (NAD) systems and nuclear NAD synthesis, and necrosis (McLean, 1970). The necrosis is zonal in the liver lobule, the particular zone affected depending on the metabolic enzymic geography of the lobule in the particular animal species, and also in man and monkey on the vascular endothelium of the central and sublobular veins.

The sequence of events of the vascular lesion has been studied by McLean et al (1964). After a single dose of Crotalaria fulva extract in the rat, centrilobular necrosis is present after the first day, but collagenous veno-occlusion of the central veins of the liver lobule only appears between 7 and 10 days later. Evidently, the necrosis of the liver cells is not secondary to venous occlusion. Centrilobular haemorrhage is seen from day 2 onwards and signs of hepatic venous outflow tract obstruction appear after 2 - 5 days (McLean & Hill, 1969).

Rappaport et al. (1967) and McLean (1969) demonstrated, through transillumination studies on rats, that the outlet end of the sinusoids is blocked by stationary columns of red cells, 16 - 24 h following administration of PAs. The reaction is typically patchy and results in stasis and extravasation of red cells spreading backwards from the centre of the lobule. For at least 3 days, no circulatory detail can be seen with transillumination. Portal pressure is significantly raised 3 days after administration of fulvine (Rappaport et al., 1967), notably before the first appearance of collagenous venous occlusion at 7 days. McLean (1969) observed that 6 - 10 days after PA administration, a new irregular pattern of vascular flow, contrasting with the uniform radial pattern of flow in the normal liver lobule, develops, which corresponds to the bypass channels represented by dilated paraseptal sinusoids, as observed in human liver biopsies (section 7.3). Segments of central vein into which the blocked sinusoids open, are gradually abandoned in favour of such by-pass routes and undergo occlusion first by oedematous connective tissue and then by fibrosis. The mechanism of closure of the sinusoids is not clear. A toxic action on the sinusoidal or venous endothelium, which swells and occludes the lumen, seems possible, as suggested by Allen & Carsten's (1968) electron microscopic studies on the monkey, and studies on children (Brooks et al., 1970). The endothelial lining of the vessels is denuded and replaced by a fibrinous and proteinaceous precipitate, which, together with the oedematous wall of the vessel, becomes organized and slowly replaced by fibrous connective tissues. The occlusion of sinusoids is further contributed by the discharge of cellular debris into the space of Disse. The lumen of the sinusoids becomes occluded simultaneously with the fibrosis occurring in the central vein. Collagen fibres extend into the space of Disse and sinusoids leading to a creeping fibrosis.

The proximate toxin that escapes from the liver is returned to the heart, after which it damages the first portal of entry into the alveolar capillaries of the lung and pulmonary arteries.

6.4.1.5 Chronic effects

The chronic hepatotoxic effects of PAs have been described in a number of studies on a variety of animals and have been reviewed by Bull et al. (1968) and McLean (1970). A notable feature is that an appropriate single dose of PA has been demonstrated by Schoental and Magee (1957, 1959) to lead to a relentlessly progressive course and eventually kill the animal, more than 18 months after administration. Schoental & Bensted (1963) demonstrated that rats receiving a single dose of PA may develop chronic liver disease and finally hepatocellular carcinoma more than 13 months after administration. The morphological changes in the liver are similar in a given species of animal, whether a single sublethal dose is administered or multiple small doses.

Schoental & Magee (1957) studied the long-term effects of a single dose of lasiocarpine on rats receiving normal and protein-deficient diets. Albino rats of Porton strain were used. In the first study, 66 rats fed a normal diet were administered a single oral dose of lasiocarpine at one of 3 dose levels (50 - 74 mg/kg, 75 - 100 mg/kg, or 101 - 150 mg/kg body weight); 24 animals served as controls. In another study, 46 young female rats were administered a protein-deficient diet. Of these, 13 and 10 animals received a single oral dose of lasiocarpine at 50 - 100 mg/kg and 50 - 75 mg/kg body weight, respectively. Each of these groups was pair-fed with an identical number of animals that did not receive any PA. Of the 66 rats fed a normal diet, very few died in the first 10 days. Thirty-one animals survived longer than 3 months. They continued to be in good health until shortly before death. The numbers of animals that survived for 13 months after administration of lasiocarpine were 8/10 males and 7/7 females (lowest dose) and 5/25 males and 11/18 females (intermediate dose). In the group that received the highest dose, neither of the 2 male animals survived longer than 35 days, and only 1/4 female animals survived longer than 3 months. In the animals that died or were killed when moribund, parenchymal damage was invariably present with prominent megalocytes, ductular proliferation, and invasion of lobules by oval or elongated cells, thought to be derived from the bile-duct epithelium or the reticuloendothelial cells. Animals that survived showed various degrees of fibrosis and nodular hyperplasia and, in some, a mild thickening of the central veins. No obliterative lesions of the veins were seen. The 31 animals that survived 13 months showed similar changes, but to a much lesser extent. In the livers of animals that had repeat liver biopsies, there was no tendency to regression of the lesions.

The above data indicate that very few animals died of acute disease. In most animals, there was a latent period of 3 - 4 weeks, during which they remained well and showed little evidence of liver cell injury, followed by a progressive course often leading to fibrosis and nodular hyperplasia.

The low-protein diet adversely affected the growth of all the rats in the control as well as the treated group. Only 3 out of 23 treated animals remained alive and in apparent good health, 8 - 11 months after the treatment. Liver biopsies taken at various intervals between 4 and 10 months showed very severe fatty changes in the liver cells. There was little fibrosis and no bile-duct proliferation. In areas where the fatty changes were less severe, characteristic megalocytes were seen. Control animals had either normal livers or showed only slight fatty changes that were not comparable in severity with those in the livers of PA-treated animals. Thus protein deficiency in the diet was shown to enhance the toxic effects of the PA.

Schoental & Magee (1959) extended these studies on young Wistar rats using several other PAs including heliotrine, retrorsine, riddelliine, seneciphylline, monocrotaline, and its $\underline{N}$-oxide in various dosages ranging from 25 to 300 mg/kg body weight; the animals were studied at death from 1 - 10 days to 18 - 30 months. Pathological changes were similar to those observed with lasiocarpine in the previous study. Notable observations were that necrosis did not necessarily precede subacute or chronic changes. The livers of some animals became severely damaged and showed nodular hyperplasia. Liver biopsy, 2 - 3 days after PA treatment, did not show pathological changes in some animals, but a repeat biopsy at 41 days showed characteristic changes. Fibrous thickening of the central veins was observed in some animals, more often with monocrotaline $\underline{N}$-oxide, but no occlusion of the hepatic veins was seen.

The studies of Nolan et al. (1966) confirmed the observations of Schoental & Magee (1957, 1959). They gave a single dose of lasiocarpine at 120 mg/kg body weight, by stomach tube, to 108 (equal numbers of both sexes) Sprague Dawley weanling rats (60 - 135 g body weight). Thirty animals of both sexes served as controls. Groups of 10 animals, each consisting of 8 treated and 2 control animals, were killed at various intervals from 1 to 123 days. Of the 80 treated animals, 28 died within 26 days. No delayed hepatic lesions were found in 59 rats between days 1 and 18. Between 19 and 123 days, delayed lesions were found in 34/49 rats. These 34 rats showed megalocytosis, but no ductular proliferation or fibrosis.

In a second study, 127 twenty-one-day-old male Wistar rats were given a single oral dose of lasiocarpine at 80 mg/kg body weight; 65 animals served as controls. Liver biopsy was carried out on day 3 in 108 and on day 9 in 15 of the treated animals. In 47 animals, additional biopsies were carried out at intervals. Of the 127 PA-treated rats, 98 died during the first 9 days, and 29 after 10 - 50 days. In contrast to the first study, 32/58 survivors exhibited delayed subacute and chronic lesions, as described by Schoental & Magee (1957). Of these, 8 animals developed cirrhosis. The observations indicated that the lesions of acute zonal necrosis, which appeared on, or before, the third day, healed without residual lesions. However, 55% of the 58 survivors developed subacute/chronic lesions that tended to be progressive after a latent period of 2 - 3 weeks. There appeared to be an intimate relationship between chronic lesions and megalocytosis, which was seen in 52/58 surviving animals. No obliterative vascular changes were observed and so the lesions could not be ascribed to impaired circulation.

Schoental (1959) demonstrated the toxic effects on the newborn of PAs administered to lactating mother rats. Wistar rats (200 - 300 g) were administered lasiocarpine, orally or by ip injection, at 25 - 40 mg, in 5 - 10 doses of 5 - 10 mg, twice weekly or more (24 rats), or retrorsine at 4 - 10 mg per dose, in 1 - 14 approximately daily doses (23 rats). The litters were left with the mothers for 24 days or more (except for 1/2 hour separation during the PA treatment). The litters were examined by biopsy at frequent intervals or at autopsy when they died or were killed in a moribund state. Litters of the lasiocarpine-treated rats showed only insignificant fibrosis or some megalocytosis. In the retrorsine-treated group, the majority of the young rats survived for about 18 days, but all rats died before reaching the age of 30 days. The milk secretion of the mothers was apparently not affected by the PA treatment. Of the 98 animals in 9 litters, 45 died by the 20th day and 45 survived 30 days. Animals dying in the first fortnight did not show gross liver lesions, but those that died at weaning time or later, all showed signs of liver disease. Animals dying at 18 - 30 days showed hydropic or fatty vacuolation of hepatocytes and haemorrhage into distended sinusoids. The change was severe in animals dying at 1 - 2 months, and some central veins showed a narrowed lumen. The lactating rats that received the alkaloids survived longer than their young, and most showed no ill effects from the treatment. This evidently indicates either a high susceptibility of suckling rats or a high concentration of PAs in milk.

In studies by Allen et al. (1963), 2 groups of 4-week-old turkeys, each consisting of 12 birds, were fed diets containing ground Crotalaria spectabilis seed at 2.5 g/kg and 5 g/kg, respectively, for 120 days. Twelve animals served as controls. At the end of the study, 11/12 birds receiving 5 g/kg seed and 6/12 receiving 2.5 g/kg seed in the diet developed cirrhosis. The minimum period of feeding required to produce cirrhosis was 75 days, provided the diet was reduced to a level that was not lethal.

Allen & Carstens (1971) induced the Budd-Chiari syndrome in monkeys by monocrotaline. Six adult female and 9 adult male Macaca speciosa monkeys, weighing 5.2 - 6.5 kg each, were divided into 2 groups, each comprising 5 control (3 males and 2 females) and 10 treated animals (6 males and 4 females). The treated group was given a subcutaneous (sc) injection of monocrotaline at 60 mg/kg body weight at monthly intervals for 3 months. Needle biopsy of the liver was carried out every month for 5 months and laparotomy, 6 months after PA treatment.

The treated animals showed marked vascular changes and various degrees of occlusion in the centrilobular and sublobular veins as well as the larger vessels. There was also characteristic haemorrhagic necrosis in the centrilobular zones and megalocytes were seen. The portal venous pressures were raised. The animals were autopsied at 6 months. The livers were markedly shrunken weighing an average of 68 g in contrast to those of control animals, which weighed 130 g. There were severe occlusive vascular changes and irregular fibrosis in the lobules. The adjacent sinusoids were dilated as a compensatory mechanism.

Swick et al. (1982a) studied the effects on guinea-pigs of long-term dietary administration of Senecio jacobaea and compared them with the toxic effects of single doses of injected Senecio alkaloids and monocrotaline. The possible protective effect of cysteine was also examined. Fifteen guinea-pigs of 250 - 300 g initial body weight were divided into 2 treated groups, being fed 10% Senecio jacobaea, or 10% Senecio jacobaea plus 1% cysteine in the diet, and a control group. The whole plant of Senecio jacobaea was air dried and powdered for incorporation into the diets. The animals were fed for 365 days. They were autopsied at death or at the termination of the study. In a second study, 7 guinea-pigs of 500 g body weight were injected intraperitoneally with either monocrotaline, jacobine, or mixed Senecio jacobaea PAs. The chronic lethal dose LD_{100} of Senecio jacobaea was 1264 g/kg initial body weight or 526% of the initial body weight with an average survival time of 279 days. No mortality was observed in control animals. This contrasts with the chronic LD_{100} of Senecio jacobaea for rats of 58% of initial body weight (Swick et al., 1979) and that of cattle equivalent to 5 - 20% body weight (Bull et al., 1968). Addition of cysteine to the diet was only slightly, but not significantly, protective. Pathological examination of the livers of the guinea-pigs fed Senecio jacobaea revealed extensive megalocytosis and severe cytoplasmic vacuolation with biliary hyperplasia and fibrosis, primarily in periportal areas. The centrilobular and midzonal areas were spared.

Monocrotaline was non-toxic at doses up to 1000 mg/kg body weight, whereas jacobine and mixed alkaloids from Senecio jacobaea were lethal at much lower levels. Similar results showing resistance to monocrotaline in guinea-pigs were also reported by Chesney & Allen (1973a). In in vitro studies, they related this resistance to lack of conversion of PA to pyrroles by guinea-pig microsomes.

A morphological peculiarity of chronic hepatotoxicity in a large variety of laboratory and farm animals is megalocytosis (Bull, 1955; McLean, 1970), i.e., the appearance of exceptionally large hepatocytes, 10 - 30 times the volume of normal cells with proportionately large nuclei. Relevant literature has been reviewed by Jago (1969), McLean (1970), and Mattocks (1986). Advanced megalocytosis was produced by Jago (1969) within 4 weeks in 2-week-old rats by administering a single dose of lasiocarpine at 76 µmol/kg body weight. Megalocytes tend to appear in the periportal and midzones of the liver lobules with normal sized cells around the central veins. The nuclear chromatin is proportionately increased, but the cells appear incapable of entering into mitosis, as only abnormal mitoses are seen. Jago (1969) demonstrated a fall in the mitotic index (from 1.61 to 0.04) in liver cells of 2-week-old rats, one day after injection of 50 µmol lasiocarpine/kg. The electron microscopic appearance also supports the above observations (Afzelius & Schoental, 1967). A striking proliferation of rough endoplasmic reticulum and multiple centrioles is seen in the cytoplasm, and the cytoplasmic organelles are disorganized, suggesting increased metabolic activity but inability of the cells to divide. Such cells may persist for the life-time of the animal (up to 2 years in the rat) and the liver never returns to normal (Mattocks, 1986). Megalocytes have also been described in the kidney (Bull et al., 1968), the lung (Barnes et al., 1964; Butler et al., 1970; Hooper, 1974), and the duodenum (Hooper, 1975c).

Data on the total chronic lethal dose of heliotrine in rats were discussed by Bull & Dick (1959) and Bull et al. (1968). For a variety of dosing rates, and with withholding periods of 10 - 20 weeks interposed, the total doses ranged from 2.2 to 7.8 LD_{50}. In Table 10, these data are extended with results for other alkaloids. The overall range of the total lethal dose is 1.2 - 10.3 LD_{50}.

6.4.2 Effects on lungs

Current literature has been extensively reviewed by Kay & Heath (1969), and Mattocks (1986). PAs have been shown to produce pulmonary hypertension with associated vascular changes in the pulmonary circulation in a number of experimental animal species including the rat, mouse, frog, turkey, pig, sheep, rabbit, and horse (McLean, 1970) as well as in non-human primates (Allen & Chesney, 1972; Chesney & Allen, 1973b) and the dog (Miller et al., 1978). The alkaloids have been administered by feeding the animals with: PA-containing

Table 10. Total chronic lethal doses in rats
(ip administration, 2 or 3 times per week)

Dose (x LD_{50})	Time to death (days)	Total lethal dose (x LD_{50})	Reference
Heliotrine (male rats, unless otherwise stated)			
0.2	58	5	Bull & Dick (1959)
0.1	123	5.1	
0.04	303	4.1	
0.02	508	4.1	
0.01		5.2 - 5.3	Bull & Dick (1960)
(with interval of 10 - 20 weeks after 21 days)			
0.11		2.2 - 4.3	Bull & Dick (1959)
0.1		7.8	Jago (1971)
(35-day-old male rats)			
0.1		4.7	Jago (1971)
(337-day-old male rats)			
0.1		5.8	Jago (1971)
(35-day-old female rats)			
0.1		4.5	Jago (1971)
(337-day-old female rats)			
Lasiocarpine (male rats)			
0.1	210	9	Culvenor & Jago (1979)
0.05	482	10.3	
0.02	676	5.7	
0.01	595	2.6	
(0.005)	(638)	(1.4)	
0.1	81 - 175	2.4 - 5	Bull & Dick (1959)
0.1	-	6.3 - 10.9	Jago (1971)
Lasiocarpine (female rats)			
0.1	108	4.6	Culvenor & Jago (1979)
0.05	274	5.8	
0.02	471	4	
0.01	487	2.1	
(0.005)	(692)	(1.5)	
0.1		2.4 - 7	Jago (1971)
Monocrotaline			
0.1		1.2 - 2.4	Bull et al. (1968)
0.05		2.5 - 4.4	
Senecionine[a]			
0.2		2 - 7.4	Bull et al. (1968)
0.1		1.7 - 5.7	
0.04		(3 survived)	

[a] Assuming LD_{50} mg/kg (c.f., Mattocks, 1986).

seeds of plants (notably Crotalaria spectabilis) (Turner & Lalich, 1965; Kay & Heath, 1966; Kay et al., 1967a) or the dried plant itself (e.g., Senecio jacobaea) (Burns, 1972), aqueous solutions of fulvine (Barnes et al., 1964; Wagenvoort et al., 1974a,b) or monocrotaline (Lalich & Ehrhart, 1962; Huxtable et al., 1977), subcutaneous injections of monocrotaline (Allen & Chesney, 1972; Chesney & Allen, 1973b) and seneciphylline (Ohtsubo et al., 1977), or intravenous injections of some pyrrolic esters and analogues of pyrrolizidine alkaloids and their metabolites (Mattocks & Driver, 1983). Lafranconi & Huxtable (1984) studied the hepatic metabolism and pulmonary toxicity of monocrotaline in in vitro perfusion studies. Some of the representative studies on the morphological effects of toxic lung injury are listed in chronological order in Table 11.

Chronic lung lesions have been produced by most compounds that produce chronic liver lesions, though higher doses were required in some instances (Culvenor et al., 1976a). However, not all PAs that are hepatotoxic are also pneumotoxic. Among the pneumotoxic alkaloids, fulvine (Barnes et al., 1964) and monocrotaline are particularly active (Mattocks, 1986). Molecular structure activity requirements are the same as for hepatotoxicity, since both are caused by the same toxic metabolites produced in the hepatocytes.

6.4.2.1 Acute effects

Pulmonary lesions produced by PAs have been extensively investigated, mostly in rats, but also in non-human primates. Monocrotaline has been the alkaloid most frequently used, but lung lesions have also been seen in rats following fulvine and seneciphylline administration. Besides pure alkaloids, PA-containing seeds of some plants, most notably Crotalaria spectabilis, have also been used.

Miller et al. (1978) gave a single iv injection of monocrotaline at 60 mg/kg body weight to 10 mongrel dogs. Toxic effects, recorded within 2 h, included ultrastructural changes in the endothelial cells of the alveolar capillaries, prominent accumulation of platelets, and the appearance of interstitial oedema (Table 11). Valdivia et al. (1967a,b) used 25 Sprague Dawley rats in their study and made similar observations on the rat lung within 4 h of a single subcutàneous injection of monocrotaline at a dose of 60 mg/kg body weight (Table 11). Interstitial oedema and elastolysis of the alveolar wall, increase in number of mast cells, and other associated changes were observed within 4 h of the injection, followed by alterations in endothelial and interstitial cells. All of the changes progressed steadily for up to 3 weeks. It was concluded that the initial changes

Table 11. Summary of experimental data on the morphological effects of toxic lung injury due to pyrrolizidine alkaloids (in chronological order)

Animal/ strain/ sex	Number of animals/ controls	Toxic agent	Dose (mg/kg)	Administration dose	Single/ multiple	Killed after or survival for	Pathological effects	References
Acute effects								
Male Sprague Dawley rat	25/5	monocrotaline	60 (body weight)	subcutaneous	single	2 – 48 h 1 – 3 weeks alterations,	interstitial oedema, endothelial cell elastolysis, etc.	Valdivia et al. (1967a,b)
Mongrel dog	10/0	monocrotaline	60 (body weight)	intravenous	single	2 h	interstitial oedema, changes in endothelial cells	Miller et al. (1978)
Male Sprague Dawley rat	5/3	monocrotaline	40 (body weight)	subcutaneous	single	0 – 21 days	microvascular leak, right ventricular hypertrophy after 2 weeks	Sugita et al. (1983a)
Chronic effects								
Female Sprague Dawley rat	35/12	monocrotaline	10 – 30 (diet)	ad libitum feeding			pulmonary arteritis (with 20 – 30 mg dose only)	Lalich & Ehrhart, (1962)
	6	alcohol-extracted seeds of Crotalaria spectabilis	2500 (diet)	ad libitum feeding		51 days	none	
	19/4	monocrotaline	10 – 75 (diet)	ad libitum feeding		26 – 232 days		Turner & Lalich (1965)

Table 11 (contd).

Female Wistar Furth rat	24/6	Crotalaria spectabilis seeds	200 - 1600 (diet)	ad libitum feeding	105 - 172 days	right ventricular hypertrophy and dilatation, pulmonary arterial hypertrophy, pulmonary arteriolar hypertrophy, endocardial fibrosis	Allen & Chesney (1972)
Female weanling Wistar rat	10/34	Crotalaria spectabilis seeds	1000 (diet)	ad libitum feeding	30 - 60 days	right ventricular hypertrophy, pulmonary arterial hypertrophy, pulmonary arteritis	Kay & Heath (1966)
Male suckling Sprague Dawley rat	22/12	monocrotaline	120 (body weight)	subcutaneous single	20 - 47 days	right ventricular hypertrophy, pulmonary arterial hypertrophy, pulmonary arteritis, fibrin thrombi	Hayashi & Lalich (1967)
Female Wistar rat	30/10	fulvine	50 (body weight)	intraperitoneal single	3 - 37 days	right ventricular hypertrophy, pulmonary arterial hypertrophy, pulmonary arteriolar hypertrophy, pulmonary arteritis, arteriolar thrombi	Kay et al. (1971a)
			80 (body weight)	intragastric single	7 - 35 days		

Table 11 (contd).

Animal/strain/sex	Number of animals/controls	Toxic agent	Dose (mg/kg)	Administration dose	Single/multiple dose	Killed after or survival for	Pathological effects	References
Monkey (Macaca arctoides) (both sexes)	12 (30 days old)	monocrotaline	30 (body weight)	subcutaneous	one followed by 3 (2, 4, and 6 months after first injection)	199 - 325 days (average, 241)	right ventricular hypertrophy and dilatation, left ventricular hypertrophy, pulmonary arterial hypertrophy, pulmonary arteriolar hypertrophy, pulmonary hypertension	Allen & Chesney (1972)
	12 (15 months old)					163 - 334 days (average, 217)	pulmonary arterial hypertension (isolated, veno-occlusive disease (liver)	
Monkey (Macaca arctoides) (both sexes)	20/6	monocrotaline	30 (body weight)	subcutaneous	as above	165 - 325 days (average, 326)	right ventricular hypertrophy and dilatation, pulmonary arterial hypertrophy, pulmonary arteriolar hypertrophy, pulmonary arteritis, endocardial fibrosis	Chesney & Allen (1973b)

Table 11 (contd).

Animal	M/F ratio	Compound	Dose	Route	Duration	Effects	Reference
Female Wistar rat	50/12	fulvine	80 (body weight) or 50 (body weight)	intragastric single intraperitoneal	1 – 6 weeks	vasoconstriction, pulmonary arterial hypertrophy, pulmonary arteriolar hypertrophy, right ventricular hypertrophy, thickening of veins and venules	Wagenvoort et al. (1974a,b)
Male Wistar rat (4 weeks old)	16/0	seneciphylline	50 – 80 (body weight)	subcutaneous single	1 – 3 weeks	pulmonary arteriolar and right ventricular hypertrophy (after 3 weeks), right ventricular dilatation in 2 animals	Ohtsubo et al. (1977)
Male Sprague Dawley rat	21/14	<u>Crotalaria spectabilis</u>	1000 (diet)	ad libitum feeding	3 – 35 days	pulmonary arterial hypertrophy and pulmonary arteritis (2/21); right ventricular hypertrophy	Meyrick & Reid (1979, 1982)

of destruction of pulmonary capillaries and the other components of the alveolar wall preceded the arteriolar hypertrophy and arteritis observed by other investigators following monocrotaline administration, and alone was sufficient to cause right ventricular hypertrophy. Sugita et al. (1983a,b) administered a single dose of monocrotaline at 40 mg/kg body weight to 5 Sprague Dawley rats and adduced further evidence, by biochemical and radioisotopic studies, of microvascular leak in the alveolar wall within the first 3 days of injury, which preceded right ventricular hypertrophy observed 2 weeks following administration (Table 11).

A histological and electron microscopic study was made by Hurley & Jago (1975) of the lungs of rats administered dehydromonocrotaline. Female black and white hooded rats weighing 80 - 100 g were used. Dehydromonocrotaline dissolved in dimethylformamide (DMF) was administered iv as a single dose at 30 mg/kg body weight to 12 rats and at 15 mg/kg body weight to 7 rats. Four control rats were administered DMF alone. A colloidal suspension of carbon black was injected iv, 6 - 18 h after injection of dehydromonocrotaline, and the animals killed 19 - 44 h after treatment.

After an interval of 6 - 8 h, there was a direct toxic effect on the endothelial cells of pulmonary capillaries and small venules. Many endothelial cells had prominent nuclei and thickened cytoplasm, which contained more RNA granules than usual. There was also an increase in the number of mitochondria. The endothelial damage did not seem to have caused permanent disruption of the small blood vessels, and, 2 days after injury, all vessels were patent. Large numbers of mononuclear cells, which appeared in the interstitial tissues of the lung 44 h after injury, seemed to be altered emigrated blood monocytes.

6.4.2.2 Chronic effects

Lalich & Erhart (1962), fed 35 Sprague Dawley rats a diet containing monocrotaline at 10 - 30 mg/kg (Table 11). Animals receiving a daily dose of monocrotaline of 20 mg/kg diet or more, showed progressive changes in the lungs after 24 days of feeding. Of the 23 animals receiving 20 - 30 mg/kg, 12 showed pulmonary arteritis, 4 of these even at the dose level of 20 mg/kg diet. Pulmonary haemorrhages were observed in 16 animals. No changes were observed in animals fed alcohol-extracted seeds or other derivatives of monocrotaline.

Identical changes were observed in similar studies by Turner & Lalich (1965) on two strains of rats, Sprague Dawley (19) and Wistar Furth (24). The first group of 19 female Sprague Dawley rats was fed a diet containing monocrotaline at an initial level of 10 mg/kg diet. Depending on the response of each individual animal, the monocrotaline level was raised to a maximum of 75 mg/kg diet (Table 11). Fourteen rats survived for more than 100 days and 8 reached the maximum dietary level of monocrotaline, the last animal dying after 232 days. The second group of 24 female Wistar Furth rats was fed a diet contaminated with Crotalaria spectabilis seeds, initially at 0.2 mg/kg diet, gradually rising to 1.6 mg/kg by increasing the levels by 0.2 mg/kg every week. All test animals survived 100 days and 15 reached the maximum levels of Crotalaria fed. The last animal died after 172 days of feeding. Animals developed signs of toxicity and right ventricular strain, e.g., cyanosis, etc. Progressive thickening of media in the muscular pulmonary arteries, progressive muscularization of arterioles, and changes characteristic of pulmonary hypertension, were seen. Some pulmonary arteries showed medial necrosis. No changes were observed in the pulmonary veins. Significant hypertrophy of the heart, as judged by the heart weight in relation to body weight, was seen in almost all animals that survived 100 days or more (32/38), and the right ventricles were dilated. The hypertrophy affected the right side of the heart only, and generally corresponded with the vascular changes. There was a marked hyperplasia of the mast cells in the mediastinal lymph nodes and around bronchi and pulmonary arteries. Similar observations were made by Barnes et al. (1964) and Valdivia et al. (1967a,b).

Kay and his group studied cardiac and pulmonary vascular changes in rats fed Crotalaria spectabilis seeds (Kay & Heath, 1966; Kay et al., 1967a,b) or administered fulvine (Kay et al., 1971a).

A group of 10 female weanling Wistar albino rats were fed a diet containing 1 g powdered seeds of Crotalaria spectabilis/kg until they died of cardiorespiratory distress, after 36 - 60 days of feeding. Thirty-four control rats were fed a normal diet. At autopsy, the atria of the heart, the right ventricle, and the left ventricle with the interventricular septum were weighed separately. The medial thickness of the muscular pulmonary arteries was measured, and expressed as percentage of external diameter. The medial thickness of the muscular pulmonary arteries increased in all test rats; acute or healing pulmonary arteritis was seen in 3 animals. Statistically significant cardiomegaly was present in all rats fed the seeds, contributed chiefly by the right ventricle. The readings from all the test rats were well outside the upper 95% confidence limit. The increase in the medial thickness of the pulmonary arteries correlated well with the weight of the right ventricle (Fig. 10). Three rats showed pulmonary arteritis (indicated by a solid triangle). It was presumed that the organic basis for increased pulmonary resistance was the abnormal muscularization of the radicles of the pulmonary arterial system. Essentially similar results were obtained in an identical study repeated on 8 test rats and 5 controls (Kay et al., 1967a). The test rats developed pulmonary hypertension in 37 days, levels of which were correlated with the medial thickness of the muscular pulmonary arteries and that of the pulmonary trunk, as well as with the weight of the right ventricle.

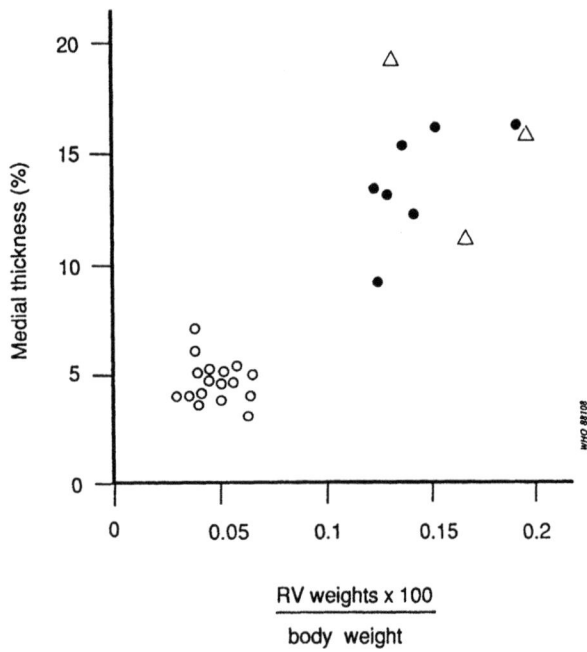

Fig. 10. The relation between the average percent medial thickness of pulmonary arteries, and the weight of the right ventricle, expressed as a percentage of the total body weight, in control and test rats (Kay & Heath, 1966). Control animals – o; test animals – o; animals showing pulmonary arteritis – Δ.

Ghodsi & Will (1981) made similar observations in Sprague Dawley rats given a single subcutaneous injection of monocrotaline at 60 mg/kg body weight. Forty rats weighing 180 - 200 g were used; of these, 20 constituted the control group. The control animals received the same volume of saline. Each week, 3 rats from each group were catheterized and pulmonary artery pressures were measured. In the treated group, 2 out of 5 animals showed a mild increase in pulmonary artery pressure at the end of 8 days. A further 4 out of 5 animals showed a mild to moderate rise in pulmonary artery pressure after 2 weeks. The highest value recorded in test rats was 56 mmHg compared with a normal upper limit of 22 mmHg. The medial thickness of pulmonary arteries was correlated with pulmonary artery pressures ($\underline{P}$ < 0.02) as was the thickness of the right ventricle. The correlation between the pulmonary artery pressures and right ventricular hypertrophy was statistically significant ($\underline{P}$ < 0.05).

Kay et al. (1971a) administered a single dose of fulvine to rats, intraperitoneally at 50 mg/kg body weight, or through a stomach tube at 80 mg/kg. Of the 30 treated rats, 17 survived 23 days. All of the animals showed changes characteristic of hypertensive pulmonary vascular disease with right ventricular hypertrophy and muscular hypertrophy of the pulmonary trunk and the muscular pulmonary arteries. Pulmonary arterioles were also muscularized and contained fibrin thrombi. Four animals showed pulmonary arteritis.

Essentially similar changes in the pulmonary arterial system were produced within 20 - 28 days by Hayashi & Lalich (1967) in 22 male suckling rats administered a single injection of monocrotaline at 120 mg/kg body weight, and within 28 days by Ohtsubo et al. (1977) in 16 male, 4-week-old rats given a single injection of seneciphylline at 80 mg/kg body weight. Hooper (1974) did not find any such effect on feeding powdered Senecio jacobaea at 100 - 200 mg/kg diet to 9 male white mice for up to 193 days.

In studies by Allen & Chesney (1972), non-human primates (Macaca arctoides) were administered 4 doses of monocrotaline at 30 - 60 mg/kg body weight by subcutaneous injection (Table 11). Twelve infant monkeys (30 days old) and 12 adults (15 months old) were studied with different results. Vascular changes, characteristic of pulmonary hypertension and resultant cor pulmonale, were observed in the infant monkeys, as described by earlier workers in the rat. Only isolated small hepatic veins were occluded. On the other hand, the adult animals showed a more severe involvement of the liver with changes characteristic of veno-occlusive disease and only an occasional pulmonary blood vessel was involved. The authors postulated that the different responses in infant and adult animals were due to the different stages of maturation of the enzyme systems of the hepatocyte in the two age groups. It is possible that the different reactions in the liver and lung in the 2 groups may be due to the fact that the enzymatic pathways responsible for producing metabolites that cause hepatic damage are poorly developed in the infant, but those responsible for causing pulmonary lesions are better developed.

Chesney & Allen (1973b) made observations similar to those of Allen & Chesney (1972) in twenty, 30-day-old monkeys in a similar study using monocrotaline injections and recorded, in addition, endocardial fibrosis of the right heart. The treated animals developed classical clinical features of cardiopulmonary distress, which was also evidenced by changes in the blood-gas parameters. The raised right heart pressures were confirmed by actual measurements of the blood pressure in the right ventricle, pulmonary artery, and descending aorta. The authors considered this study to be a good experimental model to investigate hypertensive pulmonary vascular disease, or pulmonary and endocardial fibrosis. The type of vascular changes seen in the animals were comparable with those associated with pulmonary hypertension in man in cardiopulmonary disease (Barnes et al., 1964; Kay & Heath, 1966; Kay et al., 1967a; Chesney & Allen, 1973b).

Wagenvoort et al. (1974a,b) made light microscopic and ultra-structural studies on 50 female Wistar rats, 1 - 6 weeks following a single oral dose of fulvine at 80 mg/kg body weight or an ip dose at 60 mg/kg body weight. Twelve animals served as controls. Vasoconstriction of muscular pulmonary arteries and arterioles was seen initially, one week following administration. This was evident by the coiled appearance of the muscular nuclei and excessive crenation of the internal elastic lamina. The nuclei of smooth muscle cells as well as those of endothelial cells were partly squeezed between the folds of the lamina. After 3 - 4 weeks, these blood vessels began to thicken, with muscular hypertrophy and fibrinoid necrosis of the arterial muscle. Animals surviving administration of fulvine developed right ventricular hypertrophy, proliferation of endothelial cells in the arteries and even thickening of the veins.

In a study by Meyrick and Reid (1979), 21 Sprague Dawley rats were fed a diet containing 1 g powdered seeds of Crotalaria spectabilis/kg for various periods ranging from 3 to 35 days. The earliest demonstrable change in the pulmonary arterial system of the animals was seen on day 3 and consisted of the appearance of muscle in normally non-muscular arteries of the lung. The muscular pulmonary arteries began to show hypertrophy of the media from day 7, which reached statistically significant levels on day 10 in smaller arteries and on day 14 in the larger arteries. Significant right ventricular hypertrophy was seen on day 21. These changes were confirmed by ^{3}H-thymidine uptake studies (Meyrick & Reid, 1982).

6.4.2.3 Mechanisms of toxic action

Considerable progress has been made recently in the understanding of biochemical and pharmacological changes that occur in PA-induced lung disease.

Turner & Lalich (1965) and Takeoka et al. (1962) postulated that pulmonary hypertension was mediated by the release of 5-hydroxytryptamine from the mast cells, which became hyperplastic in the mediastinal lymph nodes and around bronchi and pulmonary arteries (Turner & Lalich, 1965) (section 6.4.2.2) following administration of monocrotaline, causing vasoconstriction. On the other hand, Kay et al. (1967b) found that the number of mast cells corresponded with the severity of exudative changes in the lung and were not related to the genesis of pulmonary hyperplasia.

Besides the medial muscular hypertrophy of pulmonary arteries reported in several studies cited above, swelling and lysis of the endothelial cells, contributing to luminal narrowing and thickening of the wall with fibrosis, have been described (Allen & Carstens, 1970).

Weanling rats are more susceptible to these changes than older animals, and the changes follow a strict temporal sequence. Oral administration of monocrotaline to rats at 20 mg/litre in drinking-water produced a sequence of changes over 3 weeks, that included an increase in lung mass, which was significant by day 9, stimulation of pulmonary RNA and protein synthesis (maximal on day 10), increased pulmonary arterial blood pressure (significant by day 10), and right ventricular hypertrophy by day 14 (Huxtable et al., 1978; Lafranconi et al., 1984). The increase in the lung mass was not accompanied by change in the total collagen content and was contributed possibly by hypertrophy of endothelial cells, but the increased mass of the right ventricle was associated with a 4-fold increase in collagen content (Lafranconi et al., 1985).

An early event is inhibition of serotonin removal by pulmonary endothelium (Huxtable et al., 1978). This phenomemon, combined with the increased release of serotonin by mast cells that has been observed, may be involved in the development of pulmonary hypertension (Carillo & Aviado, 1969). Right ventricular hypertrophy is blocked by propanolol, whereas the development of pulmonary hypertension is unaffected (Huxtable et al., 1977). Novel metabolites have been found to be released by livers perfused with monocrotaline in vitro, and these metabolites block serotonin transport in vitro, when perfused through isolated lungs (Lafranconi & Huxtable, 1984). These data suggest that the slow release of metabolites from the liver into the circulation following low-level exposure to monocrotaline results in specific inhibition of endothelial cell function (Huxtable et al., 1978).

The effect of monocrotaline treatment on pulmonary angiotensin converting enzyme (ACE) activity in the rat is disputed. Hayashi et al. (1984) observed a reduction in the ACE activity of pulmonary tissue in pyrrolizidine-exposed rats in parallel with the development of pulmonary alterations, while the ACE activity of the plasma remained unchanged. However, other authors have reported that, though the specific activity of ACE falls in the isolated perfused lungs of monocrotaline-treated rats, or in lung homogenates from such animals, when activity is expressed as total activity per lung, there is no significant alteration in the lungs of treated animals compared with those of untreated animals (Huxtable et al., 1978; Lafranconi & Huxtable, 1983). Therefore, the significance of changes in ACE activity is open to question.

Molteni et al. (1984) also found evidence of endothelial cell damage by monocrotaline in their ultrastructural and biochemical studies on rats. Eighty male Sprague Dawley rats were used; half were administered monocrotaline at 20 mg/litre in the drinking-water and half were given plain water. The average daily water consumption was 35 ml/rat. Thus, the treated rats were estimated to have received 2 mg/kg per day. Five animals each from the treated and control groups were killed at intervals of 1 - 12 weeks after the start of the study. The endothelial damage was measured by ACE activity, plasminogen-activator (PLA) activity, and prostacyclin (PGI_2) production. These were correlated with pulmonary arterial perfusion and ultrastructural changes in the lung. In the treated groups, after an initial rise at 1 week, the ACE activity showed a steady decline from 1 to 6 weeks, after which it plateaued at 55% of normal. PLA activity did not change for 2 weeks, but decreased by 59 and 79% of the control value after 6 and 12 weeks, respectively. On the other hand, the PGI_2 production increased progressively reaching 140 and 270% of the control level after 6 and 12 weeks, respectively. These endothelial functional changes were not accompanied by significant changes in pulmonary arterial perfusion as visualized by ^{99m}Tc-labelled macroaggregated albumin perfusion studies. The activities of ACE and PLA and the production of PGI_2 are considered sensitive indices of endothelial function in rats. The above results indicated endothelial cell dysfunction. The ultrastructural studies also revealed oedema of capillary subendothelial, perivenous and periarterial tissues at 1 week, and interstitial inflammatory infiltrates at 2 weeks. At 6 - 12 weeks, there was thickening of the pulmonary arteries and enlargement of right side of the heart.

Stenmark et al. (1985) studied the role of alveolar inflammation and arachidonate metabolism in monocrotaline-induced pulmonary hypertension in rats. Five groups of male Sprague Dawley rats were treated as follows: (a) 20 rats received 40 mg monocrotaline/kg body weight, sc; (b) 20 rats received monocrotaline, 40 mg/kg sc plus diethylcarbamazine

(DEC) 100 mg/kg sc, every 12 h; in addition, 250 mg DEC was added to 100 ml of drinking-water. This treatment started 2 days prior to the start of the study and was continued daily for 3 weeks; (c) 12 control rats received normal saline plus monocrotaline at 40 mg/kg sc; (d) 12 rats received indomethacin at 2 mg/kg sc for 2 days, prior to receiving monocrotaline at 40 mg/kg and then daily for 3 weeks; (e) 6 animals each received a single sc injection of normal saline and served as additional controls.

One, 2, and 3 weeks after monocrotaline or saline injection, lung lavage was carried out for cell counts and assay for enzyme activity and cyclooxygenase metabolites, the degradation products of prostacyclin (PGI_2) and thromboxane A_2 (TXA_2), as 6-keto-prostaglandin ($PGF_{1\alpha}$) and TXB_2, respectively. At 3 weeks, the animals were anaesthetized, right ventricular pressures measured by catheterization and the heart removed. The 2 ventricles were separated and weighed for the determination of heart-weight ratio (right ventricle/left ventricle + septums RV/LV + S) an indicator of right ventricular hypertrophy.

The right ventricle showed hypertrophy at 2 weeks and the right ventricular pressure was increased at 3 weeks following monocrotaline administration (Fig. 11). The leukocyte count in the lavage fluid increased at 3 weeks, with a rise in the percentage of polymorphonuclear leukocytes and large, abnormal alveolar macrophages in the test animals. B-N-acetyl-D-glucoseaminidose activity was also elevated at 3 weeks, indicating activation of leukocytes. There was also a rise in the concentration of 6-keto-$PGF_{1\alpha}$ at 1 and 3 weeks, as well as in TXB_2 at 3 weeks, compared with those in control animals.

The administration of DEC inhibited both the increase in heart-weight ratio (RV/LV + S) and the increase in pulmonary artery pressure (Fig. 11) that occurred 3 weeks after monocrotaline administration, and reduced the percentage of polymorphonuclear cells, abnormal alveolar macrophages, and hexoseaminidase activity in the lavage fluid, compared with that from animals that had received monocrotaline only. The rise in the levels of 6-keto-$PGF_{1\alpha}$ was inhibited ($\underline{P}$ < 0.05) by 73% and that of TXB_2 by 74% in the lung lavage.

The administration of indomethacin did not have any effects on either the heart-weight ratio or the pulmonary arterial pressure 3 weeks after monocrotaline administration (Fig. 11), but it inhibited ($\underline{P}$ < 0.05) the rises in 6-keto-$PGF_{1\alpha}$ (by 90%) and TXB_2 (by 91%) that occurred in the lung lavage of monocrotaline-treated animals at 3 weeks.

The above studies indicate that both the cyclo-oxygenase and the lipo-oxygenase pathways of arachidonate metabolism are activated by monocrotaline as early events in its toxic effect.

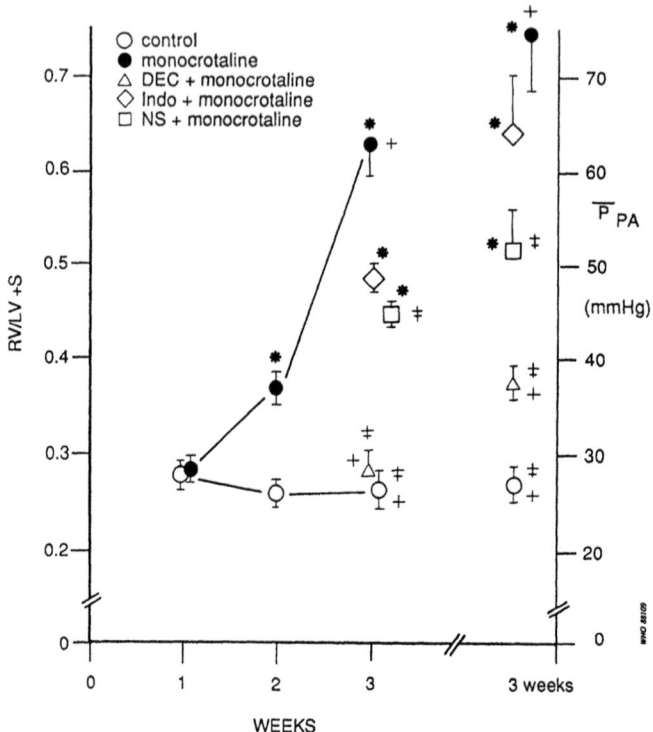

Fig. 11. Effects of monocrotaline administration on pulmonary artery pressures and right ventricular hypertrophy in rats and of the pharmacological blockade of pathways of arachidonate metabolism by diethylcarbamizine and indomethacin.

Note:

Left axis: measurements (SE) of heart/weight ratios [right ventricle/ left ventricle + septum (RV/LV + S)], 1, 2, and 3 weeks after administration at time zero of monocrotaline.
Right axis: mean pulmonary arterial pressure at 3 weeks [40 mg/kg; o, mean of 15 rats] or of saline (o, mean of 6 rats), subcutaneously.
Δ = mean values for 15 rats given monocrotaline at time zero plus daily diethylcarbamazine (DEC) for 2 days before and 3 weeks after monocrotaline.
▽ = mean values for 12 rats given monocrotaline at time zero plus daily indomethacin.
□ = mean values for 12 rats given monocrotaline plus daily normal saline injections.
* = values different ($P < 0.05$) from those in untreated controls.
† = values different ($P < 0.05$) from those in monocrotaline-treated rats.
+ = values different ($P < 0.05$) from those treated with monocrotaline and normal saline.

Source: Stenmark et al. (1985)

Activation of the cyclo-oxyenase pathway, demonstrated by increased concentrations of the prostaglandin metabolites 6-keto-$PGF_{1\alpha}$ and TXB_2 in lavage fluid, was inhibited by indomethacin, but this inhibition did not prevent the monocrotaline-induced injury. DEC attenuated both the inflammatory response and pulmonary hypertension and inhibited the formation of slow reacting substances including leukotriene D_4. Since DEC produces a pharmacological blockade of the lipo-oxygenase pathway, it seems that the latter, rather than the cyclo-oxygenase pathway, is responsible for perpetuating the pathophysiological mechanism leading to monocrotaline-induced pulmonary hypertension.

Hilliker et al. (1984) demonstrated that antibody-induced thrombocytopaenia attenuates right ventricular hypertrophy induced by monocrotaline in rats. In another study, Hilliker & Roth (1984) also produced evidence that hydrallazine, a vasodilator and inhibitor of platelet prostaglandin synthesis, dexamethason, an antiinflammatory agent and inhibitor of phospholipase, and sulfinopyrazone, an inhibitor of platelet prostaglandin synthesis inhibited monocrotaline-induced right ventricular hypertrophy in rats, supporting the hypothesis that platelets and vasoconstrictor agents play a role in monocrotaline-induced pulmonary hypertension. Likewise, prior chemical sympathectomy with 6 hydroxydopamine (100 mg/kg) or inhibition of serotonin synthesis with p-chlorophenylalanine (500 mg/kg) reduced the degree of monocrotaline-induced right ventricular hypertrophy in rats, but did not prevent or reduce pulmonary vascular muscularization (Tucker et al., 1983). Thus, the sympathetic nervous system and serotogenic mechanisms seemed to be involved in the development of right ventricular hypertrophy, but not in the development of the pulmonary vascular lesion induced by monocrotaline. Kay et al. (1985) also demonstrated that pretreatment with p-chlorophenylalanine, which inhibits 5-hydroxytryptamine (5HP) synthesis, also significantly ($p < 0.05$) reduced right ventricular systolic pressure, right ventricular hypertrophy, and medial thickness of muscular pulmonary arteries in monocrotaline-treated rats. Similar observations were made in rats exposed to hypoxia. It was therefore suggested that 5HP might play a role in monocrotaline-induced or chronic hypoxic pulmonary hypertension.

The biosynthesis of rat lung polyamines, putrescine, spermidine, and spermine, generally considered to be important regulators of cell growth and differentiation, is increased prior to the evolution of monocrotaline-induced pulmonary hypertension in rats. Continuous administration of α-difluoromethylornithine (DFMO), which is a highly specific irreversible inhibitor of ornithine decarboxylase (DDC), a rate-limiting enzyme in polyamine biosynthesis, attenuated the development of monocrotaline-induced pulmonary hypertension in rats (Olson et al., 1984). This effect was mediated by the DFMO, by inhibiting the synthesis of putrescine and spermidine, and not by blocking the hepatic metabolism of monocrotaline to pyrroles (Olson et al., 1985). Thus, it was suggested that lung polyamine biosynthesis might be essential for the expression of monocrotaline-induced perivascular oedema as well as medial thickening in the development of monocrotaline-induced pulmonary hypertension vascular disease.

On the basis of the preceding studies, mostly on the rat, the mechanism of chronic long-term injury to the lung by monocrotaline seems to be as follows. Within hours of PA administration, there is damage to the pulmonary endothelial cells accompanied by vascular leak leading to pulmonary oedema. Platelet aggregation also occurs. The endothelial damage indicated by ultrastructural and biochemical studies activates the production of prostacyclin and lipogenic products, which mediate increases in vascular permeability and inflammatory reaction. There is simultaneous production of 5 hydroxytryptamine and several polyamines. The injected monocrotaline is completely metabolized within hours, and no significant quantity is found in the body at 24 h (Hayashi, 1966) and, though some active metabolites may still be detectable by isotope studies, even at 14 days (Hsu et al., 1974), the rats do not have any lung lesions. The slow evolution of vascular changes suggests that it is not caused by monocrotaline but through biological pathways activated by the initial injury.

Methylprednisolone (MP), which reduces acute lung oedema caused by monocrotaline (MCT), has been shown to reduce MCT-induced pulmonary hypertensive vascular changes in rats and the resultant right ventricular hypertrophy (Langleben & Reid, 1985). Daily ip administration of MP at 5 mg/kg body weight, was found to be more effective than 2 large doses of MP at 30 mg/kg, 2 h before and 2 h after a single sc injection of MCT at 60 mg/kg. It was suggested that secondary changes, though triggered by the acute MCT injury, become self sustaining and are more significant for vascular structural remodelling.

Structural arterial remodelling with vasoconstriction, medial hypertrophy of the muscular pulmonary arteries, and muscularization of the pulmonary arterioles follow as late effects, resulting in pulmonary hypertension and right ventricular hypertrophy of the heart.

The results of the above studies suggest a direct toxic effect of the alkaloid on the endothelial cells of the alveolar capillaries and on the pulmonary arteries, as well as a pulmonary hypertensive effect on the heart.

6.4.3 Effects on the central nervous system

The dominant signs of pyrrolizidine poisoning in horses are neurological (Rose et al., 1957a,b; McLean, 1970). Similar signs can also occur in cattle and sheep. It has been claimed that such signs are probably non-specific secondary effects following primary liver disease resulting in hyperammonaemia (Rose et al., 1957a). However, neurological abnormalities in which animals walk in a straight line until they come to an object, and then stand with their heads pressed against the object, indicate specific lesions in the central nervous system. Spongy degeneration of the central nervous system occurs in cattle, sheep, and pigs (Hooper et al., 1974; Hooper, 1975a,b).

Trichodesma alkaloids, in particular, appear to be neurotoxic. There is a considerable body of literature in the USSR on Trichodesma intoxication of mice, rabbits, and dogs, which has been reviewed (Ismailov et al., 1970). Mice given Trichodesma alkaloids subcutaneously at 0.5 mg/kg develop paresis of the hind limbs within 12 - 17 days. Opisthotonus and clonic convulsions are also seen. Doses of 10 - 15 mg/kg of alkaloids produces death in all animals within 2 - 6 h, as the result of respiratory depression. Higher doses produce immediate death.

6.4.4 Effects on other organs

Right ventricular hypertrophy, secondary to the primary effects on the pulmonary arteries, and the resultant pulmonary hypertension in animals treated with PAs or PA-containing plants have been dealt with in section 6.4.2. Lalich & Merkow (1961) reported myocarditis consisting of focal oedema and infiltration with a minimal number of lymphocytes and mononuclear cells in some rats fed Crotalaria spectabilis seeds mixed with the diet at a concentration of 0.13 - 2 g/kg. Treated groups consisted of 11 - 24 animals each; there were 12 controls. The changes were seen in all groups of animals, but the maximum number of rats (10 out of an unspecified number of the group) showing these changes was in the group that received 0.5 g/kg diet for 20 - 31 days. Generally, there was a close correlation with the presence of pulmonary arteritis.

Renal changes have been described by a number of investigators. Hayashi & Lalich (1967) observed mild to moderate changes in renal glomeruli consisting of necrosis, capillary thrombosis, and degenerative changes in the epithelial and mesangial cells, thickening of interlobular arteries, and arterial thrombosis in suckling, male Sprague Dawley rats given a single sc dose of monocrotaline at

120 mg/kg body weight. Renal changes were seen, to some extent, in all animals surviving for 41 - 47 days.

Carstens & Allen (1970) studied the effects of feeding Crotalaria spectabilis seed on the rat kidney. Fifty male Sprague Dawley rats were fed a diet containing ground Crotalaria spectabilis seed at 0.2 - 0.8 g/kg for 8 months. The seeds were estimated to contain approximately 3.5 g monocrotaline/kg; 10 animals served as controls. Renal changes were seen in 33/50 PA-treated rats. In 22 rats, over 75% of the glomeruli were hyalinized and capsules thickened. In the less severely affected kidneys, the glomerular basement membrane was thickened and homogeneous deposits were seen in mesangial areas. Afferent arterioles and interlobular arteries were markedly thickened. In the most severely affected vessels, the internal elastic lamina was necrosed and the larger arteries showed fibrinoid necrosis.

Renal tubular megalocytosis was the dominant lesion described by Hooper (1974) in mice. Nine male white mice, 10 weeks of age, were fed Senecio jacobaea, which contained a concentration of alkaloids (jaconine, jacobine, and seneciphylline) of 2.7 g/kg and a concentration of N-oxide of 0.9 g/kg, mixed with the diet. The S. jacobaea was given at 100 g/kg diet for 9 weeks, before being raised to 200 g/kg diet. Five animals served as controls. The animals were killed from 63 to 193 days after the start of the study. All treated animals, except 2 killed on day 63, showed changes. The large cells occurred in both the proximal tubules and the loop of Henle. Similar cells were seen in the alveolar and bronchiolar epithelium. No glomerular lesions were described. The author mentioned having seen the above changes in rats given repeated sublethal injections of fulvine and spectabiline. On the other hand, Kurozumi et al. (1983) observed glomerular lesions in rats given a single injection of monocrotaline.

A variety of renal lesions has been observed in pigs, a common pathological feature being renal megalocytosis, which was observed in pigs poisoned by at least 4 different plant genera containing a variety of toxic alkaloids (Harding et al., 1964; Peckham et al., 1974) and has also been observed in wild pigs grazing in areas rich in PA-containing plants in northern Australia (Hooper, 1978). McGrath et al. (1975) described glomerular lesions in pigs given Crotalaria spectabilis seed daily for 43 days. Severe renal lesions comprising tubular dilatation, megalocytosis, and necrosis of tubular epithelial cells with casts in the lumen, interstitial and periglomerular fibrosis, and glomerular hyalinization were reported by Hooper & Scanlan (1977) in pigs fed Crotalaria retusa seeds containing monocrotaline. Renal megalocytosis has also been reported in C. retusa poisoning in horses, sheep, and mice poisoned by S. jacobaea but not by H. europaeum, and in vervet monkeys with chronic retrorsine poisoning (Van der Watt et al., 1972).

Lesions have been reported in the stomach and intestines in field and experimental animals after poisoning with pyrrolizidine alkaloids, but are difficult to identify as specific PA injury. Hooper (1975c) conducted studies on sheep, rats, and mice. In the study on sheep, 12 male cross-bred lambs, 7 - 8 weeks of age, were newly weaned on to a standard commercial calf grower diet. Lasiocarpine was administered at the rate of 15 - 20 mg/kg body weight every 2 - 4 days. Each animal was killed when in terminal coma. Survival time ranged from 4 to 17 days. In the rat study, young Wistar-Furth rats (sex not stated) weighing 150 - 200 g were used. In one group of 11 rats, each animal received an ip injection of lasiocarpine at the rate of 40 mg/kg body weight. Three animals received isotonic saline. Animals were killed or died 2 - 6 days after the injection. A second group of 13 rats received a dose of 35 mg lasiocarpine/kg body weight; 4 control animals received saline. All rats received a second injection 48 h later. They were killed 3 - 60 days after the second administration of lasiocarpine. In the mouse study, 3 mature male white mice received 6 injections each of lasiocarpine at the rate of 45 mg/kg body weight followed by 4 injections of 90 mg/kg body weight at 48-h intervals. There was one control animal.

All animals showed characteristic hepatic lesions. Sheep also showed severe oedema, haemorrhage, and epithelial necrosis in the gall bladder; lesions were also found in the central nervous system and occasionally in the kidney. All animals showed severe intestinal atrophy. There was inhibition of crypt cell mitosis leading to mitotic irregularities, abnormal large cells and syncitial cells, especially in the duodenum of sheep, and severe villous atrophy with ulceration. Lesions in the intestines were similar to those caused by radiation and radiomimetic agents. It was suggested that the local intestinal radiomimetic effect was due to local exposure to the pyrrole metabolite of lasiocarpine after excretion through the bile duct. It was proposed that a more conspicuous and rapid development of duodenal megalocytosis was due to very rapid turnover of cells in the duodenum.

Other probably secondary effects included haemolysis in sheep in association with advanced liver disease and high liver-copper levels (Bull et al., 1956), anaemias and disturbance in iron metabolism and haematopoiesis (Schoental & Magee, 1959; Schoental, 1963; Peckham et al., 1974; Hooper & Scanlan, 1977) (section 6.4.11), pancreatic oedema and fibrosis (Bras & Hill, 1956; Schoental & Magee, 1959), cerebral oedema, haemorrhage, and congestion in the rat brain (Davidson, 1935; Rosenfield & Beath, 1945).

Tumours in the different organs have been dealt with separately under carcinogenesis (section 6.4.8).

6.4.5 Teratogenicity

The teratogenic potential of PAs was demonstrated by Green & Christie (1961) who produced a variety of dose-related fetal abnormalities in the rat, with a single intraperitoneal injection of heliotrine administered during the second week of gestation. The dosages ranged from 15 to 300 mg/kg maternal body weight. Litters exposed to a dose of less than 50 mg did not show any abnormalities, but abnormalities were observed in litters exposed to higher doses, and increased in frequency and severity with increasing dose. The abnormalities included retardation of development, musculoskeletal defects, especially hypoplasia of the lower jaw, cleft palate, and other abnormalities. Doses above 200 mg resulted in the intrauterine death or resorption of many fetuses.

Similar studies were performed by Peterson & Jago (1980) who compared the effects of heliotrine with its metabolic pyrrole derivative dehydroheliotridine (DHH), when administered in a single ip injection to rats on the 14th day of gestation. Heliotrine was administered at 200 mg/kg body weight and DHH at 30 - 90 mg/kg, 14 days after conception. Effects on embryos, evaluated on the 20th day, showed that both heliotrine and DHH retarded growth and were teratogenic, but that the effects of a 40 mg/kg dose of DHH were equivalent to those of 200 mg/kg heliotrine, i.e., the metabolite was 2.5 times as effective on a molar basis. DHH produced a number of skeletal abnormalities including retarded ossification, distorted ribs, long bones, cleft palate, and feet defects. At higher doses, growth almost ceased in many tissues and the fetuses were very immature. However, the embryonic liver parenchyma did not show the antimitotic effects of DHH.

The teratogenic properties of heliotrine were also demonstrated in _Drosophila_ larvae fed low levels of the alkaloid (Brink, 1982).

6.4.6 Fetotoxicity

The subject of fetotoxicity has been reviewed by Mattocks (1986). Sundareson (1942) demonstrated the ability of pyrrolizidine alkaloids to cross the rat placenta. Twice weekly injections of the PA, starting at, or after, the 12th day of gestation, resulted in premature delivery of some litters and many were born dead. The same author showed that the alkaloids themselves and not just the pyrroles formed in the dams' livers, could pass the placental barrier by injecting senecionine into 19-day-old rat fetuses _in utero_, which produced the characteristic toxic lesions in the dams. The fetuses were also found to be more resistant to the lethal effects of the PA than the mother rats. When 4 fetuses were each administered 1.25 mg of PA, representing about 200 - 400 mg/kg body weight, which is much higher than the LD_{50} for an adult rat, 3 of them were still alive after 2 days. Green & Christie (1961) did not find any liver damage in fetuses from pregnant rats given teratogenic doses of heliotrine. Only mild liver damage was found in the embryo rats whose mothers had been injected with PAs (heliotrine, lasiocarpine, retrorsine, or monocrotaline) (Bhattacharya, 1965). In contrast, Schoental (1959) demonstrated that lasiocarpine and retrorsine, when administered to lactating rats produced little effect on the mothers, but produced acute liver lesions in the suckling infants. The lesions were most severe in 3- to 7-week-old animals. It was suggested that the infants were affected by the milk from the lactating mothers, which possibly contained the metabolic products of the PAs.

It would seem from the above studies that the embryo is relatively more resistant to the toxic effects of PAs in utero than it is after birth. Mattocks & White (1973) postulated that this could be due to the low capacity for the metabolic activation of PAs of the embryo liver, as they had shown that the ability of liver enzymes to convert retrorsine to toxic metabolites was low in rats, immediately after birth, but picked up rapidly afterwards. The susceptibilities of rats of various ages to the hepatotoxic effects of the PAs was proportional to their capacity to form and retain the pyrrolic metabolites. Twenty-day-old rats were found to be more sensitive than older animals.

The effects of fulvine administration on pregnant rats between 9 and 12 days of gestation were studied by Persaud & Hoyte (1974). Dose-related fetal resorptions were observed, but no hepatic lesions were seen in the fetuses. On the other hand, Newberne (1968) observed damage, in both the maternal and fetal livers, when lasiocarpine was administered to pregnant rats. Acute liver necrosis was observed in the livers of mothers as well as fetuses in animals that had received 100 mg lasiocarpine/kg body weight on day 13 of gestation. However, in animals that received 2 doses of 35 mg/kg body weight on days 13 and 17 of pregnancy, liver necrosis was seen in the fetal liver but not in that of the mother. It is not known why lasiocarpine acts differently from other alkaloids and has a greater effect on the fetal liver. Mattocks (1986) has postulated the possibility that fetotoxicity was caused chiefly by toxic metabolites formed in the maternal liver, and that a greater proportion of such metabolites reached the fetus from lasiocarpine than from other PAs.

6.4.7   Mutagenicity

A number of PAs that have been shown to be powerful dose dependent mutagens in Drosophila melanogaster have been listed by Mattocks (1986). All the compounds are hepatotoxic though the degree of mutagenicity is not necessarily proportional. Table 12 provides a summary of the mutagenicity tests on different PAs, related compounds and plant extracts. Clark (1959) demonstrated the mutagenic effect of heliotrine in Drosophila, in which a considerable increase in sex-linked recessive lethals was produced, apparently by interfering with the maturation of germ cells, so that as soon as the available spermatozoa were used, the males were no longer capable of breeding. The cell damage was irreversible. The mutagenic effect of feeding Drosophila males for 24 h with a medium containing 10^{-3} mol monocrotaline was comparable to about 1000 R of X rays (Clark, 1976). The Basc test with Drosophila melanogaster is considered a highly sensitive mutagenicity test for PAs (Candrian et al., 1984a).

Seneciphylline and senkirkine, known to occur in animal feeds and medicinal herbs, respectively, were tested for their ability to produce sex-linked recessive lethals in males of Drosophila melanogaster using the Basc (3-day feeding method) by Candrian et al. (1984a). Seneciphylline was found to be mutagenic at concentrations of 10^{-5}, 10^{-4}, and 10^{-3} mol, which produced 3.8% (983 chromosomes tested), 9% (708 chromosomes tested), and 15.3% (327 chromosomes tested) sex-linked recessive lethals, respectively. Senkirkine (10^{-5} mol) was found to produce 4.4% sex-linked recessive lethals (2541 chromosomes tested) against 0.17% maximum sensitivity in the late spermatid stage of spermatogenesis indicating that PAs act as indirect mutagens. Flies fed with milk from lactating rats given an oral dose of 25 mg seneciphylline/kg showed 1.2% sex-linked recessive lethals (1477 chromosomes tested) compared with 0.3% (1533 chromosomes tested) in controls.

Mutagenic properties of 7 PAs extracted from plants to Salmonella typhimurium TA100 have been demonstrated by a modified Ames method by Yamanaka et al. (1979). The PAs were clivorine, fukinotoxin, heliotrine, lasiocarpine, ligularidine, LXC201, and senkirkine. Pre-incubation of these alkaloids with liver S9 mix and bacteria in liquid medium was essential for demonstration of the property. PAs in the heliotridine and otonecine family were mutagens, while retronecine bases were inactive. Monocrotaline and heliotrine were not active mutagens to Escherichia coli WP2, even though they were quite cytotoxic (Green & Muriel, 1975). They were active in repair deficient strains. Retrorsine was active in inducing mutations on the Ames Salmonella/microsome assay (Wehner et al., 1979). Extracts from medicinal plants and noxious weeds were mutagenic towards Salmonella in the Ames assay (Pool, 1982; White et al., 1983; Koletsky et al., 1978).

Table 12. Mutagenicity tests on pyrrolizidine alkaloids, related compounds, and source plants

Compound or material	Type of test[a]	Response[b]
Clivorine	A	+
	HPC	+
Echimidine	D	+
Echinatine	D	+
Fulvine	D	+
Heliotrine	D	+
	P	+
	F	+
	CC	+
	A	+
	B	+
	TM	+
	CM/CC	+
Integerrimine	D	+
Jacobine	D	±
	P	+
Lasiocarpine	D	+
	P	+
	F	+
	A	+
	HPC	+
	CM/CC	+
	TM	+
Ligularidine	A	+
Lindelofine	A	0
Lycopsamine	A	0
Monocrotaline	D	+
	P	+
	CC	+
	B	+
	A	0
	HPC	+
	CT	+
Petasitenine (fukinotoxin)	A	+
	HPC	+
	CM	+

Table 12 (contd).

Compound or material	Type of test[a]	Response[b]
Platyphylline	D	0
Retrorsine	A	+
	D	+
	CT	+
Rosmarinine	CT	0
Senecionine	D	+
	A	0
Seneciphylline	A	0
	P	+
	D	+
Senkirkine	A	+
	HPC	+
	CM/CC	+
	D	+
Supinine	D	±
	P	+
Mixed alkaloids from Senecio jacobaea	A	0
Senecio numorensis spp.. fuchsii (extract)	CM	+
	A	0
	A	+
Senecio jacobaea (extract)	A	+
Senecio longilobus (extract)	A	0
Symphytum officinale (comfrey extract)	A	0
Retronecine bis-p-chloro-benzoate	P	+
Synthanecine A bis-N-ethyl-carbamate	CT	+
Retronecine	A	0
Heliotridine	D	0
Viridofloric acid	A	0
Heliotric (heliotrinic) acid	D	±
	HPC	+

Table 12 (contd).

Compound or material	Type of test[a]	Response[b]
Dehydroretronecine	CT A SCE	+ ± +
Dehydroheliotridine	CM	+
Pyrrole	HPC	0
2,3-Bishydroxymethyl-1-methyl-pyrrole	CT A SCE	+ ± +
2-Hydroxylmethyl-1-methyl-pyrrole	A SCE	0 ±
3-Hydroxylmethyl-1-methyl-pyrrole	A SCE	+ ±

[a] A = Salmonella ("Ames") test.
B = Other bacterial tests.
CC = Clastogenic activity in cultured cells.
CM = Mutagenicity in cultered mammalian cells.
CT = Cell transformation test.
D = Mutagenicity in Drosophila.
F = Tests in fungus (Aspergillus nidulans).
HPC = Hepatocyte primary culture/DNA repair test.
P = Chromosomal aberrations in plant cells.
SCE = Sister chromatid exchange.
TM = Transplacental micronucleus test.

[b] + = active.
± = marginally active.
0 = inactive.

From the limited data available, it seems that the carcinogenic activity of individual alkaloids parallels their mutagenic behaviour, but not their relative hepatotoxicities (Culvenor & Jago, 1979).

6.4.7.1 Chromosome damage

Pyrrolizidine alkaloids have been shown to be capable of damaging chromosomes in plants, fungi, bacteria, tissue cell cultures, and the fruit fly (Drosophila melanogaster). Literature on this topic has been reviewed by Bull et al. (1968), McLean (1970), and Mattocks (1986).

Several PAs are known for their ability to damage the chromosomes of growing plant cells (Mattocks, 1986). Similar properties have been demonstrated in leukocyte cultures from the marsupial (Potorus tridactylus) (Bick & Jackson, 1968; Bick, 1970). Bick & Culvenor (1971) found dehydroheliotridine, a metabolite of heliotrine, to be 10 times more active than the alkaloid.

Infusions of Symphytum officinale L., described in Polish pharmacopoeia as Radix symphyti, are recommended as expectorants, especially for children. Furmanowa et al. (1983) demonstrated the mutagenic effects of an alkaloidal fraction and infusion in this plant in the meristematic cells of the lateral roots of Vicia faba L. var minor. Lasiocarpine, a proven carcinogen, served as a positive control.

Chromosome damage by PAs in the hamster lung cell line was demonstrated by Takanashi et al. (1980). Stoyel & Clark (1980) used the transplacental micronucleus test in pregnant female mice and showed the chromosome damaging properties of heliotrine (225 mg/kg body weight) and lasiocarpine (86 mg/kg) within 20 h of the injection.

The genotoxicity of heliotrine, monocrotaline, seneciphylline, and senkirkine was studied by Bruggeman & Van der Hoeven (1985) using the sister-chromatid exchange (SCE) assay in V79 Chinese hamster cells co-cultured with primary chick embryo hepatocytes. Exposure to these PAs resulted in the high induction of SCEs, a more than 5-fold increase in the SCE rate with 2.5 mg heliotrine/litre, 4-fold with monocrotaline at 5 mg/litre, 8-fold with seneciphyline at 1.2 mg/litre, and more than 5-fold response with senkirkine at 2.5 mg/litre. For all compounds, a dose-response relationship was observed at concentrations that did not seriously affect survival. PAs are also known to induce DNA repair in rodent hepatocytes (Green et al., 1981; Mori et al., 1985). DNA repair synthesis was elicited by 15 alkaloids, including 11 of unknown carcinogenic potential (Mori et al., 1985).

There are also a few reports of chromosome damage by PAs in man. Martin et al. (1972) found chromosome damage in the blood cells of children with veno-occlusive disease, probably caused by fulvine. It has also been shown by Ord et al. (1985) that dehydroretronecine, is able to induce SCE in human lymphocytes. Kraus et al. (1985) studied the PAs senkirkine and tussilagine, which occur in a medicinal plant Tussilago farfara, for their ability to induce chromosome damage in human lymphocytes in vitro. They were not found to enhance the number of chromosome aberrations up to concentrations of 1000 µmol. However, heliotrine, used for comparison, induced chromosomal aberrations at concentrations of 100 µmol. In addition, heliotrine was also found to be capable of damaging unstimulated eg G_0-phase lymphocytes.

6.4.8   Carcinogenesis

Carcinogenesis has been reviewed by McLean (1970), IARC (1976, 1983), and Mattocks (1986). A number of purified PAs, purified or crude extracts of plants containing them in a mixture or the actual plant, dried and milled, and several PA metabolites or synthetic analogue compounds have been tested for carcinogenecity. However, these include only relatively few of the known cytotoxic PAs. Data relating to some of the representative studies on rats are summarized in Table 13. Studies on liver tumours found in rats given PAs and plant materials are summarized in Table 14. All experimental animal studies, with the exception of one on chickens (Campbell, 1956) and one on Syrian golden hamsters (Fushimi et al., 1978), have been carried out on rats.

The liver is the most common organ involved in experimental studies. Tumours produced are mostly of epithelial origin, but a significant number are also vascular. Lack of precision and diversity of terms used to describe similar or identical tumours makes it difficult to compare the types of carcinogenic effect in different studies. Some terms have been used interchangeably, e.g., hepatomas, hepatocellular carcinomas, haemangiogenic and cholangiogenic tumours; nodular hyperplasia, pre-neoplasma, neoplastic nodules, and hepatocellular tumours. In most studies, there are no supporting photomicrographs to draw any inference as to whether the tumours were malignant. Difficulties in the interpretation of data have been commented on by Schoental et al. (1954) and McLean (1970).

Lasiocarpine has produced the largest yield of tumours. In the studies of Svoboda & Reddy (1972), 16/18 animals surviving for more than 56 weeks after receiving ip multidoses of lasiocarpine developed malignant tumours of the liver. Of these, 10 animals had more than one tumour. Continuous feeding of rats on a regimen containing lasiocarpine resulted in all animals (24/24) developing tumours (NCI, 1978). In one study, a single oral administration of retrorsine (Schoental & Bensted, 1963) to weanling rats resulted in 7 of the 29 animals that survived for more than one year developing 11 tumours of a wide variety, at least of 5 which were malignant. It is of note that this PA is known to have caused two cases of human toxicity together with riddelline in two cases, though the total intake was proportionately lower (Stillman et al., 1977; Fox et al., 1978; Huxtable, 1980) (Table 15).

Tumours produced covered a very wide range in unrelated tissues and organs, for example, the pancreas, urinary bladder, pituitary, bone, retro-peritoneal tissues, and skin, among others. Hepatocellular carcinoma and haemangiosarcoma were the most common.

Table 13. Summary of data on the carcinogenic action of PAs and PA-containing plants in rats (in chronological order)

Material tested	Strain, number, and sex of animals	Route of administration and dosing regimen	Other treatment	Duration of treatment	Period of observation	Tumours produced	Tumour incidence in surviving animals	Reference
Alkaloids of Senecio jacobaea	Wistar rat (13 males, 12 females)	drinking-water (0.05 g/litre) followed by		1 week (males), 2 weeks (females); gap of 7 weeks	until death	only nodular hyperplasia of liver		Schoental et al. (1954)
	9 males, 1 female (surviving)	0.03 g/litre, 3 days/week		until death				
Retrorsine	10 males, 4 females	0.03 g/litre, 3 days/week		until death	until death	hepatomas	4/14	
Isatidine	8 males, 14 females	0.05 g/litre followed by 0.03 g/litre, 3 days/week		20 months	until death	hepatomas	10/22	
	3 males, 4 females	as above	choline (0.5% in drinking-water), 4 days/week	until death		hepatomas	3/7	

Table 13 (contd).

Isatidine (contd)	2 males, 3 females	single ip injection of 2 mg in 0.2 ml tricaprilyn followed by skin application of 0.5% solution, 3 days/week	15 months	hepatoma (?)	1/5	Schoental et al. (1954)
	controls (7 males, 7 females)		15 months	not mentioned		
Retrorsine	Porton Wistar weanling rat (50 males)	single intragastric dose of 30 mg/kg body weight	400 r radiation to 31/50 surviving for 100 days; 4/13 had head shielded	until death	hepatomas	5/25	Schoental & Bensted (1963)
					hepatocellular carcinoma with metastases	1/25	
					mammary tumours	2/25	
					lung carcinoma	1/25	
					renal carcinomas	2/25	
					colonic carcinoma	1/25	

Table 13 (contd).

Material tested	Strain, number, and sex of animals	Route of administration and dosing regimen	Other treatment	Duration of treatment	Period of observation	Tumours produced	Tumour incidence in surviving animals	Reference
Retrorsine (contd)	Porton Wistar rat (50 males)	as above	as above			splenic haemangio-endothelioma	1/25	Schoental & Bensted (1963)
						osteosarcoma bone	1/25	
						leukaemia	1/25	
						"spindle cell" tumour (neck)	1/25	
	95 males, 95 females (weanling) (controls)	single oral dose of 30 mg/kg body weight			until death	hepatomas	5/29	
						mammary tumour	1/29	
						lung carcinoma	1/29	
						splenic haemangio-endothelioma	1/29	
						uterine carcinoma	1/29	

Table 13 (contd).

Retrorsine (contd)	Porton Wistar rat (controls)			retroperitoneal sarcoma	1/29	Schoental & Bensted (1963)
	6 males (weanling)	no PAs	400 r radiation	squamous cell carcinoma (jaw)	1/29	
				leukaemia	2/6	
			until death	osteosarcoma	1/6	
				renal adenoma	1/6	
	10 males (weanling)	single intragastric dose (30 mg/kg body weight) (9 days after partial hepatectomy)	partial hepatectomy	hepatomas	2/9	
			until death	squamous cell carcinoma (jaw)	1/9	
Mixed PAs from seeds of Amsinckia intermedia (intermedine and lycopsamine)	Randomly bred from Porton Wistar weanling rat (15 males)	single intragastric dose at 500 - 1500 mg/kg body weight	more than 1 year ?	pancreatic islet cell adenoma[a]	1/15	Schoental et al. (1970)
				"papillary tumour"[a] of urinary bladder	1/15	
				pituitary adenoma[a]	1/15	

Table 13 (contd).

Material tested	Strain, number, and sex of animals	Route of administration and dosing regimen	Other treatment	Duration of treatment	Period of observation	Tumours produced	Tumour incidence in surviving animals	Reference
Mixed PAs from seeds of Amsinckia intermedia (contd)	as above	as above				pancreatic islet cell adenocarcinoma	1/15	Schoental et al. (1970)
						exocrine pancreatic adenoma	1/15	
Leaves and stems of Heliotropium supinum L.	2 males (weanling)	10% mixed with diet		1 month	until death (period not stated)	pancreatic islet cell adenoma	1/2	
	6 males	single intragastric dose at 200-300 mg/kg body weight of crude alkaloidal fraction			until death (longer than 1 year)	pancreatic islet cell adenoma	1/6	
	controls					not stated		
Senecio longilobus	Harlan rat 50 males, 50 females	0.75% of diet			until death within 131 days	none		Harris & Chen (1970)

Table 13 (contd).

Senecio longilobus (contd)	50 males, 50 females	0.5% of diet	until death within 200 days	non		Harris & Chen (1970)
	40 males, 40 females	0.5% of diet for 1 month alternating with normal diet for 2 weeks	1 year	liver cell carcinomas	4/23	
				peritoneal mesothelioma	1/23	
	50 males, 50 females	0.5% of diet for 1 week alternating with normal diet for 1 week	54 weeks	liver cell carcinomas	16/47	
				angiosarcoma	1/47	
	controls (10 males, 10 females)			not stated		
Lasiocarpine	Fischer rat (25 males)	intraperitoneal injection at 7.8 mg/kg body weight, twice weekly for 4 weeks, then once a week for an additional 52 weeks	till moribund or had palpable tumours	hepato-cellular carcinomas	10/18	Svoboda & Reddy (1972)
			60 – 76 weeks	cholangio-carcinoma	1/18	
				lung adenomas	5/18	
				skin squamous cell carcinomas	6/18	

Table 13 (contd).

Material tested	Strain, number, and sex of animals	Route of administration and dosing regimen	Other treatment	Duration of treatment	Period of observation	Tumours produced	Tumour incidence in surviving animals	Reference
Lasiocarpine (contd)	Fischer rat (25 males)	as above				ileal adenocarcinoma	2/18	Svoboda & Reddy (1972)
						ileal adenomyoma	1/18	
						testicular interstitial cell tumour	1/18	
	controls (25 males)					lung adenomas	2/25	
Retronecine, hydrochloride	Porton Wistar new-born rat (6 males, 6 females)	single sc injection of 300 - 1000 mg/kg body weight				spinal cord ependymoblastoma	1/10	Schoental & Cavanagh (1972)
						"pituitary tumour"	5/10	
						"mammary tumour"	1/10	
Hydroxysenkirkine	5 males (weanling)	single ip injection of 100 - 300 mg/kg body weight				brain astrocytoma	1/5	Schoental & Cavanagh (1972)

Table 13 (contd).

Heliotropium ramosissimum	5 females	5% of diet fed to 1 pregnant rat during 1st 15 days of pregnancy and from 10th day of parturition until weaning; female offspring (5) fed on experimental diet for 10 days at 6 months of age		Schwann cell tumour of spinal cord	1/5
	controls			not stated	
Petasites japonicus (young flower stalks)	ACI rat Group I: 12 males, 15 females	4% of diet for 6 months; subsequently, 8% of diet alternating weekly with normal diet	480 days or until moribund	liver cell adenomas liver haemangio-endotheliomas liver cell carcinomas	Groups I II 6/27 4/19 3/27 8/19 2/27 1/19	Hirono et al. (1973)
	Group II: 11 males, 8 females	4% of diet	until death		
	controls 7 females/ 8 females	normal diet		none	

Table 13 (contd).

Material tested	Strain, number, and sex of animals	Route of administration and dosing regimen	Other treatment	Duration of treatment	Period of observation	Tumours produced	Tumour incidence in surviving animals	Reference
Monocrotaline	Sprague Dawley rat (50 males)	gastric intubation weekly of 25 mg/kg body weight for 4 weeks, then 8 mg/kg body weight for 38 weeks			72 weeks	liver cell carcinomas	10/42	Newberne & Rogers (1973)
	50 males	As above, with a diet deficient in lipotropes				liver cell carcinomas	14/33	
Heliotrine	Porton Wistar weanling rat	intragastric						Schoental (1975)
	4 males	300 mg/kg body weight; repeated after 3 weeks to 2 surviving rats			until death (5 months)	none		

Table 13 (contd).

Heliotrine (contd).					Schoental (1975)
6 males	300 mg/kg body weight nicotinamide (500 mg/kg body weight), ip, before and after each heliotrine administration; repeated after 3 weeks	until death (5 months)	none		
4 males	230 mg/kg body weight; repeated to 2/4 surviving after 5 days	27 months	pancreatic islet cell adenoma	1/1	
			pituitary adenoma	1/1	
12 males	230 mg/kg body weight nicotinamide (350 mg/kg body weight), ip, before, plus 2 doses after each heliotrine administration; repeated 6.5 days later	until death or killed when moribund	fibrosarcoma of jaw	1/8	
			pancreatic islet cell adenomas[b]	3/6	
		up to 27.5 months	hepatoma[b]	1/6	
			urinary bladder papilloma[b]	1/6	
12 males			testicular interstitial cell tumour[b]	1/6	

Table 13 (contd).

Material tested	Strain, number, and sex of animals	Route of administration and dosing regimen	Other treatment	Duration of treatment	Period of observation	Tumours produced	Tumour incidence in surviving animals	Reference
Heliotrine (contd)	2 males	ip injection	nicotinamide at 350 mg/kg body weight		19 – 27.5 months	pituitary adenoma	1/2	Schoental (1975)
	controls (6 males)	no treatment			19 – 27.5 months	pituitary adenomas	3/6	
Tussilago farfara (coltsfoot) (pre-blooming flowers)	ACI rat (6 males, 6 females) (1.5 months old)	32% in diet for 4 days, then 16% until end of study		600 days	600 days	haemangio-endothelioma	8/12	Hirono et al. (1976)
						liver cell adenoma[c]	1/12	
						hepatocellular carcinoma[c]	1/12	
						bladder papilloma[c]	1/12	
	5 males, 5 females	8% in diet		600 days	600 days	liver haemangio-endothelioma	1/9	

Table 13 (contd).

Tussilago farfara (contd)	6 males, 5 females	4% in diet	600 days	600 days	none	Hirono et al. (1976)
	controls (8 males, 8 females)	none	600 days	600 days	none	
Dehydroretronecine	Sprague Dawley rat Group I: 75 males	sc injection bi-weekly for 4 months at 20 mg/kg body weight, followed by 10 mg/kg body weight for 8 months	partial hepatectomy on 15 animals after 4 months	12 months		Allen et al. (1975)
				10 months	rhabdomyo-sarcomas (5 with metastases)	36/60
Monocrotaline	Group II: 75 males	sc injection bi-weekly at 5 mg/kg body weight for 12 months	partial hepatectomy on 15 animals after 4 months	12 months		
				10 months	rhabdomyo-sarcomas	2/60
					hepato-cellular carcinomas	2/60
					acute myeloid leukaemia	2/60
					pulmonary adenomas	2/60

Table 13 (contd).

Material tested	Strain, number, and sex of animals	Route of administration and dosing regimen	Other treatment	Duration of treatment	Period of observation	Tumours produced	Tumour incidence in surviving animals	Reference
Monocrotaline (contd)	controls (50 males)	sc injection bi-weekly, 0.1 mol phosphate buffer, pH7	partial hepatectomy on 5 animals after 4 months	12 months	10 months	none mentioned		Allen et al. (1975)
	Sprague Dawley rat (60 males)	sc injection on alternate weeks (5 mg/kg body weight)		12 months		any tumours (several animals had more than 1 tumour)	17/60	Shumaker et al. (1976)
						pulmonary carcinomas	11/60	
						hepato-cellular carcinomas	5/60	
						acute myeloid leukaemia	3/60	
						rhabdomyo-sarcomas	4/60	
						adrenal adenomas	8/60	
						kidney adenoma	1/60	

Table 13 (contd).

Dehydroretronecine	60 males	sc injection on alternate weeks at 20 mg/kg body weight followed by 10 mg/kg body weight, alternate weeks for 8 months	12 months	12 months until moribund	rhabdomyosarcomas at site of injection	39/60	Hayashi et al. (1977)
	controls (45 males) (same group as above)				adrenal adenomas	2/45	
Monocrotaline	Sprague Dawley rat (80 males)	single sc injection of 40 mg/kg body weight	45 days	72 days	pancreatic insulinomas	16/23	
Petasitenine	ACI rat (3 males) (1 month old)	drinking-water (0.05% solution)	until death or moribund	up to 16 months	none		Hirono et al. (1977)
	5 males, 6 females	0.01% solution			haemangioendothelial sarcomas	5/10$\underline{d}$	
					liver adenomas	5/10$\underline{d}$	

Table 13 (contd).

Material tested	Strain, number, and sex of animals	Route of administration and dosing regimen	Other treatment	Duration of treatment	Period of observation	Tumours produced	Tumour incidence in surviving animals	Reference
Petasitenine (contd)	controls (10 males, 9 females)	none				fibrosarcoma (subcutaneous)	1/10	Hirono et al. (1977)
Lasiocarpine	Fischer rat (20 males)	mixed with feed at a concentration of 50 mg/kg		55 weeks	59 weeks	liver angiosarcomas	9/20	Rao & Reddy (1978)
						hepatocellular carcinomas	7/20	
						malignant adnexal tumour of skin	1/20[e]	
						malignant lymphoma	1/20[e]	
	controls (10 males)					none		

Table 13 (contd).

Petasites japonicus (flower stalks)	ddN mice 24 males 21 females	4% of diet as dried flower stalks	480 days	480 days	lung adenomas	24/39	Fushimi et al. (1978)
					lung adenocarcinoma	6/39	
					liver reticullum cell sarcoma	4/39	
					liver-haemangio-endothelial sarcoma	1/39	
					thymoma	1/39	
					leukemia	2/39	
	Control 23 males 27 females	nil	480 days	480 days	lung	1g	
					kidney haemangio-endothelial sarcoma	1g	
					spleen haemangioma	1g	

Table 13 (contd).

Material tested	Strain, number, and sex of animals	Route of administration and dosing regimen	Other treatment	Duration of treatment	Period of observation	Tumours produced	Tumour incidence in surviving animals	Reference
Petasites japonicus (contd)	Swiss strain mice 20 males 20 females	4% diet as dried flower stalks		480 days	480 days	lung adenoma	5/26	Fushimi et al. (1978)
						leukemia	1/26	
	controls 23 males 20 females	nil		480 days	480 days	liver haemangio-endothelioma	1g	
						breast carcinoma	1g	
						lung adenoma	3g	
						leukemia	2g	
	C57BL/6 mice 20 males 20 females	4% diet as dried flower stalks		480 days	480 days	no tumours		
	controls 20 males 20 females	none		480 days	480 days	lung adenoma	1g	

Table 13 (contd).

Petasites japonicus (contd)	Syrian golden hamsters 13 males 17 females	4% diet as flower stalks	480 days	480 days	adrenal cortical adenoma & breast carcinoma	1/25	Fushimi et al. (1978)
	controls 12 males 9 females	none	480 days	480 days	no tumours		
Symphytum officinale (leaves or root)	ACI rat (1 - 1.5 months old) Group I.1: 11 males, 8 females	33% of diet as leaves	480 days	until death or moribund	liver adenomas	5/19	Hirono et al. (1979b)
					urinary bladder papilloma	1/19	
					urinary bladder carcinomas	2/19	
					(rats with tumours)	5/19	
	Group I.2: 10 males, 10 females	33% of diet as leaves	600 days	until death or moribund	liver adenomas	11/19	
					urinary bladder papillomas	2/19	
					(rats with tumours)	11/19	

Table 13 (contd).

Material tested	Strain, number, and sex of animals	Route of administration and dosing regimen	Other treatment	Duration of treatment	Period of observation	Tumours produced	Tumour incidence in surviving animals	Reference
Symphytum officinale (contd)	group II: 11 males, 10 females	16% of diet as leaves		600 days	until death or moribund	liver adenomas	7/21	Hirono et al. (1979b)
						haemangio-endothelial sarcoma	1/21	
						urinary bladder papillomas	2/21	
						urinary bladder carcinoma	1/21	
						lymphatic leukaemia	1/21	
						colonic adenoma	1/21	
						pituitary adenoma	1/21	
						(rats with tumours)	7/21	

Table 13 (contd).

Symphytum officinale (contd)	Group III: 14 males, 14 females	8% of diet as leaves	600 days	until death or moribund	liver adenoma	1/25	Hirono et al. (1979b)
	Group IV: 12 males, 12 females	8% of diet as root	until death		liver adenomas	19/22	
					urinary bladder papilloma	2/19	
					(rats with tumours)	19/22	
	Group V: 24 males, 24 females	4% of diet as root reduced by stages to basal diet after 180 days	until death	until death or moribund	liver adenomas	16/42	
					urinary bladder carcinoma	1/42	
					adrenal cortical adenomas	2/42	
					(rats with tumours)	16/42	
	Group VI: 12 males, 12 females	2% of diet as root for 190 days reduced by stages to basal diet	280 days		liver adenomas	10/23	

Table 13 (contd).

Material tested	Strain, number, and sex of animals	Route of administration and dosing regimen	Other treatment	Duration of treatment	Period of observation	Tumours produced	Tumour incidence in surviving animals	Reference
Symphytum officinale (contd)	Group VII: 15 males, 15 females	1% of diet as root for 275 days and subsequently a basal diet alternating with 0.5% at 3-week intervals		until death		liver adenomas	16/24	Hirono et al. (1979b)
						liver haemangiosarcomas	4/24	
						cholangiocarcinoma	1/24	
						adrenal cortical adenomas	1/24	
						(rats with tumours)	17/24	
	Group VIII: 15 males, 15 females	0.5% of diet as root		entire study		liver adenomas	8/30	
						liver haemangiosarcomas	9/30	
						pituitary adenoma	1/30	
						uterine adrenocarcinoma	1/30	

Table 13 (contd).

Symphytum officinale (contd)					
	controls (65 males, 65 females)	none	till the end	gonadal stromal tumour (rats with tumours) urinary bladder papilloma subcutaneous fibrosarcoma caecal adenoma mammary fibroadenoma retroperitoneal teratoma	1/30 10/30 1 1 1 1 1	Hirono et al. (1979b)
Senkirkine	ACI rat (20 males)	ip injection of 22 mg/kg body weight, twice weekly, then once weekly	4 weeks, 52 weeks 650 days or till moribund	liver adenomas myeloid leukaemia testis interstitial tumour	9/20 1/20 1/20	Hirono et al. (1979a,b)

Table 13 (contd).

Material tested	Strain, number, and sex of animals	Route of administration and dosing regimen	Other treatment	Duration of treatment	Period of observation	Tumours produced	Tumour incidence in surviving animals	Reference
Symphytine	20 males	ip injection of 22 mg/kg body weight, twice weekly, then once weekly	4 weeks, 52 weeks		650 days or till death or moribund	liver adenoma	1/20	
						liver haemangio-sarcomas	3/20	
	controls (20 males)					myeloid leukemia	1/20	
						fibroma-soft tissues	1/20	
						testis-interstitial tumour	1/20	
Clivorine	ACI rat (6 males, 6 females)	drinking-water (0.005%)		340 days	480 days	liver haemangio-sarcomas	2/12	Kuhara et al. (1980)
	ACI rat (6 males, 6 females)	drinking-water (0.005%)				hepatic neoplastic nodules	6/12	
						testicular interstitial cell tumour	3/12	
	10 males, 10 females					pituitary adenoma	1/20	

Table 13 (contd).

Clivorine (contd)				testicular interstitial cell tumour	3/20[f]	Kuhara et al. (1980)
				adreno-cortical adenoma	1/20[f]	
				pancreatic aunar cell adenoma	1/20	
Crude alkaloidal extract from Senecio numorensis fuchsii	Sprague Dawley rat (20 males, 20 females)	intragastric dose of 8 mg/kg body weight, 5 times per week	104 weeks	114 weeks	liver tumours (all types)	13/40 (2 males, 11 females)	Habs et al. (1982)
				other tumours (all types)	11/40 (5 males, 6 females)	
	20 males, 20 females	intragastric dose of 40 mg/kg body weight, 5 times per week	104 weeks	114 weeks	liver tumours (all types)	34/40 (5 males, 29 females)	
				other tumours (all types)	10/40 (7 males, 3 females)	
	controls (20 males, 20 females)			liver tumour	1/40 (male)	
				other tumours (all types)	10/40 (4 males, 6 females)	

Table 13 (contd).

Material tested	Strain, number, and sex of animals	Route of administration and dosing regimen	Other treatment	Duration of treatment	Period of observation	Tumours produced	Tumour incidence in surviving animals	Reference
Farfugium japonicum (leaves and stalks)	ACI rat (15 males, 14 females)	diet (20%)		480 days	till death or end of study	liver haemangio-sarcomas	6/29	Hirono et al. (1983)
						liver adenomas	7/29	
						adrenal-cortical adenomas	7/29	
						adrenal phaeochromo-cytoma	1/29	
						urinary bladder papillomas	2/29	
	ACI rat (15 males, 14 females)	diet (20%)				testicular interstitial tumours	2/29	
						ileal adeno-carcinoma	1/29	

Table 13 (contd).

Senecio cannabifolius (leaves and stalks)	15 males, 15 females	diet (8%)	none survived > 240 days		Hirono et al (1983)
	14 males, 14 females	diet (4%)	none survived > 240 days		
	12 males, 12 females	diet (1%)	480 days	till dead	
				liver haemangio-sarcoma	1/23
				liver adenomas	13/23
				adrenal cortical adenomas	5/23
				urinary bladder papilloma	1/23
				testicular interstitial tumour	1/23
	12 males, 12 females	diet (0.2%)	480 days	till death	
				liver haemangio-sarcomas	8/24
				liver cell adenomas	3/24
				adreno-cortical adenomas	3/24

Table 13 (contd).

Material tested	Strain, number, and sex of animals	Route of administration and dosing regimen	Other treatment	Duration of treatment	Period of observation	Tumours produced	Tumour incidence in surviving animals	Reference
Senecio cannabifolius (contd)						adrenal phaeochromocytoma	1/24	Hirono et al. (1983)
						testicular interstitial tumours	2/24	
						pituitary adenoma	1/24	
						caecal fibrosarcoma	1/24	
	controls (25 males, 24 females)	basal diet		560 days		cortical adenomas of adrenal	3/49	

Table 13 (contd).

Senecio cannabifolius (contd)	controls (25 males, 24 females)	basal diet			Hirono et al. (1983)
			testicular interstitial tumours	3/49	
			pituitary adenomas	2/49	

a These tumours were present in the same animal.
b Each coexisted in one animal (seen in animals surviving 22 - 27.5 months).
c Together with haemangioendothelioma of the liver.
d Two animals had both tumours.
e Found in the same animal with angiosarcoma.
f In the same animal.
g No data available on number of surviving animals in the control groups.

Note: Figures in column 2 indicate the number of animals used at the start of the study. In column 8, animals developing a tumour out of the number surviving up to the end of study/or dying with tumour are indicated. Different terms have been used in animals developing more than one tumour are indicated separately in a footnote. Different studies to describe the same tumour in the liver, e.g., haemangiosarcoma, angiosarcoma, haemangioendothelioma, haemangioendothelial sarcoma, and haemangiogenic sarcoma.

Table 14. Liver tumours found in rats given PAs and plant materials (condensed results from various authors)[a]

Alkaloid or plant given	Alkaloid type		Route of administration	Number of rats at autopsy	Number and types of liver tumours found	Reference
	Necine	Type of ester				
Clivorine	otonecine	macrocyclic diester	oral	12	2 haemangiosarcomas 6 neoplastic nodules	Kuhara et al. (1980)
Lasiocarpine	heliotridine	"open" diester	ip oral	18 20	11 hepatocellular carcinomas 9 angiosarcomas 7 hepatocellular carcinomas	Svoboda & Reddy (1972); Rao & Reddy (1978)
Monocrotaline	retronecine	macrocyclic diester	oral	75	24 hepatocellular carcinomas	Newberne & Rogers (1973)
Petasitenine	otonecine	macrocyclic diester	oral	11	5 haemangiosarcomas 5 adenomas	Hirono et al. (1977)
Retrorsine	retronecine	macrocyclic diester	various	14	4 "hepatomas"	Schoental et al. (1954); Schoental & Head (1957)
Isatidine	retronecine	macrocyclic diester	various	7	4 "hepatomas" and "nodular hyperplasia"	Schoental et al. (1954); Schoental & Head (1957)
Senecionine (+ fuchsisenecionine)	retronecine (platynecine)	macrocyclic diester, monoester	oral	80	19 hepatocellular tumours 16 cholangiogenic tumours 12 "haemangiogenic" tumours	Habs et al. (1982)

Table 14 (contd).

Senkirkine	otonecine	macro-cyclic diester	ip	20	9 adenomas	Hirono et al. (1979a)
Symphytine	retronecine	"open" diester	ip	20	1 adenoma 3 haemangiosarcomas	Hirono et al. (1979a)
Senecio longilobus (seneciphylline, retrorsine, etc.)	(retronecine)	macro-cyclic diester	oral	47	16 hepatocellular carcinomas 1 angiosarcoma	Harris & Chen (1970)
Petasites japonicus (petasitenine)	(otonecine)	macro-cyclic diester	oral	46	11 haemangiosarcomas 10 adenomas 3 hepatocellular carcinomas	Hirono et al. (1973)
Tussilago farfara (senkirkine)	(otonecine)	macro-cyclic diester	oral	12	8 haemangiosarcomas	Hirono et al. (1976)
Symphytum officinale (various)	(retronecine)	mono-ester + "open" diester	oral	175	81 adenomas 3 haemangiosarcomas	Hirono et al. (1978)
Farfugium japonicum	senkirkine		oral	29	7 adenomas 6 haemangioangiosarcomas	Hirono et al. (1983)
Senecio cannabifolius	petasitenine		oral	21	16 adenomas 9 haemangioangiosarcomas	Hirono et al. (1983)

[a] From: Mattocks (1986).

Crude extracts of plants or whole plants have also been demonstrated to produce a variety of tumours. For example, Senecio longilobus has been shown to produce tumours in a significantly large proportion of experimental animals (Harris & Chen, 1970). Hirono et al. (1973, 1976, 1978, 1979b) demonstrated the carcinogenic properties of a number of plants, used as food in Japan, in the rat.

A summary of the relevant experimental data follows.

6.4.8.1 Purified alkaloids

Kuhara et al. (1980) administered clivorine in the drinking-water at a concentration of 0.05 g/litre to 12 rats of both sexes, continuously for 340 days followed by plain water. There were 20 control rats. All the treated rats survived 440 days. Eight of the 12 animals developed liver tumours including 2 haemangioendothelial sarcomas and 6 that were described as "neoplastic nodules". No liver tumours were seen in the control animals (Table 13).

Schoental (1975) tested the carcinogenic properties of heliotrine, with and without prior administration of nicotinamide. Nicotinamide protects against liver necrosis and so may enhance tumour yeild, as shown by Rakietin et al. (1971), who studied pancreatic tumours in rats given streptozootocin.

Heliotrine was administered intragastrically to 4 groups of 26 male weanling rats in 1 or 2 doses of 230 mg/kg and 300 mg/kg body weight; 2 groups also received nicotinamide at 350 - 500 mg/kg body weight, administered ip 10 - 15 min before, and 2.5 h after administration of heliotrine, as per the dosing regimen shown in Table 13. There were 8 controls. All animals administered the higher dose of heliotrine (300 mg/kg body weight) died within 5 months of the PA administration. The livers showed lesions characteristic of PA toxicity. No tumours were seen. Only 1 out of 4 animals receiving heliotrine alone, survived 27 months and it showed an islet cell adenoma as well as adenoma of pituitary, but so did 3 of the 8 controls. In the group receiving 230 mg heliotrine/kg and also treated with nicotinamide, 4 animals died within 5.5 months (one with a fibrosarcoma and the other in a moribund condition), chiefly from toxic liver disease, and two more had to be killed. Of the 6 animals surviving more than 22 months after heliotrine treatment, islet cell adenoma was seen in 3 together with other tumours as shown in the table. This tumour is stated to be extremely rare in the animal strain used. Hepatoma was seen in only one animal. The role of nicotinamide in this study is not clear.

Svoboda & Reddy and their group have carried out 2 studies using lasiocarpine (Svoboda & Reddy, 1972, 1974; Rao & Reddy, 1978). Svoboda & Reddy (1972, 1974) gave repeated ip injections to 25 rats at a dose of 7.8 mg/kg body weight (0.1 LD_{50}) for 56 weeks as per regimen shown in Table 13. Three rats died of acute liver necrosis in the initial 4 weeks. Eighteen rats survived 56 weeks, by which time each animal had received an average cumulative dose of 125 mg lasiocarpine. Of these, 16 animals developed a variety of tumours 60 - 76 weeks after the beginning of the study (Table 13). Ten of the 16 animals had more than 1 tumour. Hepatocellular carcinoma was the most common. The squamous cell carcinoma of the skin was found to be transplantable. When the same PA, mixed with the diet at the rate of 50 mg/kg (Rao & Reddy, 1978) (Table 13), was administered to 20 rats for 55 weeks, 17 animals developed tumours. Angiosarcoma of the liver emerged as the most common tumour (9/20 animals), even though hepatocellular carcinomas were also frequently seen (7/20 animals); squamous cell carcinoma of the skin was not found, but there was a malignant adnexal tumour of the skin. The average cumulative dose of the alkaloid was estimated to be 190 - 200 mg per rat.

Monocrotaline has been studied for its carcinogenic activity in rats by Schoental & Head (1955), Newberne & Rogers (1973), Allen et al. (1975), Shumaker et al. (1976), and Hayashi et al. (1977), using different routes of administration.

Allen et al. (1975) studied the long-term effects on rats of repeated sc injections of monocrotaline or its major detectable metabolite, dehydroretronecine. Male Sprague Dawley rats were given biweekly injections of monocrotaline, at 5 mg/kg body weight (75 animals) for 12 months, or dehydroretronecine at 20 mg/kg body weight (75 animals) for 4 months followed by 10 mg/kg body weight for 8 months. Fifty control animals received phosphate buffer (Table 13). Partial hepatectomy was performed on 15 animals in each of the treated groups and 5 in the control group. They were observed for 10 months following cessation of the injections. Of the 60 animals surviving in each of the treated groups, those receiving monocrotaline showed rhabdomyosarcoma at the injection site (2 animals), hepatocellular carcinoma (2 animals), acute myelocytic leukaemia (2 animals), and pulmonary adenoma (2 animals). In the group receiving dehydroretronecine, 36 animals developed rhabdomyosarcomas, and 5 of these animals developed metastases. None of the control group developed tumours. Tissues obtained from partial hepatectomies showed that both compounds caused inhibition of mitotic division in regenerating liver. The results of the study illustrate the dual alkylating and antimitotic properties of these agents, commented on by Culvenor et al. (1969).

In a similar study by Shumaker et al. (1976), rats were administered monocrotaline at 5 mg/kg body weight on alternate weeks or its metabolite dehydroretronecine at 20 mg/kg body weight, on alternate weeks for 4 months, followed by a dose of 10 mg/kg on alternate weeks in the succeeding 8 months (Table 13). Of the 60 rats receiving monocrotaline, 17 developed one or more tumours, but not until after the treatment was discontinued. The time interval was not stated. The most common tumour was carcinoma of the lung (11 animals) followed by hepatocellular carcinoma (5 animals). A wide variety of other tumours was also seen. A notable feature was that the metabolite dehydroretronecine did not produce any tumours by systemic action, but only at the site of injection, where significant numbers of rhabdomyosarcomas (39/60 animals) were seen. The marked difference in tumour sites is explained by the fact that the parent alkaloid monocrotaline has to be metabolized before it becomes a carcinogen. For this reason, the tumours are distributed in several organs of the body, whereas dehydroretronecine is itself carcinogenic and so acts at the site of injection. When monocrotaline was delivered in a higher dose, but as a single subcutaneous injection (40 mg/kg body weight) (Hayashi et al., 1977) to 40 rats there were no malignant tumours but only adenomas of the islets in the pancreas in 16/23 surviving animals. The results of the above studies indicate that monocrotaline is tumorigenic, but the type of tumour and the malignancy both depend on the route of administration and the dosage used.

Hirono et al. (1977) studied petasitenine, the pure alkaloid isolated from the flower stalks of the plant, <u>Petasites japonicus</u> Maxim, which has been found to be carcinogenic for rats (Hirono et al., 1973). Two groups of ACI rats of both sexes, Group I of 3 animals and Group II of 11 animals, were given the alkaloid in in the drinking-water, at concentrations of 0.5 g/kg and 0.1 g/kg, respectively (Table 13). There were 19 controls. All 3 animals in group I died within 72 days showing marked hepatocellular damage; no tumours were seen. In Group II, 10 out of 11 rats survived for more than 160 days. Eight of the 10 animals developed tumours - liver cell adenomas (5) and haemangiosarcomas (5) (Table 10). Two animals had both types of tumours. The authors concluded that the carcinogenicity of the plant was due to petasitenine.

Retrorsine and its $\underline{N}$-oxide (isatidine) have been used in several studies. Schoental et al. (1954) administered retrorsine at a concentration of 3 mg/litre in the drinking-water to 14 rats and isatidine at concentrations of 5 mg/litre followed by 3 mg/litre to 22 rats until death. Twenty-five rats were administered the alkaloid mixture from <u>Senecio jacobaea</u> Lin at a concentration of 500 mg/litre followed by 300 mg/litre in the drinking-water. Dosing regimens are shown in Table 13. One group of 7 animals receiving isatidine received supplementary 0.5% choline in the drinking-water and another group of 5 animals received isatidine 2 mg as a single ip administration in 0.2 ml of tricaprilyn, followed by skin application of alkalids as 0.5% solution for 3 days/week.

The animals receiving mixed alkaloids of the plant showed extensive liver damage followed by marked nodular hyperplasia; no tumours developed. The nodules were earlier interpreted as hepatomas, but later only as an early stage in the progression from hyperplasia to neoplasia.

The retrorsine group showed extensive liver damage associated with cirrhosis and nodular hyperplasia. In 4 rats, they were interpreted as hepatomas.

In the isatidine group, 10 out of the 22 rats developed hepatomas. Tumours were present in 3 out of 7 animals receiving isatidine plus choline indicating that the latter had no protective role. One out of the 5 animals receiving the alkaloid ip and then through dermal application developed a tumour, which was also interpreted as a hepatoma.

In another study (Schoental & Bensted, 1963), 95 weanling rats were administered a single dose of retrorsine at 30 mg/kg body weight, by stomach tube (Group II). One comparable group of 50 weanling rats (Group I) received 400 r radiation in animals surviving 100 days after retrorsine administration (31/50). A third group (Group III) of 6 animals received 400 r radiation alone. Another group of 10 weanling rats received the PA, 9 days after partial hepatectomy (Group IV) (Table 13). The additional treatments were given to study whether they would act as co-carcinogens and induce neoplasia in hepatocytes, which are known to show injurious effects for long periods following PA treatment. In Group I, 19 of the 50 animals receiving a single dose of the PA died before radiation could be given. Of the 31 remaining animals that received radiation after the PA, 25 survived 12 months. Among these, 19 tumours of a wide variety were seen (Table 13). Of the 6 tumours in the liver, only 1 was malignant, having metastasized. Most of the other tumours were malignant, including those of the breast, one of which had also metastasized.

In Group II, 29 out of 95 animals that had received one dose of PA and no radiation survived for more than a year, with a mean survival time of 23 months. Among these, 7 animals developed tumours of a wide variety (Table 13). Five tumours in the liver were benign. Most of the others were malignant. Two tumours, a cystic tumour of the breast and a carcinoma of the uterus, were present in one animal. Tumours seen in Groups III and IV, which were found in animals surviving 12 months or more after the start of the study, are shown in the Table 13. The 2 tumours of the liver in Group IV were also benign. The number of control animals, if any, were not indicated.

The authors concluded that the above studies did not provide definite evidence of synergistic action in the carcinogenicity of retrorsine by whole body radiation or partial hepatectomy.

Hirono et al. (1979a) studied the carcinogenic properties of senkirkine extracted from the dried milled buds of <u>Tussilago farfara</u> (coltsfoot) and symphytine extracted from dried milled roots of <u>Symphytum officinale</u> (comfrey). Both <u>Tussilago farfara</u> (Hirono et al., 1976) and <u>Symphytum officinale</u> (Hirono et al., 1978) had earlier been demonstrated to have carcinogenic properties.

Sixty inbred ACI strain male rats were divided into 3 groups of 20 animals each and received repeated ip injections of senkirkine at 22 mg/kg body weight or symphytine at 13 mg/kg body weight as per schedule given in Table 13. All animals treated with senkirkine survived 290 days. Nine out of 20 rats developed liver adenomas mostly after 350 - 450 days from start of the study. Cirrhosis of liver was frequently observed. Of the symphytine group, all animals survived 330 days after the start of the study. Three animals developed haemangioendothelial sarcoma and one, liver adenoma. The sarcomas were noted at least 518 days after the start of the study.

Schoental & Cavanagh (1972) used two alkaloids, retronecine and hydroxysenkirkine isolated from <u>Crotalaria laburnifolia</u>, which was injected ip in single doses ranging from 100 to 300 mg/kg body weight in 5 weanling Porton Wistar rats (Table 13). One animal that had received the 300 mg/kg dose developed astrocytoma of the brain after 14 1/2 months. Retronecine hydrochloride was administered in doses ranging from 300 to 1000 mg/kg body weight by single subcutaneous injection to 10 newborn rats. One male rat, which was found to be paraplegic 6.6 months after receiving a dose of 600 mg PA/kg, was also found to have ependymoblastoma of the spinal cord. Among the litter mates of this rat, 1 male that had received a dose of 1000 mg/kg died. The remaining 6 females and 2 males were killed within 22 months of being dosed. Of the females, 5 had pituitary tumours and 1 had a mammary tumour (type not stated). Two males did not show any significant abnormalities.

One group of 5, 6-month-old female rats, born to dams that had been fed on a diet containing 50 g dried and powdered <u>Heliotropium ramosissimum</u>/kg diet, a tribal remedy, used during pregnancy and parturition, were then themselves fed on the same diet at 6 months of age, as per the regimen indicated in Table 13. One female rat was found to be paraplegic at 7 months of age. A tumour, possibly of Schwann or satellite cell origin, was found.

6.4.8.2 Plant materials

A number of plant materials have been tested for their carcinogenicity by administering either a mixture of extracted alkaloids (Cook et al., 1950; Schoental et al., 1954) or, more often, the dried and milled plant mixed with the diet. The only study in which the plant alone was fed was that of Campbell (1956), who produced tumours in chickens by feeding dried and milled Senecio jacobaea plant. It is notable that the plants tested are almost all those that have been reported to be used as herbal medicines and/or food, some of which have been reported to cause human toxicity.

Schoental et al. (1970) used mixed PAs (intermedine and lycopsamine) extracted from seeds of Amsinckia intermedia, known to cause livestock losses in the USA, and leaves and stems from Heliotropium supinum L., known to be used by women in East Africa as a herbal medicine, after childbirth. The dosing regimen and mode of administration are indicated in Table 13. Of the 15 male weanling rats that had received a single treatment with the Amsinckia PAs and survived for more than 1 year, 3 rats showed one adenoma, one adenocarcinoma of the islet cells, and one adenoma of the exocrine pancreas, respectively (Table 13). In addition, one of the animals also developed a pituitary adenoma and papillary tumour of the urinary bladder. Eight animals received the treatment with Heliotropium supinum L. (Table 13). One out of the 2 weanling rats fed on the plant with the diet and one out of the 6 animals that received a single intragastric dose of the crude alkaloid fraction developed islet cell adenoma of the pancreas. The number of control animals, if any, used in the study is not stated.

Harris & Chen (1970) tested the carcinogenicity of Senecio longilobus, which has been associated with cases of human toxicity (Stillman et al., 1977; Huxtable, 1980; Fox et al., 1978). Harlan rats were divided into 4 groups (equal numbers of both sexes) and fed diets containing dried and powdered stems and leaves of the plant in the proportion of 5 - 7.5 g/kg. The number of animals in each group and the feeding regimen are shown in Table 13. Continuous feeding did not produce any tumours, presumably because of the comparatively low survival rate of animals. Significant results were obtained when the animals were fed contaminated diet alternating with normal diet (Group IV). Of the 100 animals fed on this regimen for 54 weeks, 47 survived for more than 200 days. Seventeen of these animals developed malignant tumours in the livers - hepatocellular carcinomas in 16 and an angiosarcoma in 1 animal after a minimum feeding period of 217 days. In Group III, fed a contaminated diet for 1 year, 23 rats lived for more than 200 days. Four rats (3 males and 1 female) developed hepatocellular carcinomas and one, a peritoneal mesothelioma, after a minimum feeding period of more than 428 days. The tumour-bearing animals were predominantly male. The authors emphasized the relative rarity of liver tumours in the strain of animals used. Results demonstrated that S. longilobus is carcinogenic for rats.

For a study of Schoental & Cavanagh (1972), using dried and powdered Heliotropium ramosissimum, see section 6.4.8.1.

Hirono et al. tested the carcinogenic effects of 3 widely-used herbs containing PAs. Hirono et al. (1973) studied the possible carcinogenic effects of the young flower stalks of Petasites japonicus Maxim, which has long been used in Japan as food or herbal remedy as well as the PA, petasitenine, isolated from it. The flower stalks of the plant were dried, milled, and fed to young ACI rats of both sexes mixed with the basal diet in the proportion of 40 - 80 g/kg, as indicated in the feeding regimen in Table 13, until the animals were moribund or dead. One group of 27 rats was fed the plant mixed in the proportion of 40 g/kg diet for 6 months followed by 80 g/kg diet on alternate weeks. The second group of 19 rats was fed a contaminated diet (40 g/kg) continually. Three animals in Group I died of pneumonia. Eleven out of the remaining 24 rats in this group developed liver tumours after 15 - 16 months of feeding. There were 2 hepatocellular carcinomas, 6 liver cell adenomas, and 3 haemangiosarcomas, of which 2 had metastasized. In Group II, 2 animals died from non-tumorous causes. Of the remaining 17, 8 developed haemangiosarcomas, 4 liver cell adenomas and 1 had a hepatocellular carcinoma. The incidence of haemangiosarcoma was statistically higher in Group II.

In a similar study on mice and hamsters (Fushimi et al. (1978), groups comprising 20 - 24 male and 20 - 21 female 6-week-old ddN, Swiss, and C57BL/6 mice, and 13 male and 17 female hamsters, were fed a diet combining 4% young, dried, and milled flower stalks of Petasites japonicus Maxim for 480 days. All surviving animals were killed at the end of the study. Lung adenomas and adenocarcinomas were found in 30/39 surviving male and female ddN mice combined (compared with 1/50 in the respective controls) in addition to other tumours (Table 13). No significant differences in tumour incidence were observed between treated Swiss and C57BL/6 mice and hamsters, and the corresponding controls. No data were given on the tumour incidence according to sex, in treated and control animals or on survival in the controls.

In studies on rats (Hirono et al., 1976), the dried and powdered flower buds of Tussilago farfara (coltsfoot) were mixed with the diet. Forty-nine inbred ACI strain rats were divided into 4 groups and fed diets containing 40 - 320 g T. farfara/kg for up to 600 days. The number of animals in each group and the dosage and feeding regimen are given in Table 13. Two groups receiving a diet containing 80 - 320 g/kg on a continual basis developed tumours in the liver of the same types as animals in the studies using Petasites japonicus, e.g., haemangioendotheliomas, liver cell adenomas, and hepatocellular carcinomas (Table 13). All 12 animals in Group I survived for more than 380 days after the start of the study. Of these, 8 (5 males, 3 females) developed haemangio-endothelial sarcoma of the liver. In addition, 3 of the 8 rats developed simultaneously hepatocellular adenoma, hepatocellular carcinoma, or urinary bladder papilloma. In Group II receiving coltsfoot at 80 g/kg in the diet, 9 out of 10 animals that survived for more than 420 days developed a haemangioendothelial sarcoma.

Hirono et al. (1978) fed Symphytum officinale, similarly dried and milled, to the ACI strain of inbred rats of both sexes at different levels ranging from 80 to 330 g/kg as leaves or 5 - 40 g/kg as root, for 280 days or more. Eight groups of animals of 19 - 48 animals each were used on different regimens of feeding. The feeding regimen, the number of animals in each group, and the duration of treatment are given in Table 13. Sixty-five males and 64 females served as controls. Tumours were induced in all groups receiving leaves or roots. The most common tumour was liver adenoma. Haemangiosarcomas were observed but infrequently (3 animals in the whole study). No carcinomas of the liver were seen, but there was a wide variety of other tumours. It is noteworthy that, in Group VII, 14 of 15 animals fed the lowest dose of 10 g/kg for 275 days followed by 5 g/kg or basal diet alternating every 3 weeks, developed tumours, 2 of which were malignant (haemangiosarcomas). In the control group, single animals each had papilloma of urinary bladder, caecal adenoma, subcutaneous fibrosarcoma, mammary fibroadenoma, or retroperitoneal teratoma. The livers of animals that did not have the tumours showed other features commonly encountered in animals administered PAs, e.g., megalocytosis, liver cirrhosis, hyperplastic nodules, etc., suggesting that they were induced by the PAs contained in the plant. The authors concluded that the carcinogenic activity of this plant was weaker than that observed in animals fed Petasites (coltsfoot). Hirono et al. (1979b) have summarized the above studies on PAs found in edible plants in Japan.

Habs (1982) and Habs et al. (1982) tested a crude alkaloid extract from the plant Senecio numorensis sp. fuchsii containing fuchsisenecionine (500 g/kg) and senecionine (10 g/kg). The extract was administered intragastrically in 2 doses of 8 mg and 40 mg/kg body weight, respectively, to 2 groups of rats of both sexes, comprising 40 animals each, 5 times a week for 104 weeks (Table 13). A large, dose-related number of liver tumours was produced, originating in the liver cell and the sinusoidal system, and predominantly affecting the female animals. The tumour incidence was higher in Group II (dose, 40 mg/kg) with 34 tumours compared with 13 in Group I. In Group I (dose, 8 mg/kg), 2 tumours were found in 20 males compared with 11 tumours among 20 female rats. Similarly, in Group II, only 6 tumours were found in 20 males compared with 29 among 20 female rats. The tumours included 19 of "hepatocellular origin", 16 of "cholangigenic origin", and 12 of "haemangiogenic origin" (Table 14). Besides the liver tumours, 12 males and 9 females in the treated groups and 4 males and 6 females in the control group developed a variety of extra-hepatic tumours. Senecionine is known to be hepatotoxic and capable of being converted in the rat liver into cytotoxic metabolites. Fuchsisenecionine is a saturated PA, not previously known to be cytotoxic. It is therefore likely that senecionine was the hepatotoxic component and that perhaps the two PAs acted synergistically with each other, though there is no actual evidence for synergism. However, this needs confirmation. Moreover, there appeared to have been other unknown components in the mixture tested (Mattocks, 1986).

Hirono et al. (1983) studied the carcinogenicity of 2 more plants (Farfugium japonicum and Senecio cannabifolius) of the tribe Senecioneae in the family Compositae, the leaves and stalks of which are used in Japan as human food. Fresh leaves and stalks of the plants were dried, milled, and mixed with the basal diet. Inbred strain ACI rats of both sexes, with preponderance of females, 1.5 months old, were divided into 6 groups. They were fed diets containing various proportions of the dried plant materials, as indicated in Table 13. The study was terminated at 480 days, except for one group, which was studied for 560 days. Besides the groups shown in the tables, 2 more groups of 30 and 28 animals of equal numbers of both sexes were fed 8% and 4% of Senecio cannabifolius, respectively. None of these animals survived more than 177 days and all died of hepatotoxicity. A wide range of tumours was observed, mostly in the liver, as shown in Table 13, the most common being haemangiosarcomas, which were not encountered in the control group.

The carcinogenicity of Farfugium japonicum is considered to be due to senkirkine and petasitenine. Senecio cannabifolius contains senecicannabine, a new macrocyclic PA, seneciphylline, and jacozine. It is probable that the carcinogenicity of Senecio cannabifolius is due to these PAs (Hirono, et al., 1983).

Hayes et al. (1985) studied the biological mechanisms by which PAs initiate carcinogenesis in male Fischer 344 rats. Lasiocarpine (single or double injection of up to 80 μmol/kg body weight) and senecionine (single or double injection of up to 160 μmol/kg) were inactive as initiators of Y-GT-positive nodules in rats exposed to 2-acetylaminofluorene and partial hepatectomy. Administration of lasiocarpine or senecionine 12 h after partial hepatectomy resulted in the development of very few nodules. Lasiocarpine given in a single or double dose (up to 80 μmol/kg) delayed hepatic regeneration by at least 8 weeks after partial hepatectomy, and pre-treatment with this PA reduced the initiating capacity of diethylnitrosamine and N-nitrosomethylurea in rats subsequently selected with 2-acetylaminofluorene and partial hepatectomy. Resistant nodules selected with lasiocarpine also had the typical resistant nodule phenotype (positive for Y-GT and epoxide hydrolase) and also lacked PA-induced megalocytosis. Lasiocarpine treatment also resulted in small regenerative nodules that were distinct from resistant nodules, because they were negative for Y-GT and epoxide hydrolase.

6.4.8.3 Pyrrolizidine alkaloid metabolites and analogous synthetic compounds

The subject has been reviewed by Mattocks (1986). The pyrrolizidine alkaloids are converted into pyrrolic esters in the hepatocytes. These may then be hydrolysed into pyrrolic alcohols, which are more water soluble and less active than the esters. The esters may be widely distributed throughout the body. Some of these compounds and their analogues have been tested for carcinogenicity.

(a) Pyrrolic esters

(i) Dehydromonocrotaline (monocrotaline pyrrole)

Mattocks & Cabral (1979) tested dehydromonocrotaline on the skin of mice. In their first study on male BALB/c mice, they applied the compound on the back at 1- to 2-week intervals. Thirty-three applications of 1 μmol did not produce any skin tumours in 16 mice, but 2 developed lung adenomas; no tumours occurred in 14 control mice treated with the solvent (acetone). In the second study (Mattocks & Cabral, 1982), 11 female LACA mice each received 47 applications of 2.5 μmol dehydromonocrotaline; one animal developed a malignant skin tumour. A second batch of 10 mice was given similar treatment and subsequently received 61 twice-weekly applications of croton oil at the same site. Half of these animals developed tumours. In the control group of 10 mice, only 1 tumour was seen. The results of these studies indicate that dehydromonocrotaline requires the action of a promoter to manifest carcinogenic potential.

(ii) Dehydroretrorsine

This supposed pyrrolic metabolite of retrorsine was applied to the skin of male BALB/c mice at 1- to 2-week intervals; 33 treatments of 0.5 or 1 µmol failed to produce tumours in 15 mice that survived for up to 60 weeks (Mattocks & Cabral, 1979).

(iii) 1-Methyl-2,3-bistrimethylacetoxymethylpyrrole

Mattocks & Cabral (1979, 1982) made 2 studies on this compound. In the second study, which was more significant, 22 female LACA mice received 47 dermal applications of 0.5 µmol each. This caused marked skin damage with ulceration and scarring. Malignant skin tumours developed in 19 out of 21 surviving animals. Two hydrolysis products of this ester, pivalic acid and 1-methyl-2,3-bishydroxymethylpyrrole, were similarly tested (Mattocks, 1986). Tumours were produced, though fewer.

The results of the above studies suggest that the intact pyrrolic ester is a carcinogen.

(b) Pyrrolic alcohols

Studies have been conducted using dehydroretronecine (retronecine pyrrole) and dehydroheliotridine, which are secondary metabolites of monocrotaline and heliotridine-based PAs, respectively. Originally they were regarded as (+)- and (-)- forms, respectively, of dihydro-7-hydroxy-1-hydroxymethyl-5H-pyrrolizine. Kadzierski & Buhler (1985, 1986) showed that the metabolite from monocrotaline is racemic and concluded that the product from all heliotridine and retronecine esters is the same (±)- form.

Johnson et al. (1978) painted the skin on the back of 16 female Swiss mice with dehydroretronecine, each dose equalling 20 mg/kg body weight or about 5 µmol per mouse, once a week for 4 weeks and then twice more after 6 months. Six mice developed skin tumours. Subcutaneous injections of the same dose yielded tumours in 13 out of 21 mice. Twenty-eight out of 55 mice given both topical applications and sc injections developed skin tumours.

When the same study was repeated on 34 BALB/c mice, only one skin tumour developed (Mattocks & Cabral, 1982). When 17 animals were given the same dose at more frequent intervals, e.g., 65 weekly doses, no tumours developed.

Similar results were obtained in a study by Shumaker et al. (1976), already desecribed in section 6.4.8.1, in which 39/60 rats given repeated subcutaneous injections of dehydroretronecine developed rhabdomyosarcomas at the site of injection. The compound appears to be a direct-acting carcinogen for rats and mice, though the susceptibility of various strains of mice varies.

Peterson et al. (1983) gave 9 ip injections (60 - 76.5 mg/kg body weight) of dehydroheliotridine to rats over a 32-week period. A large variety of tumours was produced. It was concluded that this metabolite may be responsible for the carcinogenicity of its parent PA.

6.4.8.4 Molecular structure and carcinogenic activity

Mattocks (1986) has reviewed the present position concerning the relationship between molecular structure and carcinogenic activity. Data available at present are not adequate for any strict correlation to be established between the molecular structure of PAs and the types of tumours produced by them in the rat. However, the common determinants in the molecular structure of all carcinogenic PAs are that they are macrocyclic or "open" diesters, in which the aminoalcohol moiety is retronecine, heliotridine, or otonecine. These are all esters of unsaturated necines and are capable of being metabolized to pyrrolic esters in the mammalian liver. Studies on pyrrolic esters, the toxic metabolic product of PAs, have yielded equivocal results. Monocrotaline has been shown to bind covalently to DNA, which is associated with the carcinogenic activity of the pyrrolizidine alkaloids (Robertson, 1982). Dehydromonocrotaline, the primary metabolite of monocrotaline has been found to be an incomplete dermal carcinogen (Hooson & Grasso, 1976; Mattocks & Cabral, 1982), whereas the synthetic compound 1-methyl-2,3-bistrimethylacetoxymethylpyrrole, which is chemically similar to dehydromonocrotaline, is clearly carcinogenic (Mattocks & Cabral, 1982). Dehydroretronecine (DHR), a second metabolite of monocrotaline and possibly other retronecine-based PAs, is also carcinogenic for the skin and is considered a proximate carcinogenic metabolite, since, unlike monocrotaline, it acts at the site of application directly and not at remote sites.

6.4.9 Antimitotic activity

Literature on this phenomenon has been reviewed by Jago (1969), McLean (1970), and Mattocks (1986). The most characteristic feature of the chronic hepatotoxicity of the PAs is the presence in the liver of megalocytes (Bull & Dick, 1959; McLean, 1970), which are generally enlarged hepatocytes containing large, hyperchromatic nuclei (section 6.4.1.5). These appear to be the result of a combined action of PAs on the hepatocytes, a stimulus to regenerate following parenchymal cell injury, and the powerful antimitotic action of the pyrrole metabolites of the PAs (Mattocks, 1986; Jago, 1969). This property has served as the basis for using a PA (indicine-N-oxide) as a chemotherapeutic agent for cancer (Letendre et al., 1981, 1984). Peterson (1965) showed that the number of mitoses following partial hepatectomy was reduced to 50% or less of normal values by prior administration of hepatotoxic alkaloids, and that the effect was dose dependent. The hepatocytes seemed to continue to grow without dividing. The effect can be produced by a single sublethal dose of the alkaloids (Schoental & Magee, 1957) or can be a cumulative effect of small doses (Bull & Dick, 1959). The lesion appears within a few weeks and may persist for the lifetime of the animal (Mattocks, 1986). It was characteristically described in the liver of the rat, but has also been reported in a number of other animals, e.g., mouse, sheep, horse, and pig, and in some other organs, e.g., kidney and lung (McLean, 1970). It has not been observed in human livers (Tandon, H.D. et al., 1978; Mattocks, 1986), but it has been observed that cultured human fetal liver cells become enlarged when exposed to PAs (Armstrong et al., 1972) indicating a susceptibility to the antimitotic effect of PAs. The lesion is not specific for PAs but has been reported to be produced by a number of toxins (McLean, 1970), including semisynthetic derivatives of PAs but not non-hepatotoxic PAs, such as platyphyline (Jago, 1970). The ultrastructural features of megalocytes are controversial. However, consistent with other functional and metabolic features, the cells show morphological characters suggesting increased metabolic activity with increased exchange of material between the nucleus and cytoplasm.

This unique reaction of the hepatocytes to PA has been used by Jago (1970) to develop a method for assessing hepatotoxicity and by Culvenor et al. (1976a) for screening 62 PAs for acute and chronic hepatotoxicity and pneumotoxicity.

Mattocks (1986) suggested a scheme for the antimitotic action of PAs in vivo, consistent with observations of the phenomenon in animals. The PA is irreversibly metabolized by the hepatic microsomes into a pyrrolic ester, which can be hydrolysed to a pyrrolic alcohol. The latter is the agent that inhibits mitosis. The more reactive primary metabolite may also do this, by reacting with tissue constituents to give products identical to pyrrolic alcohols, inhibiting mitosis. On the other hand, it can also produce acute injury to the cells, which stimulates regeneration. Thus administration of the PA or the pyrrolic ester can induce megalocytosis to a much greater extent than a secondary metabolite alone.

Antimitotic activity does not seem to be directly related to inhibition of DNA synthesis, since the latter recovers within a week while mitotic inhibition continues for up to a period of 4 weeks (Peterson, 1965; Armstrong & Zuckerman, 1970). Mattocks & Legg (1980) have shown that the level of DNA synthesis is reduced in cells that do not divide, but is not totally inhibited. Samuel & Jago (1975) investigated the position in the cell cycle of the antimitotic action of lasiocarpine and of its pyrrolic metabolite, dehydroheliotridine. Their studies indicated that the alkaloid acts during the late S or early G_2 phase of the cell cycle.

6.4.10 Immunosuppression

Dehydroheliotridine (DHH), a pyrrolic metabolite, has a significant immunosuppressant activity on the primary response in young mice; when injected ip shortly before the antigenic stimulus (Percy & Pierce, 1971). The secondary response to antigenic stimuli; as measured by the reduction in the number of 7S and 19S specific antibody-synthesizing cells of the spleen, was suppressed when DHH was administered at the time of secondary stimulus, but not when it was given 24 or 36 h after the antigenic stimulus. It was suggested that dehydroheliotridine selectively destroys or inactivates cells involved in the initial stages of antigen recognition and processing.

6.4.11 Effects on mineral metabolism

Aberrations of mineral metabolism have been observed in several species of animals. The most notable among them relate to haemolysis and copper metabolism. Anaemia has been reported to occur in rats following PA poisoning (Schoental & Magee, 1959; Schoental, 1963) and the kidney and liver show haemosiderosis (Hayashi & Lalich, 1967). Besides the haemolysis, PAs have been found to exert a direct inhibitory effect on haematopoiesis in the livers of new-born rats (Sundaresan, 1942). Disturbances of iron metabolism and haematopoiesis have also been demonstrated in rats fed Senecio jacobaea and supplementary copper (Swick et al., 1982d) and

they have been found to develop raised copper levels in the
liver and spleen, when fed on this plant (Swick et al.,
1982b). Miranda et al. (1979) reported elevated levels of
iron in the liver and spleen and of copper in the liver in
rats fed on tansy ragwort. Studies with radioactive iron also
indicated a specific inhibitory effect of PAs on haematopoiesis. High copper levels in the liver associated with
haemoglobinuria have been reported in sheep grazing on
heliotrope (Bull et al., 1956), signs of disease closely
resembling those of chronic copper poisoning. St. George-
Grambauer & Rac (1962) reported a similar outbreak of fatal
jaundice due to haemolytic crisis of chronic copper poisoning
in sheep that had grazed Echium plantagineum over 2 or more
seasons. The pathological changes in the liver were
indistinguishable from those of Heliotropium, and the livers
had a high copper content. Studies of Miranda et al. (1981b)
indicate that dietary copper can enhance PA hepatotoxicity in
rats. It has been suggested that the hepatotoxic effects of
some PAs may interfere with the excretion of copper (Bull &
Dick, 1959; Farrington & Gallagher, 1960). Similar effects
have been observed in pigs fed Senecio (Harding et al.,
1964). White et al. (1984) did not observe any rise in
hepatic copper levels in sheep fed Senecio jacobaea.

6.4.12 Methods for the assessment of chronic hepatotoxicity and pneumotoxicity

Pyrrolizidine alkaloids produce acute as well as chronic
liver damage. The acute effect is seen as extensive necrosis
(Schoental & Magee, 1957), while the chronic effect in rats is
manifested characteristically by the presence of greatly
enlarged parenchymal cells (Schoental & Magee, 1957; Bull &
Dick, 1959), which persist long after a single exposure
(Schoental & Magee, 1959). The latter effect of megalocytosis
may manifest without the liver cells going through the process
of necrosis (Schoental & Magee, 1959). This property has been
used by Jago (1970) to develop a method for the assessment of
the relative chronic hepatotoxicity of different alkaloids.
It has since been used by other investigators (Culvenor et
al., 1976a). It consists of the intraperitoneal administration
of a single dose of between 0.025 and 3.2 µmol of the
alkaloid per kg body weight in 0.2 ml aqueous solution to
14-day-old suckling hooded Wistar rats of both sexes. The
litters are randomized among the mothers, one day before the
administration of the toxin, and then weaned at 28 days.
Animals are killed 4 weeks after the injection. The relative
acute toxicity is indicated by the dose levels that cause
death within approximately a week and chronic hepatotoxicity

by those that produce hepatic megalocytosis within 4 weeks of the injection. With this method, it is not only possible to evaluate the hepatotoxicity of a given alkaloid but also to compare the hepatotoxic effects of different compounds in relation to each other on a molar basis. Chronic effects on the lungs can also be assessed by the same method. Culvenor et al. (1976a) found this method satisfactory for compounds of medium to high hepatotoxicity but failed to detect toxicity in certain compounds of known, low hepatotoxicity.

6.5 Effects on Wild-life

6.5.1 Deer

There is very little information on the consumption of PA-containing plants by non-domesticated animals, birds, and other wild-life. In one instance, the deaths of white-tailed deer in coastal marshes in Louisiana in 1967, was ascribed to the consumption of Crotalaria and/or Heliotropium species in a period of feed scarcity (Seger et al., 1969). The animals had thin and watery blood, abnormal bone marrow, and serious atrophy of cardiac and mesenteric adipose tissue. In a 9-year-old doe, the liver was dull and somewhat granular and evinced megalocytosis of the hepatocytes. In another 4-year-old animal, the liver appeared normal but with microscopic evidence of early megalocytosis, with considerable vacuolization in the centrilobular hepatocytes. Plants of the genera Crotalaria and Heliotropium were abundant and there were some Senecio species. There were signs of ingestion of the plants by the deer.

Senecio jacobaea (tansy ragwort) has been fed experimentally to black-tailed deer (Odocoileus hemionus columbianus) in Oregon to determine their susceptibility to poisoning (Dean & Winward, 1974). The ragwort was given ad libitum together with different levels of basal ration (85% alfalfa, 10% barley, 5% molasses) to captive deer. The ragwort was eaten, only when the basal ration was inadequate. One group, not given any basal ration for 6 days, began eating ragwort and consumed 5.4 kg dry weight per animal in 42 days. This represented 24% of the animal body weight. The animals did not show any toxic signs, and blood levels of SGOT and bilirubin were normal.

6.5.2 Fish

The effects of S. jacobaea alkaloids on rainbow trout (Salmo gairdneri) fingerlings has been investigated (Hendricks et al., 1981). Duplicate groups of 80 fingerlings were fed for up to 12 months on diets containing 20 or 100 mg alkaloid/kg. The alkaloid comprised 91% jacobine, 3% jacazine, 2.5% senecionine, and 2.5% seneciphylline. The 100 mg/kg diet resulted in severe growth depression and mortality, which began at 3 - 4 months. Both levels of PAs produced severe hepatic lesions. The livers from these fish were shrivelled, mottled yellow or whitish in colour, nodular, fibrous, and sometimes haemorrhagic. Microscopically, there was megalocytosis, severe fibrosis, and bile-duct proliferation. Characteristic veno-occlusive changes were seen in the centrilobular and hepatic veins, which, in the case of fish receiving the 100 mg/kg diet, appeared after only 2 months on the diet.

6.5.3 Insects

There are no reports of the toxicity of PAs for insects, but there is substantial literature on the use of PAs by certain insect families that have evolved with the ability to store the alkaloids as defensive chemicals and to convert them into pheromones and other signalling chemicals (see recent reviews by Brown (1984) and Boppré (1986), and an earlier complementary paper by Edgar (1982)). In some species, such as moths of the family Arctiidae, the larvae feed on PA-containing plants. In other families, such as Nymphalid butterflies of the sub-families Danainae and Ithomiinae, the larvae of most species live on other plants, but the adult males seek out PA-containing species and contrive to ingest alkaloids from wilting, dead, or damaged plant material or from nectar. The alkaloids so acquired have a functional role as defensive chemicals against predators and, in some species, are also converted into pheromones and other signalling chemicals involved in mating. The alkaloid derivatives may be pyrrolic compounds related to dehydroretronecine or derivatives of the esterifying acids. In one Arctiid genus, Creatonotus, the alkaloids have a morphogenetic or hormonal effect, determining the size of the pheromone-disseminating organ. Thus, for some insect species, PA-containing plants may be necessary for survival.

7. EFFECTS ON MAN

The toxic effects of pyrrolizidine alkaloids are principally on the liver. The toxic disease, produced by consuming PAs derived from certain plants, is called veno-occlusive disease (VOD), the pivotal and pathognomonic lesion being the occlusion of the central and sublobular hepatic veins in the liver. The larger hepatic vein tributaries are characteristically unaffected in contrast with the findings in Budd-Chiari syndrome (Bras, 1973).

7.1 Clinical Features of Veno-Occlusive Disease (VOD)

There are several good clinical accounts of the disease, mostly in the earlier reports from Jamaica, in children (Jelliffe et al., 1954a,b), and adults (Stuart & Bras, 1955), which have been summarized by Stuart & Bras (1957). Maksudov (1952) has described the clinical features among children in outbreaks in the USSR, where it was called toxic hepatitis with ascites, and Ismailov (1948a,b), Mnushkin (1949, 1952), and Zheltova (1952) have described them among adults. Srivastava et al. (1978) also described the clinical findings among children in a large outbreak. Children seem to be particularly vulnerable as is evident from the report of Stuart & Bras (1957) and that of Mohabbat et al. (1976) concerning a large outbreak, though the disease is rare before the age of one or two years. Frequently, more than one member of the family becomes affected (Stuart & Bras, 1957; Mohabbat et al., 1976; Tandon et al., 1976; Arora et al., 1981) within days or weeks of each other.

The disease generally has an acute onset, characterized by rapidly developing and progressing symptoms of upper abdominal discomfort, dragging pain in right hypochondrium, ascites, and sometimes oliguria and oedema of the feet. Nausea and vomiting may be present. Jaundice and fever are rare. There is generally gross, tender, smooth hepatomegaly often accompanied by massive pleural effusion, and sometimes slight splenomegaly and minimal ankle oedema. Liver function tests may show only mild disturbance. The acute disease is associated with high mortality and a subacute or chronic onset may lead to cirrhosis. Death often occurs after oesophageal haemorrhage.

Stuart & Bras (1957) summarized the clinical data of 84 patients ranging in age from 6 months to 53 years. The highest incidence occurred between the ages of 1 and 3 years (39%), and 26% of patients belonged to the 3- to 6-year age group. Thus, children up to 6 years accounted for 65% of total cases. Although the VOD was relatively uncommon at the 2 extremes of age, early infancy and adult life, mortality was highest in these groups, being 60% and 54%, respectively. Hepatomegaly and some degree of ascites were invariably present in acute cases; in 48 out of 64 patients, ascites was acute enough to require paracentesis. In 38 patients, hepatomegaly was grossly severe, reaching more than half way down to the umbilicus. Jaundice was relatively uncommon. Among the liver function tests, the most significant and consistent changes were found in the serum-cholinesterase (t = 2.67, $0.02 > P > 0.01$) and serum-albumin levels (t = 2.82, $0.01 > P$). Mortality rates associated with signs of parenchymal liver damage were 74% with clinical jaundice or high levels of serum-bilirubin, 62% with diminished serum-cholinesterase, and 58% with considerable anorexia and apathy. The mortality rates among cases with acute, subacute, or chronic disease were reported to be 27%, 17%, and 57%, respectively. Death was mostly due to hepatocellular failure (71%) in the acute phase and haemetemesis in the chronic phase (75%). More recent publications (Lyford et al., 1976; McGee et al., 1976; Tandon, B.N. et al., 1977; Datta et al., 1978a,b) also describe the haemodynamic data of the hepatic blood flow and the results of portovenographic studies that suggest outflow tract obstruction in the liver at the post-sinusoidal level, and irregularity and obstruction/distortion of the hepatic venous radicles, respectively.

The clinical course of the disease has been shown schematically by Stuart & Bras (1957) (Fig. 12). However, the temporal relationships in the different phases of the disease are not precise, and their account, at best, represents a trend.

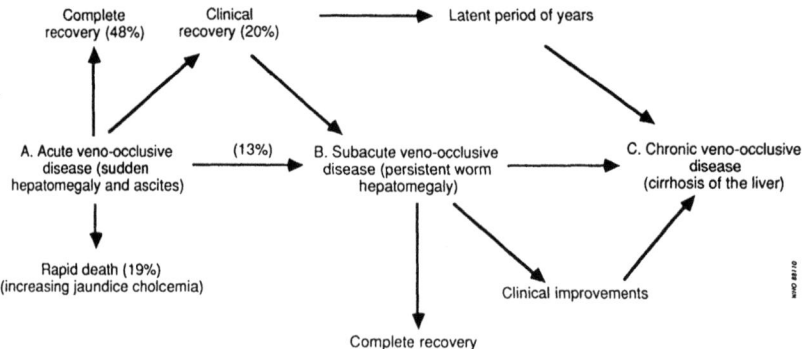

Fig. 12. Clinical natural history of veno-occlusive disease of the liver. B and C may be present with no clinical history of preceding illness. From: Stuart & Bras (1957).

The onset of the disease may be sudden (acute) or insidious (subacute or chronic). The acute disease may recover completely or result in death. A few patients may go on to the subacute phase, with almost none or very few symptoms, but a persistent hepatomegaly. The patient may subsequently recover completely, or may, after or without apparent clinical improvement, go on to the chronic phase of disease, mostly ending up in cirrhosis. Some patients with the acute disease may go on to the subacute phase, even after clinical recovery, or as postulated by the authors, after a latent period of several years, progress to cirrhosis.

Braginskii & Bobokhadzaev (1965) related an experience concerning the evolution of this disease that was similar to that observed in the USSR. About 50 - 60% of cases made a full recovery and 35% made a partial recovery with continuing hepatomegaly. About 2% of cases developed persistent ascites with "loss of working capacity". It has been suggested that these cases may develop cirrhosis.

7.2 Salient Pathological Features of Veno-Occlusive Disease

The pathological features of the disease at different stages have been described in detail (Terekhov, 1939, 1952; Dolinskaya, 1952; Bras et al., 1954; Bras & Watler, 1955; Bras & Hill, 1956; Stuart & Bras, 1957; Stirling et al., 1962; Aikat et al., 1978; Tandon, B.N. et al., 1978; Tandon, H.D. et al., 1978). The following description of Bras & Watler (1955) characterizes the morphological changes at different stages of the disease described by most investigators.

Morphological features of the liver at autopsy in 19 patients from Jamaica who died at various stages of VOD have been described. The ages ranged from 10 months to 45 years. Fourteen patients were below the age of 14 years. Of the 14 patients, 9 had developed cirrhosis, 3 died of acute disease, and 7 had various levels of fibrosis superimposed over acute VOD. In the acute stages, there was acute centrilobular congestion. The centrilobular and sublobular veins showed different degrees of thickening of the wall and occlusion of the lumen, mainly due to subintimal swelling composed of loose reticular tissues and a few cells including endothelial cells, occasional histocytes, lymphocytes, and polymorphs, suggesting an acute exudative process. In addition, there was a small amount of fibrin. Organization of this exudate gradually led to collagenization and thickening of the wall. The centrilobular congestion resulted in compression and even disappearance of liver cell cords. The reticular framework of the liver lobule was frequently preserved but was ruptured at places. Clearly recognizable thrombi occurred sporadically but were not commonly seen. The portal veins and hepatic arteries were normal.

Stirling et al. (1962) in their description of the early lesion of VOD of the liver also emphasized that thrombosis of the hepatic veins was not an important histogenetic factor in the evolution of the lesion. In the later stages of the disease, hepatic venous occlusion was chronically established. Perivenous reticular collapse and the resultant condensation and reduplication of the reticular framework resulted in non-portal fibrosis and later cirrhosis. Macroscopically, the cirrhosis looked like Laennec's cirrhosis. The cirrhotic process, which was non-portal to begin with, involved portal tracts which also became incorporated in the connective tissue septa. The presence of oesophageal varices was a common finding. Extra-hepatic collateral vessels and congestive splenomegaly were frequently seen. Within the liver, intra-hepatic collateral circulation was established with the coalescence of sinusoids. The larger hepatic veins and inferior vena cava did not show any thrombi or other pathological change. The subintimal swelling observed in the hepatic veins was not seen in any other vessels in the body. The authors concluded that acute VOD gradually leads to Laennec's cirrhosis, but, in the initial phases, it is non-portal.

No clear dose or temporal relationships between the liver and parenchymal and vascular changes in the liver lobules are evident from the available human case reports.

It is notable that megalocytosis, which is the hall-mark of chronic PA toxicity in experimental animals and a morphological manifestation of the antimitotic effect of PAs, has not been observed in human subjects (Tandon, H.D. et al., 1978) (section 6.4.9).

Tandon H.D. (personal communication) has analysed all published and unpublished observations including liver biopsies and autopsies derived from 3 outbreaks of VOD of which two occurred in India (Tandon, B.N. et al., 1976, 1977; Tandon, R.K. et al., 1976; Tandon, H.D. et al., 1977) and one occurred in Afghanistan (Tandon, B.N. et al., 1978; Tandon, H.D. et al., 1978). He commented on the frequency with which the characteristic veno-occlusion may not be seen in the needle biopsy in the acute phase of the disease, though it is invariably observed in the autopsy material. Haemorrhages may persist for up to 1 year after subsidence of acute symptoms. There is no significant inflammation accompanying the veno-occlusive changes. Cirrhosis was reported to have developed within 3 months in one patient, after an initial biopsy had shown no fibrosis (Stuart & Bras, 1957).

Brooks et al (1970) made an ultrastructural study of the liver in VOD in Jamaican children. They found extensive damage in the sinusoidal epithelium resulting in entra-vasation of red cells in Disse's space and between hepatocytes. The closed structure of the vessels, the absence of fenestrations, the existence of a basement membrane, and the presence of collagen in the wall contribute to the resistance to cellular debris and erythrocytes which track up Disse's space and tend to narrow the lumen of the sinusoid where it enters the vein. Fibrin may be present in this location and occasionally in the sinusoids, but not within hepatic veins themselves. The venous block, therefore, does not appear to be the result of thrombosis.

Pancreatic changes similar to those in Kwashiorkor were stated to be commonly observed (Bras & Hill, 1956; Stuart & Bras, 1957).

7.3 Human Case Reports of Veno-Occlusive Disease

Available data on human cases are summarized in Table 15, with the countries they have been reported from in alphabetical order. The first report of this disease in man and its relationship with consumption of wheat flour contaminated by seeds of a plant of the genus Senecio was made by Albertjin in 1918 in a report made to the government of South Africa, as quoted by Wilmot & Robertson (1920), who gave the first account of this disease in scientific literature from that country. It is possible that it may have existed even earlier, as its occurrence in farm animals had been described since the beginning of the century in veterinary literature (Bull et al., 1968). Willmot & Robertson (1920) recorded the occurrence of 'bread poisoning' in South Africa for a period of about 10 years, during which 80 cases had been observed, mostly resulting in death. The detailed account of clinical data and pathological findings in 11 cases was consistent with what is known as VOD today. The flour from which the bread was prepared was found to be contaminated with the flower heads and seeds of Senecio spp.

Isolated case reports are available in the literature (Hashem, 1939; Wurm, 1939) describing changes in childrens' liver characteristic of veno-occlusive disease; however, in these studies, no mention was made of possible etiological agents.

A pattern of disease described in the Soviet literature as dystrophy of the liver, and later as toxic hepatitis with ascites has been known in the Central Asian republics of the USSR especially Uzbekistan and Tadjikistan for a long time (Terekhov, 1952). The local inhabitants called the disease, characterized by rapidly filling ascites, the "camel-belly" syndrome and are reported to have long distrusted a weed with fine seeds, known locally as "Kharmyk", as a source of the

Table 15. Pyrrolizidine alkaloid poisoning and human reports of veno-occlusive disease (VOD) (in alphabetical order of the countries reported from)

Country/ number cases	Age group	Suspected vehicle of intoxication	Name of plant	Alkaloid	Alkaloid concentration and intake	Nature of lesion	Outcome (cirrhosis/ died)	Reference
Afghanistan								
8000 (approximately)	<14–80 years	contaminated wheat flour	Heliotropium popovii gillianum	mainly heliotrine, in addition to 1 or 2 alkaloids similar to lasiocarpine (75 – 100% as N-oxides)	1.32–1.49 g/kg (in the seeds) mostly daily intake, 2 mg; total consumption, up to 1.46 g	all stages; acute to subacute	many died	Tandon & Tandon (1975); Mohabbat et al. (1976)
Barbados								
3	2–4 years	herbal infusion	Crotalaria retusa	not analysed		acute lesions	no details	Stuart & Bras (1956)
China, People's Republic of								
2	49–57 years	herbal infusion	Gynura segetum (Compositae)	not analysed		acute lesions	died	Hou (1980)
Ecuador								
1	35 years	herbal infusion	Crotalaria juncea	not analysed		acute lesions	recovered	Lyford et al. (1976)

Table 15 (contd).

Egypt						
3		not known				Hashem (1939)
59	1-12 years	plant seed decoction in 17	not identified	acute lesions	cirrhosis in one	Safouh & Shehata (1965)
Federal Republic of Germany						
8	75-116 days	not known (alimentary toxaemia)		acute lesions	all died	Wurm (1939)
Hong Kong						
4	23-28 years	herbal infusion	Heliotropium lasiocarpum pyrrolizidine alkaloids	herbs-alkaloid base 0.42g/kg; N-oxides 1.4g/kg; daily intake, 12 mg base, 18 mg N-oxide; total intake, 570 - 1350 mg up to development of symptoms in 3 patients over 19-45 days, 1300 mg up to death in 1 patient. 630 mg in one case who remained asymptomatic.	acute VOD 1 died	Kumana et al. (1983, 1985); Culvenor et al. (1986)

Table 15 (contd).

Country/ number cases	Age group	Suspected vehicle of intoxication	Name of plant	Alkaloid	Alkaloid concentration and intake	Nature of lesion	Outcome (cirrhosis/died)	Reference
India
2	25 and 35 years	herbal decoction and pills	not identified	unknown		acute	recovered	Gupta et al. (1963)
25	12-38 years (6 patients studied)	possibly contaminated cereal	not identified	not analysed		acute lesions; cirrhosis in one	12 died; 1 cirrhosis	Tandon, R.K. et al. (1976); Tandon, H.D. et al. (1977)
108	3-60+	contaminated cereal	Crotalaria nana	crotananine and crotaburmine	0.81 g/kg	various stages of disease	up to 63% died	Tandon, B.N. et al. (1976, 1977)[a]; Siddiqui et al. (1978a,b)
35	1- > 25 years	contaminated cereal	Crotalaria	as above	5.3 g/kg seed (0 - 1.9% of seeds in cereal); daily intake, 40 mg	acute	as above	Krishnamachari et al. (1977)[a]; Siddiqui et al. (1978a,b)

Table 15 (contd).

Country	Age	Source	Plant	Alkaloid	Dose	Pathology	Outcome	Reference
India (contd)								
6	20-70	herbal medicine in 4 cases	*Heliotropium eichwaldii* by 3 patients	heliotrine as N-oxide	12 – 20 g/kg; total intake, 4 g and 10 g in patients who died of fulminant disease	acute lesions; cirrhosis in one; organizing thrombi in hepatic veins in 4 cases	3 died (2 with fulminant disease)	Aikat et al. (1978); Datta et al. (1978a,b)
Iraq								
9	3-10 years	possibly wheat flour	possibly *Senecio*	not identified		acute	1 died	Al-Hasany & Mohamed (1970)
South Africa								
11 (only 5 described)	11-19 years	wheat flour as bread	*Senecio ilicifolius*; *S. burchelli*	not identified senecionine?		centrilobular haemorrhages; central vein "dilated"	"majority died"	Wilmot & Robertson (1920)
12		wheat flour	not identified; "possibly *Senecio*"	not identified		centrilobular haemorrhages; organizing thrombi in hepatic veins; 3 cases of VOD		Selzer & Parker (1951)

Table 15 (contd).

Country/ number cases	Age group	Suspected vehicle of intoxication	Name of plant	Alkaloid	Alkaloid concentration and intake	Nature of lesion	Outcome (cirrhosis/ died)	Reference
South Africa (contd)								
12	5 - 30 months	herbal medicines admitted in 4 cases; food not analysed	not identified	none found in herbal medicines		"acute" in 5 cases; "subacute" in 7 cases; no detail	3 died of disease; no follow-up on 5	Stein & Isaacson (1962) Stein (1957)
United Kingdom								
1	26 years	herbal infusion	Maté-Paraguay tea	identified only as pyrrolizidine alkaloids	"trace amounts"	acute with veno-occlusion going on to chronic with fibrosis	died	McGee et al. (1976)
1	13 years	herbal infusion	<u>Symphytum officinale</u>	not analysed	not known	"thrombotic" variant of veno-occlusive disease		Weston et al. (1987)

Table 15 (contd).

USA							
1	2 months	herbal infusion	Senecio longilobus	riddelline and N-oxides of retrorsine, seneciphylline, and senecionine	15 g/kg as alkaloids; total intake, 66 mg	acute lesions	died	Fox et al. (1978)
1	6 months	herbal infusion	Senecio longilobus	as above	3 g free alkaloid plus 10.5 g N-oxide per kg; total intake, 70 – 147 mg of combined alkaloid and N-oxide	acute lesions	cirrhosis	Stillman et al. (1977); Huxtable (1980)
1	49 years	herbal infusion	Symphytum sp.	symphytum pyrrolizidine alkaloids	14.1 µg/kg body weight per day; total intake, 85 mg	acute lesions	recovered after short surgery	Ridker et al. (1985); Huxtable et al. (1986)
USSR							
1000 – 1500	3-50 years	contamination of wheat flour	Heliotropium lasiocarpum	heliotrine, lasiocarpine		acute lesions	13 – 15% died; cirrhosis in many	Mirochnik (1938); Terekhov (1939)
"large number"		contamination of wheat flour	Heliotropium lasiocarpum	heliotrine lasiocarpine				Dubrovinskii (1947). Isr

Table 15 (contd).

Country/ number cases	Age group	Suspected vehicle of intoxication	Name of plant	Alkaloid	Alkaloid concentration and intake	Nature of lesion	Outcome (cirrhosis/died)	Reference
Venezuela								
1	5 years	"earth and plant eating habits"	not identified			acute lesions	died	Grases & Beker (1972)
West Indies								
11	14 months to 9 years	not known, possibly herbal infusions				acute to cirrhosis	data unclear; cirrhosis in some	Jelliffe et al. (1954b)
5	23 months to 18 years	not known, possibility of "bush teas"	not identified				1 cirrhosis	Bras et al. (1954)
4	16-45 years	history of herbal infusion in one case	not identified			acute to acute-on-chronic	3 died	Stuart & Bras (1955)
84	0.5-53 years	probably "bush teas"	not identified; (Senecio and Crotalaria)				11% had cirrhosis; 27% died	Stuart & Bras (1957)
23		possibly bush teas	not stated	not analysed		acute to "cirrhosis" (no breakup)		Bras & Hill (1956)

Table 15 (contd).

Country/ number cases	Age group	Suspected vehicle of intoxication	Name of plant	Alkaloid	Alkaloid concentration and intake	Nature of lesion	Outcome (cirrhosis/ died)	Reference
West Indies (contd)								
23	1-50 years		no precise identification; "Crotalaria fulva and possibly other toxic factors"	no analysis		cirrhosis		Bras et al. (1961)
3	"children"	"toxic substances not identified in herbal remedies"				acute	died	Stirling et al. (1962)

a Pertains to the same outbreak.

disease (Dubrovinskii, 1952) (it is notable that, in the Afghan outbreak, which occurred close to the area in Uzbekistan where the disease had been known, the toxic seeds were known as "Charmak" among the local population). The "camel-belly" disease was associated with the use of bread with a bitter taste (presumably caused by the mixing and grinding of toxic seeds with the wheat grains). Two waves of outbreaks occurred, the first between 1931-35 (Terekhov, 1939) and the second between 1945-46 (Dubrovinskii, 1947; Ismailov, 1948a,b). In the first wave of outbreaks, approximately 1000 - 1500 subjects were estimated to have been affected with an overall mortality rate of 13 - 15%. The age of the affected subjects ranged from 3 to 50 years. In the second wave of outbreaks, 60 - 70% of the population in agricultural areas were estimated to have been affected (Ismailov, 1948a,b). Domestic animals also suffered from the disease (Dubrovinskii, 1947). During the second wave of outbreaks, the disease was found to have been caused by contamination of food crops with seeds of Heliotropium lasiocarpum (Dubrovinskii, 1947; Khanin, 1948). The etiology was confirmed by studies on a variety of experimental animals using the suspect seed (Khanin, 1948; Kampanzev, 1952). Pathological features of the disease in children have been described by Dolinskaya (1952) and in all age groups by Terekhov (1939, 1952). First an acute form of the disease was recognized, followed by hepatic cirrhosis, which was found to result as a consequence of the initial disease (Ismailov, 1948a,b; Savvina, 1952). The disease has since been largely eradicated by the agricultural and public health measures taken.

Selzer & Parker (1951), from South Africa, described 12 cases, 10 of whom came from 3 families that had eaten bread made of flour of "imperfectly winnowed wheat". Five cases who were autopsied showed characteristic occlusion of the central vein of the liver lobules.

Attention on veno-occlusive disease as an entity followed a spate of reports, mostly from Jamaica, on the occurrence of this disease, mainly among children (Bras et al., 1954, 1961; Jelliffe et al., 1954a,b; Bras & Watler, 1955; Bras & Hill, 1956; Stirling et al., 1962) but also among adults (Stuart & Bras, 1955, 1957). The youngest patient reported was a 3-month-old infant (Stein & Isaacson, 1962), but the disease was stated even to have been observed in the newborn (Stein, 1957). Very often, several members of the family were affected by disease, which presented primarily as rapidly filling ascites, within a matter of days, sometimes accompanied by fever, and leading to hepatic failure. It was considered an important cause of cirrhosis among children in Jamaica (Jelliffe et al., 1954a,b). On the basis of evidence of an almost identical disease occurring among grazing animals (Hill, 1960), and a prevalent practice among the Jamaicans of using herbal infusions for treating a variety of ailments as a home remedy, herbs, notably of the Senecio and Crotalaria groups, were suspected of being a contributory factor (Bras & Hill, 1956; Bras et al., 1957, 1961; Stuart & Bras, 1957).

Hill (1960) gave a comprehensive account of current knowledge up to that time of the world-wide distribution of seneciosis in man and animals in which mention is made of this disease only in Egypt, the Federal Republic of Germany, South Africa, and the West Indies. Pyrrolizidine alkaloids were identified as the active element in the toxic factor in the plants.

However, there are earlier reports of this disease from elsewhere. Bras (1973) cited a number of reports from Europe from 1905 to 1949, including reports by Wurm (1939) and Teilum (1949), of an entity in infants and children that was called Endophlebitis hepatica obliterans. He reported having studied some tissue sections of liver in Austrian children aged 7 months - 1 year, who had died with clinical and histological patterns of VOD. Some of these conformed to the Budd-Chiari syndrome whereas others were consistent with VOD. The case of a 36-year-old woman reported by Teilum (1949) does not, however, fall in the usual category of cases accepted as veno-occlusive, since thrombotic changes were also seen in the larger hepatic veins and several, in the peripheral parts of the body, as well as the portal vein.

A comprehensive account of all available published reports of this disease to date from different parts of the world is given in Table 15. A clear distinction between the acute and chronic effects of exposure is not always possible from the published data, since, in most patients, the history of onset of disease with intake of the toxic substance is not generally volunteered or available, and so a precise temporal relationship is difficult to establish. However, this information is available in a reasonably accurate form in more recent reports (Stillman et al., 1977; Datta et al., 1978a,b; Fox et al., 1978; Kumana et al., 1985).

Gupta et al. (1963) from India described 2 cases who had drunk some herbal infusions for the treatment of skin disorders. The liver biopsies confirmed changes typical of acute VOD. However, the herb was not identified and no analysis was made of the infusion for alkaloids.

The cases of 59 children in Egypt who had symptoms of rapidly developing abdominal distension and hepatomegaly were reported by Safouh & Shehata (1965). Their ages ranged between 1 and 12 years, 47 being below 4 years of age. A dietary survey carried out on 17 patients indicated ingestion of drinks made by boiling some common plant seeds, but these

were not identified. Wedge biopsies of liver were performed
in 6 patients and 16 postmortem examinations were made. In
all the autopsy cases, there was thrombotic obliteration of
the main hepatic veins and their ostia. The central and sublobular veins were not uniformly thickened as they were
frequently dilated or disrupted and some contained fresh
thrombi. There was centrilobular necrosis of the liver
lobules. One patient developed decompensated cirrhosis in 3
years. The authors observed that the clinical and pathological
features closely resembled those of VOD in Jamaica, but
thrombosis was an unusual feature of the latter disease.

Al-Hasany & Mohamed (1970) described a short outbreak in
Iraq, occurring during a 3-month period, affecting 9 children,
all except one being below 12 years of age, and belonging to 3
Bedouin families living outside Baghdad. One of the children
died. Autopsy of this case and biopsies on the others showed
changes characteristic of VOD. Poisoning through the wheat
flour contaminated by seeds of some PA-containing plant was
suspected, but no analysis of the food was carried out.

Three of the largest outbreaks of the disease have been
reported from South Asia, two from the same site in central
India (Tandon, B.N. et al., 1976, 1977; Tandon, R.K. et al.,
1976; Krishnamachari et al., 1977; Tandon, H.D. et al., 1977)
and one from North-West Afghanistan (Tandon & Tandon, 1975;
Mohabbat et al., 1976; Tandon, B.N. et al., 1978; Tandon, H.D.
et al., 1978). The first Indian outbreak was reported by
Tandon, B.N. et al. (1976) and Tandon, R.K. et al. (1976) and
Tandon, B.N. et al. (1977) and Tandon, H.D. et al. (1977). It
occurred in a group of 5 tribal villages in central India in
1972-73. The epidemiology of the outbreak was later described
by Arora et al. (1981). Out of a total population of 2060 in
these villages, 71 households with 366 members were investigated. Among these, 39 cases had developed and 19 had died
before commencement of the investigations. The incidence rate
was 1.1% and the case fatality rate was 50%. All cases
occurred among 20 households. In many households, several
members were affected. In one household, 4 out of 5 cases
died. Six sick patients were investigated in detail with
repeated liver biopsies. One patient died 17 months after
onset and was autopsied. The clinical onset was characteristic of disease. Haemodynamic and radiographic studies
suggested a combined post and perisinusoidal block. Liver
biopsy studies showed features characteristic of acute
centrilobular haemorrhagic necrosis with progressive fibrosis,
hepatic veno-occlusion, and non-portal cirrhosis in one case
who survived 17 months after the acute onset. The etiological
factor of this outbreak was not established, though dietary
contamination with pyrrolizidine alkaloids was considered.

A second outbreak occurred at the same site in 1975 and was studied and reported independently by Krishnamachari et al. (1977) and Tandon, B.N. et al. (1976, 1977). A total population of 486 was affected, 67 cases were reported, of whom 28 (46%) had died. There was a strong family history (Tandon, B.N. et al., 1976). In a later survey, 108 patients were studied and the mortality rate was estimated to be 63% (Tandon, B.N. et al., 1977). This time the etiological factor was identified as the plant Crotalaria nana Burm, which had been growing in the fields of millet (Panicum miliare), their staple food crop. The seeds of this plant became mixed with the cereal grain during harvesting. The toxic seeds contained pyrrolizidine alkaloids that were identified as a macrocyclic ester closely similar to monocrotaline. The total alkaloid content was estimated to be 5.3 g/kg of seed, expressed as monocrotaline. The levels of contamination of the millet with seeds were reported to be 0 - 3.4 g/kg in the unaffected and 0 - 19 g/kg in the affected households (Krishnamachari et al., 1977). A precise identification could not be made, but the same material, independently studied by Siddiqui et al. (1978a,b), were reported to contain 2 new alkaloids, cronaburmine and crotananine. The seed contained an alkaloid level of 26 g/kg. The levels of contamination of the millet were up to 20 g/kg and the amount of alkaloid ingested was estimated to be up to 40 mg per day.

The largest outbreak reported to date occurred in the Gulran district of Herat Province in northwest Afghanistan, close to the border of the USSR. Tandon & Tandon (1975) identified the plant as being causative factor of the disease, which was surveyed and reported in detail by Mohabbat et al. (1976). The outbreak, was estimated to have affected a population of approximately 35 000 in 98 villages. Examination of 7200 inhabitants of the affected villages showed evidence of disease in 22.6%, which was more serious in 15%. Thus, it was estimated that approximately 8000 subjects suffered from the disease including 5000 who were seriously affected. All age groups were affected, but 46% of subjects were below 14 years of age. However, no sign of disease was found in children below 2 years of age. A detailed report concerning the pathological material obtained from 14 liver biopsies and 8 autopsies was made by Tandon, B.N. et al. (1978) and Tandon, H.D. et al. (1978). The time interval between the onset of symptoms and the biopsy/autopsy was not indicated. Pathological findings were characteristic, ranging from acute disease with characteristic veno-occlusion to non-portal cirrhosis, which was observed in 5 of the above 22 cases. The outbreak was ascribed to massive contamination of wheat, the staple food crop, following 2 years of drought, with the seeds of Heliotropium popovii H. Riedl subsp. gillianum H. Riedl, which had been growing profusely among the wheat crop. The seeds contained pyrrolizidine alkaloids at concentrations reported by 2 laboratories to be 7.2 and 13.2 - 14.9 g/kg, identified as mainly as the N-oxide of heliotrine (74%) (Mattocks, personal communication), and one or two other compounds similar in character to lasiocarpine. Samples of wheat from several villages contained an average of 40 seeds/kg, i.e., 0.03% by weight. It was estimated that an adult consumed at least 700 g flour/day, containing approximately 2 mg alkaloid (based on a mean of the seed analyses). There is some uncertainty about the estimate, since Mohabbat et al. (1976) also stated that the samples of the wheat flour were assayed and contained 0.186 - 0.50% alkaloid. This analysis and the 0.72% result for alkaloid in the H. popovii seed were from the same laboratory in Kabul (R.N. Srivastava, personal communication), and together, imply 13 - 36% seed in the wheat. If correctly reported, the result for the flour conflicts with the estimated proportion of the seed in the wheat and can scarcely have been representative.

Tandon & Tandon (1975) stated in their report of the survey during which the causative factor of the Afghan outbreak was discovered, that such cases had always been observed by Government physicians posted in this area in the past, sometimes in significant numbers, but neither the population nor the physicians remembered that the disease had occurred in the form of an outbreak.

There was no mention of VOD in the hepatic lesions observed by Sobin et al. (1969) among the 121 specimens of liver obtained at medico-legal autopsies or by needle biopsies at Kabul, though 6.6% of the 89 autopsy specimens showed "acute passive congestion with necrosis". The ages of these subjects was not stated and the authors did not discuss the cause of the lesion.

Following these outbreaks, a number of isolated cases have been described following the use of herbal medicines. McGee et al. (1976) reported a case from the United Kingdom of a 26-year-old woman who had consumed herbal tea containing pyrrolizidine alkaloids. The one sample examined contained only trace amounts of alkaloids. Neither the plant nor the alkaloids were further characterized. The patient had been taking very large quantities of the tea for about 2 years and it is possible that some of the earlier batches may have been more heavily contaminated. Maté or Paraguay tea (Ilex species), which she was drinking, is stated to be a popular drink in Brazil and is not believed to contain pyrrolizidine alkaloids. It has been stated (Huxtable, 1980) that possibly she ingested the pyrrolizidine alkaloids from some other unidentified source. The clinical course of the disease progressed rapidly. Three biopsies carried out within one month and the autopsy showed characteristic changes including centrizonal fibrosis, but no cirrhosis. It is notable that some involvement of muscular pulmonary arteries was also seen.

Lyford et al. (1976) reported the case of a 35-year-old woman from Ecuador, who had ingested herbal tea prepared from <u>Crotalaria juncea</u>. She had consumed 1 - 2 litres of this infusion daily for 6 weeks, but no qualitative or quantitative analysis for pyrrolizidine alkaloids was made. She had had arthralgias for 3 years, for which she had received treatment with indomethacin and phenylbutazone. The liver biopsy showed characteristic changes of acute VOD from which she recovered completely as proved by a repeat biopsy carried out one year later.

The occurrence of acute disease was reported in 2 infants in Arizona in the USA, aged 6 months and 2 months, respectively, following ingestion of infusions of a herb, locally called the Gordolobo Yerba by the Mexican-American population and identified as <u>Senecio longilobus</u> (Stillman et al., 1977; Fox et al., 1978; Huxtable, 1980). The plant from which this infusion had been prepared was found to contain pyrrolizidine alkaloids identified as riddelline and N-oxides of retrorsine, seneciphylline, and senecionine (Huxtable, 1980), in a concentration of 3 g free alkaloid and 10.5 g N-oxides/kg (Stillman et al., 1977). It was estimated that, during a period of 2 weeks, the 6-month-old infant received a total dose of between 70 and 147 mg of combined alkaloid and N-oxide derivative. The liver biopsy showed characteristic features of acute disease, which had progressed to extensive central, portal, and sinusoidal fibrosis (Stillman et al., 1977). The child subsequently developed cirrhosis over the next 8-month period. The 2-month-old infant was administered an infusion of the same herb for 4 days, after which he became progressively more ill and stuporous (Fox et al., 1978). On admission he was diagnosed as a case of Reye's syndrome, but subsequently developed jaundice and possibly ascites and died. The sample of herb contained a concentration of alkaloids of 15 g/kg. It was estimated that the infant had received a total of 66 mg of mixed alkaloids over the 4-day period. At autopsy, extensive centrilobular haemorrhagic necrosis of the liver was seen, which is characteristic of the acute disease. However, no occlusive lesions of the central vein of the lobules were described, and no obstructive lesions were seen in the larger hepatic veins or inferior vena cava. No mention was made of ascites. The basal ganglia showed kernicterus.

Datta et al. (1978 a,b) reported 6 cases that occurred between 1974 and 1977. All of the patients had taken herbal medicines, identified in 1 case as <u>Heliotropium eichwaldii</u>, which contained N-oxides of heliotrine. Two patients took the herb as an extract of the whole plant, which contained an alkaloid concentration of 20 g/kg, for 20 and 50 days, respectively, and developed symptoms after a time lag of 45 and 90 days, respectively. They were both estimated to have consumed 200 mg of heliotrine per day, the total alkaloid intake being 4 g and 10 g, respectively. They had taken the herb for treatment of epilepsy, and were admitted with acute onset of symptoms of abdominal pain, ascites, jaundice, hepatic encephalopathy, and gastrointestinal bleeding, which suggested fulminant viral hepatitis. They died within 2 - 12 weeks of the onset of symptoms. Only a brief description of the main autopsy findings in the liver was given, which indicated that there were changes characteristic of acute veno-occlusive disease of liver, including marked centrilobular haemorrhagic necrosis of the liver lobules, and occlusive lesions of the central and sublobular veins. It is interesting to note that both patients had also been on long-term anticonvulsant phenobarbitone therapy. The remaining 4 patients had a chronic insidious onset of disease suggesting cirrhosis of the liver in 3, and alcoholic liver disease in one. One of the former had been taking some indigenous powder, presumably prepared from a herb, the indication and nature of which are not known. The patient died from hepatic encephalopathy. No detailed description of the autopsy findings was given, but it was stated that the central and sublobular veins of the liver showed chronic occlusive changes. The inferior vena cava and large hepatic veins were patent. There was non-portal cirrhosis. A notable feature was that the arsenic levels in the liver tissue were high (500 µg/kg; normal, 1 µg/kg). There is no mention of arsenic in the report on the analysis of the indigenous powder taken by the patient. Two of the remaining 3 patients with chronic disease, had taken herbal medicine for vitiligo, and one for diabetes mellitus. The herb was identified as Heliotropium eichwaldii in the case of one of the vitiligo patients, who had taken it for 10 days and in whom the onset of symptoms occurred within 10 days. The herb was taken in the form of seed with an alkaloid content of 12 g/kg. The daily intake was estimated to be 500 mg of alkaloid and the total intake, 5 g. This patient was admitted with a clinical diagnosis of cirrhosis of the liver. The liver biopsy showed changes characteristic of acute VOD. Follow-up data are not known. The herb taken by the other 2 patients was not identified. The diabetic patient was being treated with oral hypoglycaemic drugs and was also known to be an alcoholic. The indigenous powder being taken by him contained a high concentration of arsenic (5 mg/kg). The results of haemodynamic studies suggested hepatic venous outflow tract obstruction of the intra-hepatic post-sinusoidal type in the smallest hepatic veins. Liver biopsy showed characteristic centrilobular haemorrhages. Central veins could not be recognized. There were mild changes in the liver cells, but no alcoholic hyaline was seen. The liver biopsy of the third patient showed characteristic features of acute disease with veno-occlusion.

Two cases have been reported from China (Hou et al., 1980). Both were adults who were taken ill after taking medicinal infusions prepared from Gynura segetum of the family Compositae (tribe, Senecioneae). The presenting symptoms and the cause of disease, as well as the pathological findings, were characteristic, except that one patient had jaundice and also portal vein thrombosis, which is not a usual feature of the disease. No chemical analysis of the plant was made and the alkaloid was not precisely identified. Furthermore, the total intake of alkaloid was not calculated. It has been stated that this was the first report of such a case, but that it was possible that the disease might occur among adults more frequently without being reported.

Ghanem & Hershko (1981) reported 3 cases of Arabs, one 3-year-old child and 2 adults, who were diagnosed as having VOD of the liver, on the basis of the clinical findings and morphological features of liver biopsies. One more patient had occlusive lesions of both the small and large hepatic vein radicles. They were among 29 patients with clinical features of hepatic vein thrombosis (HVT) found on a retrospective analysis of data from 9 major hospitals in Israel. Of these patients, 15 were Jews and 14 were Arabs. Notable features were that all Jewish patients were adults, whereas the majority of Arab patients were children below 10 years of age and primary HVT was 2.4 times more common among the latter. No analysis of the diet was made for PAs and their etiological role was suspected only on a presumptive basis. A survey of stored wheat grain in 9 villages showed that 2 samples were contaminated with seeds of Lolium, belonging to the Graminae family, which were found to contain 2 PAs (loline and norloline). However, these PAs are not known to be hepatotoxic. The authors argued that, even though in a classical case of VOD there should not be thrombosis of the larger hepatic vein radicles, the difference in the anatomical appearance of VOD and that of primary HVT of the near-east type is not due to a different etiological agent but rather to a difference in the dose and rate of absorption of the ingested toxic compounds.

A further report of the disease from Hong Kong, by Kumana et al. (1985), described it in 4 young Chinese women with psoriasis who took infusions of a herbal remedy, the toxic component of which has since been identified as Heliotropium lasiocarpum (Culvenor et al., 1986). They developed symptoms 19 - 45 days after starting the herbal treatment, and were examined 61 - 68 days after its initiation. The condition of patient No. 2, who continued taking the herb for 16 days after the onset of symptoms, deteriorated and she died of hepatic failure and was autopsied. The liver biopsies and autopsy confirmed the presence of acute disease in all patients. Patient No. 4 stopped taking the herb after 21 days, on account of a new rash. When assessed 77 days later, she had mild hepatomegaly only. A detailed analysis of the alkaloid content was carried out for each case. The pyrrolizidine alkaloids were quantified as if senecionine based. The herb contained 0.42 g alkaloid/kg and 1.4 g N-oxide/kg. The daily intakes of alkaloid base and N-oxide were estimated to be 12 ± 1 mg and 18 ± 4 mg, respectively. The respective cumulative doses of alkaloid (base and N-oxide) consumed by patients Nos 1, 2, and 3, up to onset of symptoms, were calculated to be 1350 mg over 45 days, 900 mg over 30 days, and 570 over 19 days, respectively. Patient No. 4 who had irrefutable histological evidence of disease but did not develop symptomatic disease, must have consumed 630 mg over 21 days. Patient No. 2, who died, was estimated to have taken a total amount of 1380 mg alkaloid over 46 days. The estimated cumulative intake per kg body weight before the development of symptoms for patients Nos. 1, 2, 3, and 4 was 26, 15, 12, and 15 mg/kg, respectively. It should be noted that, in patient No. 2, who died, the cumulative dose until the onset of symptoms was the same as in patient No. 4, who was asymptomatic. Moreover, the total intake by patient No. 2, was 23 mg/kg, which was lower than the intake by patient No. 1, who survived. The authors compared the intake data of their patients with those of the Arizona children reported by Stillman et al. (1977) and Fox et al. (1978). The 6-month-old baby, who survived but developed cirrhosis, and the 2-month-old baby, who died, are estimated by comparison to have taken cumulative doses of 12 - 25 and 11 mg/kg, respectively. The above data suggest marked variation in susceptibility among individual subjects. It is also known from experimental animal studies that the young and new-born animals are particularly vulnerable (Jago, 1970).

A case reported from the USA is ascribed to the consumption of comfrey (Symphytum sp.) powder, sold as a digestive aid (Ridker et al., 1985). A 49-year-old woman presented with classical symptoms and signs of VOD. The haemodynamic data showed a hepatic vein wedge pressure of 3.07 kPa (23 mmHg) with a sinusoidal pressure of 2.27 kPa (17 mmHg). Hepatic venograms showed near obliteration of the smaller radicles of the hepatic veins during balloon distension of one of the intra-hepatic venous tributaries, and there was extra-vasation of the dye into the hepatic parenchyma. A porta-caval shunt was carried out and the operative findings confirmed the presence of a post-sinusoidal block. The liver biopsy showed marked centrilobular necrosis and congestion with dilatation of the central veins and sinusoids, consistent with hepatic venous outflow tract obstruction. According to the clinical history, the patient had been a heavy consumer of food supplements. Apart from several vitamins and minerals, she had been drinking 3 cups of camomile tea per week and for 6 months before admission had consumed 1 g/day of a commercially available herbal tea. For 4 months before admission, she had taken 2 capsules of "comfrey-pepsin" pills with each meal. The herbal tea and the pills were analysed for PAs. Pyrrolizidine alkaloids and their $\underline{N}$-oxides were found, but the compounds were not precisely identified. On the basis of the analysis of the PA content, the patient was estimated to have consumed a total of at least 85 mg of PA (Huxtable et al., 1986) (14.1 µg/kg body weight per day, Ridker et al., 1985). The authors emphasized that the total PA consumption was relatively low. It was possible that the patient had other sources of exposure and probably she had been consuming PA-containing supplements for longer than the periods stated by her in the clinical history.

The latest is the report of a 13 year old boy from the U.K. who is stated to have developed symptoms of toxicity following administration of herbal tea prepared from comfrey leaf (Symphytum officinale) for treatment of inflammatory bowel disease for two or three years (Weston et al., 1987). The exact quantity of leaves consumed and frequency of administration were not known. The liver biopsy is stated to have shown a "thrombotic variant" of veno-occlusive disease, though the inferior vena cava and the major hepatic veins were patent on Doppler ultrasound and percutaneous phlebography. He had earlier been treated with predinisolone and sulphasalazine. The case is unusual in so far as the thrombosis of the central veins of the liver lobules, which is not a usual feature of veno-occlusive disease of the liver.

7.4 VOD and Cirrhosis of the Liver

Jelliffe et al. (1954b) were the first to draw attention to VOD being an important cause of cirrhosis among Jamaican children. Prior to this, Hashem (1939), while reviewing the records of all cases of cirrhosis admitted to a children's hospital in Egypt since 1933, described 3 cases of a special type of cirrhosis that was rare in adults. The clinical features and pathological findings were similar to those in cases of VOD. They speculated that the cause was some metabolic toxins of gastrointestinal origin. Royes (1948) suggested that the cirrhosis in the Jamaican children was very like the disease described from India and Egypt.

The clinical and pathological features of 100 cases of VOD among Jamaican children were described by Bras et al. (1954). Sixty-five of the cases were below 12 years of age. None of the 100 patients had cirrhosis initially, but 5 showed occlusive lesions in hepatic veins and features of non-portal fibrosis. Four of these 5 cases later developed cirrhosis. The authors concluded that VOD contributed to a substantial number of all cases of cirrhosis in this age group. Stuart & Bras (1957) studied 84 patients with VOD including 64 acute, 6 subacute, and 14 chronic cases. Twenty-three patients were followed up, some for up to 5 years. Autopsy performed on 21/26 cases showed cirrhosis in 11 cases. Notable features were that 1 of the 6 cases of acute disease described in detail, developed cirrhosis. Of the 2 cases with chronic disease, 1 developed cirrhosis within 3 months of a liver biopsy for acute disease, at which time the liver had shown hepatic venous occlusive lesions but no fibrosis. Autopsy findings were described by Bras & Watler (1955) in 19 patients aged 10 months - 45 years in different stages of the disease. Nine patients had cirrhosis that was non-portal to begin with but finally became indistinguishable from Laennec's portal cirrhosis. Rhodes (1957) studied the pattern of liver disease among Jamaican children. A total of 193 liver biopsies was studied derived from 39 children who had one biopsy and 59 who had more than one at intervals of 1 week - 3 years. Of the 14 autopsies on cases of VOD, 12 had cirrhosis. A notable feature was that the disease could occur asymptomatically with hepatomegaly. The autopsy material from the University College Hospital, Jamaica was analysed by Bras et al. (1961). Of the 1560 autopsies, 28.5% cases concerned infants of less than one year old, mostly from poor, predominantly black families. Cirrhosis was seen in 77 autopsies. Approximately 30% of these 77 cases were diagnosed as post-VOD. The authors postulated that they might have resulted from the ingestion of Crotalaria fulva or some other toxic substances.

In a follow-up study of 61 patients who developed the disease in an outbreak in the USSR in 1958, 28 developed "Hepatoleinal syndrome" (Braginski & Bobokhadzaev, 1965). Two of these cases, in whom the disease in the initial stages was not particularly severe, developed cirrhosis within 4 years.

Tandon, H.D. (personal communication) has analysed the pathological data derived from the Afghan and Indian outbreaks on the basis of 61 liver biopsies and 17 autopsies, including repeat biopsy studies on 15 patients who were followed-up for 1 month - 3 years after onset in the Indian outbreaks. Three of the 11 patients, followed up for 16 months or longer for persistent clinical evidence of disease, ended up with cirrhosis and 2 more had marked fibrosis with equivocal changes of cirrhosis in the biopsy. Notable features of the study were that the disease progressed to cirrhosis in patients who were put on a normal diet, free from alkaloids, after appearance of symptoms of acute disease and did not receive any subsequent exposure. Impact of alcohol intake was excluded. There was a poor correlation between the clinical and pathological severity of disease. Centrilobular haemorrhages, which are a sign of acute disease, were seen to persist for over one year in patients, some of whom were apparently well. At needle biopsy, characteristic hepatic venous occlusions were not seen in many patients in the acute phase of the disease, though they were seen in all livers at autopsy and most of them showed persistent centrilobular haemorrhages. Biopsy findings in cirrhotic livers were often not histopathologically characteristic for any specific form of cirrhosis. Features that might suggest the veno-occlusive etiology of cirrhosis at biopsy included paraseptal dilatation of sinusoids and persistent haemorrhages or haemosiderin in the septa.

In studies by Aikat et al. (1978) and Datta et al. (1978a,b), 6 cases of VOD were reported following ingestion of herbal medicines containing PAs. Four of the patients had symptoms of chronic disease and one of them developed non-portal cirrhosis.

One of the 2 infants from Arizona, USA, who suffered from VOD following the administration of PA-containing herbal medicine, developed cirrhosis during the 8-month period following the appearance of acute symptoms (Stillman et al., 1977; Fox et al., 1978; Huxtable, 1980). Huxtable (1980) mentioned the death from cirrhosis of liver of a 62-year-old woman, who had consumed the same herb as the 2 infants for 6 months prior to her death. However, there was no confirmation of the diagnosis of cirrhosis.

7.5 Differences between VOD and Indian Childhood Cirrhosis (ICC)

A type of cirrhosis of liver, peculiar to the people of Indian origin, Indian Childhood Cirrhosis (ICC), has been ascribed to PA toxic etiology (Bras et al., 1954; Rhodes, 1957), owing to the observation of occlusive changes in the central and sublobular veins of the liver, by some investigators (Radhakrishna Rao, 1935; Prabhu, 1940; Jelliffe et al., 1957; Ramalingaswami & Nayak, 1969), though this is not a characteristic feature of the disease. The presence of copper-positive granules in the hepatocytes (Salaspuro & Sipponen, 1976) has added to such a conjecture (Tanner & Portmann, 1981), because of the reported aberration of copper metabolism in experimental animals exposed to PAs (section 6.4.11). However, ICC and VOD are clinically and pathologically distinct. ICC, confined to infants and children, often affects siblings. Jaundice is a common sign and hepatosplenomegaly is a common feature. The disease is almost invariably fatal due to rapidly developing hepatocellular failure. Liver parenchymal changes are characterized by marked ballooning degeneration of hepatocytes, prominent deposits of alcoholic hyaline, severe cholestasis, and aggressive, pericellular fibrosis (Nayak et al., 1969), all features not characteristic of VOD. Moreover, occlusive changes in the hepatic veins are very rare and were not observed by any member of the liver diseases subcommittee of the Indian Council of Medical Research (1955), who made a critical study of the disease.

7.6 Chronic Lung Disease

Heath et al. (1975) reported the case of a 19-year-old African man who had died of congestive cardiac failure, and who was suspected of having ingested a herbal remedy containing the seeds of Crotalaria laburnoides. Histopathological examination of the lungs showed characteristic vascular changes of severe primary pulmonary hypertension. Powdered seeds of the plant were fed in the diet to Wistar albino rats for 60 days (Table 11). Characteristic features of pulmonary hypertension including hypertensive vascular changes in the lung and right ventricular hypertrophy of the heart were produced in the animals showing that the seeds contained an agent capable of inducing pulmonary hypertension in rats. Apart from this indirect evidence, there was no proof of such a causal relationship with the pulmonary hypertensive disease in the patient.

A brief mention is made by McGee et al. (1976) of finding changes "somewhat similar but rather more mature" (than those seen in the hepatic veins) involving some branches of the pulmonary artery in the lower lobe of the left lung, in a case of veno-occlusive disease of the liver caused by ingestion of PA-containing herbal teas. The alkaloids were not further characterized. The changes were also stated to be similar to those produced in experimental animals by PAs. Apart from these 2 cases, there is no mention of involvement of the pulmonary arterial system in any of the case reports available.

The possibility that diet-mediated agents might induce pulmonary hypertension in man has been discussed at length by Fishman (1974). A parallel was drawn with the epidemic of pulmonary hypertension that occurred in Austria, the Federal Republic of Germany, and Switzerland between 1966 and 1968 (Kay et al., 1971b), which was suspected of being caused by

Aminorex, a compound that resembles epinephrine and amphetamine in chemical structure. Although the etiological role of Aminorex could not be conclusively proved on epidemiological or experimental grounds, there continues to be a suspicion that agents taken by mouth can evoke pulmonary hypertension in man. Levine et al. (1973) reported the cases of 3 children aged 5 1/2 - 13 years, and 11 months, respectively, with portal hypertension, who developed progressive pulmonary hypertension resulting in cor pulmonale and death. In all 3 cases, there was evidence of extra-hepatic portal vein obstruction confirmed at autopsy, and the symptoms and signs of portal obstruction had appeared in early childhood. They developed symptoms of cardio-respiratory failure. Studies of pulmonary and cardiovascular function including haemodynamic studies of pulmonary circulation in 2 of the children suggested pulmonary vascular obstructive disease. At autopsy, no primary parenchymal lung disease was found. There were vascular changes of advanced pulmonary hypertension (plexiform lesions), but no evidence of thrombo-embolism was found. No factor responsible for pulmonary vasoconstriction was identified. In one case, the liver was stated to show coarse nodularity at autopsy, but the microscopic examination showed only patchy areas of portal fibrosis and regeneration. In the other 2 cases, there were only non-specific changes, with fibrosis in one. However, centrilobular congestion was present in 2 cases. It is possible that some metabolites of toxic agents, such as PAs, which are metabolized in the liver, might have blocked a metabolic pathway that ordinarily exerts a pulmonary antihypertensive effect, or that, by damaging the liver, vasoactive substances, such as histamine, serotonin, and catecholamines, might escape metabolic pathways to reach the lungs and injure the pulmonary vessels. However, no such agents were identified.

Kay et al. (1971b) made a plea that, on the basis of the experimental data available, including the ability of several agents to produce pulmonary hypertension in experimental animals, and, in spite of the fact that pulmonary vascular disease has never been demonstrated in human cases of veno-occlusive disease, careful enquiries should be made on the possibility of patients presenting with unexplained pulmonary hypertension having ingested a plant product. A similar plea was made by Heath et al. (1975).

7.7 Trichodesma Poisoning

The disease "Ozhalanger encephalitis", which occurred in the Samarkand region of Uzbekistan, USSR in the period 1942-51, is believed to have been caused by contamination of food grain with the seeds of <u>Trichodesma incanum</u> (Shtenberg & Orlova, 1955; Ismailov et al., 1970), which contain 1.5 - 3.1% alkaloids, mainly trichodesmine and incanine (Yunusov & Plekhanova, 1959). Clinical features of the disease have been described by Ismailov et al. (1970). This outbreak differed from the others described above in that the primary symptomatology was extra-hepatic. Over 200 patients were affected, not including children below the age of 10, or the breast-fed infants of the affected mothers. Following exposure, there was a latent period of about 10 days, then vertigo and recurring headaches developed leading to nausea and vomiting. This was followed by generalized malaise, which progressed to delirium and loss of consciousness. Physical signs included pathological reflexes in 59% of the patients and paresis of the extremities and the facial nerve. Death was stated to have been caused mostly by respiratory depression. Of the 200 patients affected, 44 died. Autopsy findings were relatively non-specific degenerative and necrotic lesions scattered in several organs including the central nervous system. No report of such a disease is available from outside the USSR.

7.8 Relationship Between Dose Level and Toxic Effects

In some recent human case reports, the PAs consumed have ben identified and estimates made of the daily and total intakes (Table 16). The relationship between these intake levels and the known toxicity of the alkaloids in rats has been discussed by Culvenor (1983) and Mattocks (1986).

Discussion of the relationship between the dose level and toxic effects in human cases is complicated, because the poisoning is generally due to a mixture of alkaloids found in naturally-occurring plant products, consumed as herbal remedies or food, and different plant species. There are wide differences in the acute toxicities of the alkaloids, which are the best available measure of comparative effects due to long-term intake as well. Furthermore, estimates of intake can, at best, be approximate. When a large population is affected through the contamination of a food crop, as happened in the Afghan and Indian outbreaks (section 7.3), no precise estimates are possible regarding the extent of contamination in different households, the amount of the contaminated grain consumed, the length of exposure resulting in the appearance of symptoms or signs of toxicity, or death, in the cases studied. There may also be various other contributing factors that are not apparent, e.g., food or cooking habits, on account of which no conclusive generalizations regarding the causative role of the toxic agent can be made. Reports on chemical analysis for the toxic agent and, hence, the amounts ingested, may not always be reliable (refer to the controversy in the Afghan outbreak in section 7.3). In cases, such as the one reported by Ridker et al. (1985), when the total dose of PAs estimated to have been received by the patient before the disease developed, was only a fraction of that of other alkaloids causing episodes of human toxicity (Table 16), it is not certain whether that was the only toxic alkaloid agent that the patient was exposed to, or whether there were other contributing factors, particularly as the patient was stated to be a heavy consumer of herbs, vitamins, and natural food supplements.

In the known instances of human toxicity the principal alkaloids involved are heliotrine from <u>Heliotropium</u>, echimidine and related alkaloids from <u>Symphytum</u>, riddelline and retrosine from <u>Senecio longilobus</u>, and crotananine and cronaburmine from <u>Crotalaria nana</u> (Table 16). Approximate acute toxicity values (LD_{50} in rats) for these alkaloid mixtures are 300, 500, and 50 mg/kg, respectively. For the mixture from <u>C. nana</u>, for which there are no experimental data, the acute toxicity was assumed to be similar to that of monocrotaline, 100 mg/kg. These relativities need to be taken into account in discussing dose-effect relationships for the PAs as a group. This has been done by discussing first the dose estimates for heliotrine, since it was the main alkaloid in 1 epidemic and in 6 case reports (Table 16). Then in discussing the other alkaloids, reference is made to a heliotrine-equivalent dose as well as the actual dose. The heliotrine equivalent is:

$$\text{actual dose} \times \frac{LD_{50} \text{ of heliotrine}}{LD_{50} \text{ of alkaloid mixture concerned}}$$

The estimated daily intake in poisoning by heliotrine ranges from 0.033 mg/kg body weight in the Afghan epidemic (Mohabbat et al., 1976), which after a period of about 180 days and a total intake of about 6 mg/kg caused fatalities, to 3.3 mg/kg, which led, in 2 cases in India (Datta et al., 1978a,b), to death after 20 and 50 days and total intakes of 67 and 167 mg/kg, respectively. In between are 4 cases in Hong Kong with estimated daily intakes ranging from 0.49 to 0.71 mg/kg (Kumana et al., 1983, 1985; Culvenor et al., 1986). Three cases were non-fatal at total doses of 11 - 27 mg/kg body weight and one was fatal at a total dose of 23 mg/kg. These heliotrine cases imply that daily intakes are cumulative down to 0.033 mg/kg and may be fairly rapidly fatal above 0.5 mg/kg. Above a total dose of 6 - 15 mg/kg, VOD may become evident and sometimes fatal.

In the 2 cases due to riddelline and retrorsine in <u>Senecio longilobus</u> (Stillman et al., 1977; Fox et al., 1978; Huxtable, 1980), the estimated daily intakes were 0.8 - 0.17 and 3 mg/kg

Table 16. Estimated intakes of PAs in human beings

Principal alkaloids(s) [a]	Age of subject (years)	Daily intake (mg/kg) [b]	Period (days)	Total dose mg	Total dose mg/kg	Toxic effect [c]	Reference
1. Heliotrine	various [d]	0.033	180 [e]	360	6 [e]	VOD, death	Mohabbat et al. (1976)
2. Heliotrine	(a) 20 (b) 23	3.3 3.3	20 50	4000 10000	67 167	death death	Datta et al. (1978a,b)
3. Riddelline, retrorsine	0.5	0.8–1.7 [b]	14	70–147	12–25	VOD	Stillman et al. (1977); Huxtable (1980)
4. Riddelline retrorsine	0.17	3.0 [b]	4	66	12	death	Fox et al. (1978)
5. Crotananine, crotaburmine	various [d]	0.66	c. 60 [e]	2400	40	VOD, death	Tandon, B.N. et al. (1976); Tandon, R.K. et al. (1976); Krishnamachari et al. (1977)
6. Heliotrine	(a) 28 (b) 26 (c) 23 (d) 27	0.59 0.49 0.60 0.71	45 46 19 21	1350 1380 570 630	27 23 11 15	VOD death VOD VOD	Kumana et al. (1983, 1985); Culvenor et al. (1986)
7. Echimidine	49	0.015	120	94	1.7	VOD	Ridker et al. (1985)

[a] The principal alkaloid is recorded as the free base, even if there was evidence of its presence in the plant mainly as N-oxide.
[b] Calculated for a 60-kg adult, unless definite information available. The 0.5-year infant was said to be 6 kg, and the 0.17-year infant was assumed to be 5.5 kg, the average weight for males of these ages.
[c] When mentioned, death was a consequence of severe liver damage.
[d] Epidemic.
[e] Estimate based on unpublished information available to the Task Group.

body weight (equivalent to 3 and 1 mg heliotrine/kg, respectively, and the total doses were 12 - 25 and 12 mg/kg (equivalent to 72 - 150 and 72 mg heliotrine/kg, respectively). These levels are comparable with the highest reported intakes of heliotrine and, in infants, led to the rapid development of VOD and, in one case, death.

In the epidemic due to mixed crotananine and cronaburmine in Crotalaria nana (Tandon, B.N. et al., 1976; Tandon, R.K. et al., 1976; Krishnamachari et al., 1977), the estimated daily intake of 0.66 mg/kg and the total intake of 40 mg/kg (equivalent to 2 and 120 mg heliotrine/kg, respectively) also corresponded to the highest intake of heliotrine.

In the case of Symphytum poisoning (Ridker et al., 1985), the estimated daily intake of echimidine and related alkaloids was 0.015 mg/kg and the total dose 1.7 mg/kg (equivalent to 0.009 and 1.0 mg/kg heliotrine, respectively). The dose levels are lower in equivalent terms than the lowest estimates in cases due to heliotrine by a factor of about 4 for daily intake and 6 for total intake. The estimates were based on questioning of the patient and assay of the material concerned. It seems prudent to conclude that a daily intake of pyrrolizidine alkaloid as low as the equivalent of 0.01 mg/kg heliotrine may lead to disease in humans.

There is substantial overlap between intake rates and total intakes for fatal and non-fatal poisoning. This presumably reflects the influences of a number of factors, such as individual sensitivity, age, nutritional status, and general health, but it is also due to the progressive nature of pyrrolizidine toxicity and the effects of time. In the epidemics, in which some people died, only an estimated average intake is available and some who were alive at the time of investigation may have died later. Comparing the total intakes for human toxicity with the total doses up to death observed in the long-term administration of PAs to rats, 1.2 - 10.9 times the LD_{50} dose, equivalent to 360 - 3270 mg heliotrine/kg (Table 10, section 6.4.1.5), it is evident that human beings are more susceptible to the acute and chronic effects of the alkaloids than rats, sometimes markedly so.

These considerations of the toxic effects in human beings of various intake levels could provide a basis for some assessment of the likely hazard from other types of exposure to PAs. For example, the consumption of comfrey root tea, estimated by Roitman (1981) to contain 8.5 mg alkaloid per cup, at the rate of 3 cups per day, or the ingestion of comfrey leaf at the rate of one leaf per day, could lead to alkaloid ingestion rates of 0.40 and 0.016 mg/kg. These rates are respectively, much greater than, and equal to, the lowest daily rate causing veno-occlusive disease. Lower levels of exposure arising from such sources as the consumption of milk from cows eating PA-containing plants or of honey derived from such plants, seems unlikely, in practice, to cause acute or subacute liver disease. However, care should be exercised. In an experimental situation in which cows were fed Senecio jacobaea, the milk was reported to contain up to 0.84 mg alkaloid/litre. A 30-kg child drinking 0.5 litre/day of this milk could ingest 0.014 mg/kg alkaloid, equivalent to 0.028 mg heliotrine/kg (assuming an LD_{50} of 150 mg/kg for S. jacobaea alkaloid). This is above the lowest daily rate leading to veno-occlusive disease and the lowest estimated total toxic dose would be achieved in 36 days. This level of contamination of milk is undoubtedly extreme and there is no knowledge of any contamination of commercial milk supplies. Honey derived from Echium plantagineum has been reported to contain up to 1 mg alkaloid/kg (Culvenor et al., 1981). A 30-kg child consuming 30 g/day of honey (a high consumption rate) would ingest 0.001 mg alkaloid/kg body weight. The lowest estimated total toxic dose (1.7 mg comfrey alkaloid/kg, very similar to Echium alkaloid) would be achieved in 1700 days. Although it seems likely that consumption of contaminated milk and honey would lead to acute or subacute liver disease, the possibility remains that they may contribute to chronic liver disease or liver tumours.

The possibility of carcinogenic effects due to long-term exposure to PA-containing plants has been discussed by Culvenor (1983). Some of the PAs involved in instances of human poisoning have been found to be carcinogenic in experimental animals (Table 13). Data from some of the significant experimental studies were summarized by Culvenor (1983) with approximate estimates of PA dosages administered to rats in terms of mg/kg body weight per day (Table 17). The dose rates that were carcinogenic for rats (Table 17) ranged from 2 to 6 mg/kg per day for an initial period and 0.2 - 3 mg/kg per day for a remaining period of about 12 months, except in one study in which a dose of 10 mg/kg per day was used. It can be seen that, in all except two instances of human poisonings summarized in Table 16, the estimated daily rates of intake ranging from 0.015 to 3.3 mg/kg body weight per day are within close range of those known to induce tumours in rats. In other reports, the consumption rates are above and below this range.

Epidemiological studies to assess the carcinogenic role of PAs for man are not available. In countries with a high incidence of primary liver cancer, it is possible that PAs may have an additive effect with those attributed to aflatoxin (Newberne & Rogers, 1973) and hepatitis B virus. The total evidence now available warrants long-term studies of the survivors of poisoning outbreaks, especially where a substantial number of people were affected, as in the Afghanistan outbreak.

Table 17. Rates of administration of PAs leading to tumours in rats[a]

Alkaloid	Dosing schedule	Approximate equivalent rate[b] (mg/kg per day)	Number of rats developing tumours	Reference
Lasiocarpine	(a) 7 mg/kg diet, 24 months	0.70	23/24	National Cancer Institute (1978)
	(b) 15 mg/kg diet, 24 months	1.50	24/24	National Cancer Institute (1978)
	(c) 7.8 mg/kg, ip, 2/week for 4 weeks and 1/week for 52 weeks	2.2 for 4 weeks, then 1.1 for 52 weeks	16/18	Svoboda & Reddy (1972)
	(d) 50 mg/kg diet, 55 weeks	5.0	18/20	Rao & Reddy (1978)
	(e) 0.39 mg/kg, ip, 3/week, to death	0.2	2/7	Culvenor & Jago (1979)
Monocrotaline	(a) 25 mg/kg, ip, 1/week for 4 weeks, and 8 mg/kg, for 38 weeks	3.5 for 4 weeks, then 1.1 for 38 weeks	10/50	Newberne & Rogers (1973)
	(b) 5 mg/kg sc, once per 2 weeks for 52 weeks	0.36	43/60	Shumaker et al. (1976)

Table 17 (contd).

Retrorsine	(a) 30 mg/kg, ip, single dose	–	7/29	Schoental & Bensted (1963)
	(b) 30 mg/litre in water, 3 days/week, to death	1.3	4/14	Schoental et al. (1954)
	(c) 30 – 50 mg N-oxide/litre in water, 3 days/week for 20 months	1.3 – 2.0	10/22	Schoental et al. (1954)
Petasitenine	0.1 g/litre in water, up to 16 months	10	8/10	Hirono et al. (1977)
Senkirkine	22 mg/kg, ip, 2/week for 4 weeks and 1/week for 52 weeks	6 for 4 weeks, then 3 for 25 weeks	11/24	Hirono et al. (1979a)
Symphytine	13 mg/kg, ip, 2/week for 4 weeks and 1/week for 52 weeks	3.7 for 4 weeks, then 1.9 for 52 weeks	5/24	Hirono et al. (1979a)

[a] From: Culvenor (1983).
[b] Where necessary, estimates assume a daily rat food intake of 100 g/kg body weight, and water intake 100 ml/kg body weight. Injected doses are given pro rata, for daily administration.

7.9 Pyrrolizidine Alkaloids as a Chemotherapeutic Agent
       for Cancer

The PA, indicine N-oxide derived from Heliotropium indicum, a widely used indigenous drug in Ayurvedic medicine, has been found to have an antitumour activity and has been used in clinical trials as a chemotherapeutic agent for leukaemia (Letendre et al., 1981, 1984; Cook et al., 1983) and solid tumours (Kovach et al., 1979a,b; Nichols et al., 1981; Ohnuma et al., 1982; Taylor et al., 1983). Dosing schedules typically were 5 consecutive intravenous doses of 0.15 - 3 g/m² body surface area (approximately 2.5 - 5 mg/kg body weight) repeated at 4- or 6-week intervals (Kovach et al., 1979a; Letendre et al., 1981; Ohnuma et al., 1982). Hepatic toxicity, as judged by increased SGOP levels, was infrequent and mild. However, subsequent trials with this agent have indicated more serious hepatotoxicity. In a more recent report by the same workers (Letendre et al., 1984), 5 out of 22 cases of refractory acute leukaemia, treated with indicine N-oxide, had severe hepatotoxicity, presumably induced by the drug. One of these patients had been treated for 18 months with methyl-testosterone, 4 months prior to receiving indicine N-oxide. Symptoms of severe hepatocellular failure appeared in 3 patients after the initial course of treatment. This occurred after 4 daily doses of 3 g/m² surface area in one patient and after 5 daily doses of 3.75 g/m² surface area in 2 patients. Two other patients who had received 3 g/m² surface area daily for 5 days developed hepatocellular failure after the second course of treatment, one at 3.3 g/m² and the other at 3.75 g/m² surface area, daily, for 5 days. In each patient, the onset of hepatic disease was rapid and the course was downhill. Livers of 4 of these patients examined at post-mortem showed severe centrilobular vascular congestion with necrosis of parenchymal cells, and, in one patient, a few sublobular veins were found to be occluded.

Miser et al. (1982) reported severe hepatotoxicity in 4 of 45 children treated with indicine N-oxide for refractory leukaemia or advanced solid tumours. Similarly, Cook et al. (1983) reported the case of a 5-year-old child with acute myeloid leukaemia who developed severe hepatic failure within 3 days of starting the treatment. Autopsy showed massive hepatic necrosis.

However, it should be noted that no hepatic failure was reported in more than 100 adults with solid tumours, treated with the same agent (Kovach et al., 1979a,b; Nichols et al., 1981; Taylor et al., 1983). No hepatotoxic effects were reported by Ohnuma et al. (1982) among 37 patients who received this drug for solid tumours. The major toxic effect was myelosuppression (Kovach et al., 1979b).

7.10 Prevention of Poisoning in Man

At present, prevention of poisoning can be achieved only by reducing or eliminating ingestion of the alkaloids. The two effective procedures are control of PA-containing plants in agricultural areas and educational programmes directed to the populations at risk.

The control of plant populations for this purpose has been carried out only in Uzbekistan, USSR, following the epidemics of human disease due to contamination of grain by seeds of <u>Heliotropium lasiocarpum</u> and <u>Trichodesma incanum</u>. The following measures were introduced and have been effective in preventing further outbreaks:

1. A state standard was set for the quality of seed grain, which must be certified by a State Seed Inspectorate. Current standards prohibit the sowing of wheat, rye, barley, or oats contaminated by seed of <u>Heliotropium lasiocarpum</u> or <u>Trichodesma incanum</u>.

2. A state standard was set for the quality of grain stored for food. The limits for <u>Heliotropium lasiocarpum</u> and <u>Trichodesma incanum</u> seeds are 0.2% and zero, respectively.

3. Agricultural (agritechnical) measures to ensure minimum contamination of crops and harvested grain, including specification of the most suitable methods and timing of cultivation, use of clean seed for sowing, weeding of crops prior to maturing of the grain (towards the end of May), and mechanical cleaning of grain.

4. Methods for monitoring levels of contamination of flour, bread, and similar products.

5. Publication of educational booklets describing the biological, environmental, and morphological characteristics of the toxic weeds, their pathways of distribution and the causes of the toxicoses experienced.

6. Promotion of weed control by governmental authorities and provision of legislation to enforce the control measures.

In other countries, the control of some PA-containing weeds in crops is practised by cultivation and herbicide treatment, in order to maximize yield and the general quality of the grain. In pastures, animal management and herbicide treatments are used to increase pastures and reduce poisoning of animals. Specific treatment methods differ according to the plant species and the circumstances. General references were not available to the Task Group.

In Australia, where *Heliotropium europaeum* and *Echium plantagineum* are widespread weeds in wheat-growing areas but where normal agricultural practices prevent all but occasional minor contamination, relevant tolerance standards for wheat delivered at storage silos are not specific. *Heliotropium europaeum* seed is rarely seen and would form part of the "unmillable material" the seed component of which can be up to 1% of the volume of the wheat. Seed of *Echium plantagineum* is occasionally seen in delivered grain at levels of up to 10 seeds per half litre, the tolerance level for this seed fraction being 50 seeds per half litre.

8. BIOLOGICAL CONTROL

Biological control methods have been investigated for several PA-containing plant species, notably Senecio jacobaea, Heliotropium europaeum, Echium plantagineum, and Trichodesma incanum. The effectiveness of such methods is variable and good results may be confined to certain regions where favourable conditions exist. For example, in control programmes against S. jacobaea in Australia, Canada, New Zealand, and the USA, using 3 insect species, results varied from virtually nil to nearly 100% control (Julien 1982). The effects of the introduction of the cinnabar moth (Tyria jacobaea L.) on S. jacobaea in these countries have been summarized as in Table 18.

Table 18. Results of the attempted control of Senecio jacobaea (ragwort) with the cinnabar moth

Country or region	Result
Australia	Establishment precluded by predation, parasitism and disease
Western Canada	Moth populations stabilized below that required for control
Eastern Canada	Establishment and subsequent notable reductions in ragwort levels
New Zealand	Marginal establishment, moth population limited by predation and parasitism, little impact on ragwort
USA	Widespread establishment, ragwort levels sometimes reduced at localities near the limits of its distribution

Several agents are being tested in Australia for the control of H. europaeum and one species, a flee beetle Longitarsus albineus, has been released (Julien, 1982; Delfosse, 1985). There are good prospects in this country for the biological control of Echium plantagineum and 2 other Echium spp., with 8 insect species approved for release, when legal restrictions are lifted (Delfosse & Cullen, 1985a,b). Preliminary studies have been made on the biological control of Amsinckia and other Senecio species (e.g., Pantone et al., 1985).

Given adequate funding, PA-containing plants are a suitable target for biological control.

9. EVALUATION OF HUMAN HEALTH RISKS AND EFFECTS ON THE ENVIRONMENT

9.1 Human Exposure Conditions

9.1.1 Reported sources of human exposure

The two main sources of exposure of human beings to toxic PAs that have led to major outbreaks of poisoning with high mortalities as well as to individual cases in several countries are:

(a) the contamination of cereal grains, such as wheat and millet, with the seed or other parts of plants containing alkaloids; and

(b) the consumption for medicinal or dietary purposes of herbs containing the alkaloids, either as the plant itself or as infusions.

Consumption of contaminated grain is more likely to occur in regions where food is in short supply, and particularly when drought favours infestation of the grain crop by PA-containing weeds. A qualitative field test for detecting the presence of toxic pyrrolizidine alkaloids in plant materials, using simple laboratory methods, is now available (section 2.2.2.5).

9.1.2 Plant species involved

Plant species containing toxic PAs occur throughout the world and are known in 47 genera in 6 plant families. As many as 6000 species are potentially PA-containing. The most important genera responsible for human and animal disease are Senecio and other genera of the tribe Senecioneae (family Compositae), Crotalaria (family Leguminosae) and Heliotropium, Trichodesma, and other genera of the family Boraginaceae (sections 3.1 and 3.2).

Approximately 150 different toxic PAs have been isolated from about 360 plant species that have been investigated and contain this type of alkaloid. Of these, about 12 have been involved in instances of human toxicity (section 3.1). The molecular structures of almost all of these alkaloids are known and the main outlines of structure-toxicity relationships have been established (section 2.1).

The alkaloids may occur in all plant parts and are often present as the N-oxide derivatives, which are also toxic when ingested orally. Alkaloid contents vary from low (0.1 g/kg dry weight) to very high (40 g/kg up to the maximum recorded of 180 g/kg in Senecio riddelli). Levels vary with stage of growth, locality, and other circumstances. In some, but not all, species, the alkaloid is partly decomposed during the drying or storage of the plant. The decomposition is largely enzymic and once the plant material is dry, the alkaloid is fairly stable.

9.1.3 Modes and pathways of exposure

9.1.3.1 Contamination of grain crops

Large outbreaks of poisoning have occurred through contamination of wheat crops in Afghanistan, India, and the USSR. In particular, 3 species of Boraginaceae (Heliotropium lasiocarpum, H. popovii, and H. europaeum) are well adapted to vigorous growth under the climatic conditions in which wheat is usually grown. Contamination can be effectively controlled in wheat produced using modern harvesting techniques and grain seed that is inspected and controlled for weed seed contamination, but control of contamination is more difficult where these conditions cannot be met. Contamination of staple food grain is of particular concern, since entire populations are exposed, and control may not be possible, if the people are not aware of the hazard that PA-containing weeds present.

9.1.3.2 Herbal medicines

Herbal preparations containing PAs are used as tonics, treatments, preventatives, and food supplements. Such usages are so widespread that they are nearly universal. Many are traditional, while others reflect a rejection of, or lack of access to, standard health care services (section 3.3.2).

Veno-occlusive disease was first recognized as a clinical entity in Jamaica as a result of the medicinal use of PA-containing herbs prepared from Crotalaria. Crotalaria-containing herbs have also been responsible for human poisonings in Barbados, Equador, and other locations in the West Indies. Heliotropium herbs have been reported to cause poisoning in Hong Kong and India. Symphytum- and Senecio-containing herbs have given rise to case reports in the USA. Other reports of PA poisoning are known in which the herbs used were not botanically identified. PA-poisoning has been associated with both home-prepared and commercially available herbs, the latter including prescriptions by herbalists (Weston et al., 1987).

Various other genera of PA-containing plants in the families Boraginaceae, Compositae, and Leguminosae are also widely used as herbs. No case reports are available for these genera.

Reported cases of PA poisoning due to the use of the herbs are geographically widespread, but few in number. However, the scale on which PA-containing plants are used as herbs, the typically delayed effects of long-term exposure, and the difficulties of diagnosis led the Task Group to conclude that there is every indication of under-reporting of intoxications from the use of such herbs. Symphytum root preparations, in particular, represent a hazard, and certain user groups are routinely exposed to levels of Symphytum alkaloids that are higher than those at which intoxications have been reported.

The risks associated with the use of PA-containing herbs are accentuated by the difficulties of controlling this use.

9.1.3.3 PA-containing plants used as food and beverages

Some PA-containing plants are used for food or the making of beverages in many countries, including developed countries. The following species are known to be used (though many other plants are also probably used in this way): Cacalia yatabei, Symphytum species, Ligularia dentata, Petasites japonicus, Senecio burchellii, S. inadequidens, S. pierotti, Syneilesis palmata, Crotalaria anagyrodies, C. brevidens, C. juncea, C. laburnifolia, C. pumila, C. recta, and C. retusa. No information is available on the extent to which the different types are consumed (section 3.3.3).

9.1.3.4 Other foods contaminated by PAs

Several species of Boraginaceae are nectar and pollen sources for bees. Echium plantagineum, in particular, is a widespread weed in some countries and a substantial source of honey containing a low level of alkaloid. Senecio species are also visited by bees and yield alkaloid-containing honey, though Senecio-derived honey is not known to be produced in quantity for sale. Thus, some regional and local populations are exposed to a low-level intake through the presence of PAs in honey, and surveillance may be desirable in countries producing honey (section 3.3.4).

Under experimental conditions, PAs are transmitted from the feed of dairy cows and goats into the milk. Some PA-containing species, such as Senecio jacobaea, S. lautus, and Echium plantagineum, are weeds in dairy pastures in some countries and are eaten by cattle under certain conditions. There are no published reports of alkaloid in milk supplies for human consumption (section 3.3.5).

The Task Group was not aware of any reported cases of pyrrolizidine toxicity that had been ascribed to either honey or dairy products.

No information was available to the Task Group on the possible presence of alkaloids or their metabolites in meat from animals that had consumed PA-containing plants shortly before slaughter. The results of metabolic studies in rats have indicated that the alkaloid is rapidly cleared from the body and, therefore, the levels of PAs in meat are expected to be very low. However, there is no information on the possibility of alkaloid accumulating in storage sites.

9.1.4 Levels of intake

Reliable estimates of levels of intake of PAs, especially in outbreaks of disease caused by the contamination of cereal crops with the seeds of toxic plants, are extremely difficult to make. Sampling of the contaminated grain may not be strictly representative, since the extent of the contamination may vary in different sites and households, as is evident from the estimates of PA intake in the Indian and Afghan outbreaks reported in section 7.3. Furthermore, no accurate record is possible of the amount of contaminated food consumed over an uncertain length of time. No records of the levels of toxic PA intake are available in the earlier reports of human toxicity. Where available, estimates of intake in outbreaks caused by the contamination of staple food crops have been made on the basis of random sampling of the contaminated grain in food stores, and rough estimates of daily consumption by average adults. Food-on-the-plate analyses have not been made. The estimated lengths of exposure, and hence the amount of total intake, are also approximate.

The contamination of cereal grains with the seed of PA-containing plants has caused major epidemics of human poisoning, though, in the two instances where estimates of alkaloid intake are available, the intakes were lower than in some exposures due to the use of herbal medicines. The estimated intakes are summarized in Table 16. In an outbreak in India, millet contaminated with Crotalaria nana seed had an average alkaloid content of 0.5 g/kg, and the estimated daily intake by the population was 0.66 mg/kg body weight. In a larger outbreak in Afghanistan, due to the seeds of Heliotropium popovii in wheat, the level of contamination was probably variable; representative samples of wheat contained alkaloid at 0.04 g/kg. The estimated daily intake was 0.033 mg/kg body weight. These intakes, sustained for periods of approximately 2 and 6 months, respectively, resulted in typical acute and subacute veno-occlusive disease.

The highest intake rates have been associated with the use or misuse of medicinal herbs and have resulted in acute liver damage and death. In two occasions, the consumption of Heliotropium eichwaldii as a treatment for epilepsy led to an estimated intake of 3.3 mg/kg body weight daily for 20 or 50 days, and the use of extracts of Senecio longilobus as medicine for young children led to estimated intakes of 3 and 0.8 - 1.7 mg/kg body weight. The highest intake led to extensive liver necrosis. However, it is possible that, in the above case of poisoning by Heliotropium eichwaldii, the toxicity was enhanced due to simultaneous administration of phenobarbitone, which has a potentiating effect on the microsomal enzymes in the liver cells that convert the PAs to toxic metabolites.

The use of Heliotropium lasiocarpum as a component of a herbal treatment for psoriasis involved somewhat lower daily intake rates of 0.49 - 0.71 mg/kg body weight in 4 patients, who, after periods of 19 - 46 days, developed veno-occlusive disease. The patient with the longest intake period and a total intake of 1.4 g alkaloid or 23 mg/kg body weight died.

The lowest ingestion rate leading to a case of veno-occlusive disease was also due to medicinal herbal treatment or, more specifically, to the use of a digestive aid containing a preparation of comfrey root. Commercial herb and food supplement preparations containing comfrey leaf or root are on sale in many countries. Limited assays of one comfrey-pepsin preparation prepared from comfrey root indicated a PA content of 2.9 g/kg. Another preparation made from comfrey leaf contained up to 0.27 g alkaloid/kg. The consumption of these preparations led to an estimated daily intake of 0.015 mg/kg body weight. Veno-occlusive disease was diagnosed after a 4- to 6-month period.

The consumption of Symphytum officinale (comfrey) and S. x uplandicum (Russian comfrey) in the form of food, infusions, or other preparations is widespread, though the full extent cannot be estimated. A high level of consumption as salad appears to be about 5 - 6 leaves per day and consumption as comfrey tea probably reaches a similar level. Limited assays indicate that the average alkaloid content of the leaf is about 1 mg/leaf, the concentration being higher in the younger, smaller leaves. The alkaloid intake from comfrey leaves could therefore vary from a low value, up to about 6 mg/day; or 0.1 mg/kg body weight for an adult, an intake within the range producing veno-occlusive disease. However, the Task Group noted that some people claim to have consumed comfrey at such a rate without suffering any disease.

Overall, the estimates of intake of PAs by human beings (Table 16) indicate that ingestion rates above 0.015 mg/kg body weight for the mixture of echimidine and related alkaloids in comfrey may lead to acute or subacute liver disease. If expressed in terms of equivalent doses of heliotrine (section 7.8), the estimated total doses in the known outbreaks or cases of veno-occlusive disease range from 1 to 167 mg/kg body weight. There is little real difference in the ranges of estimated total doses in non-fatal cases (1 - 120 mg/kg body weight) and those leading to death (6 - 167 mg/kg). These figures, when compared with the total lethal dose of several PAs in rats, i.e., 1.2 - 10.9 times the LD_{50} dose (equivalent to 360 - 3270 mg heliotrine/kg) (Table 10), would seem to indicate that man is markedly more sensitive than the rat to the toxic effects of PAs with regard to the development of acute and chronic effects on the liver. It should be noted that these estimates are based on limited raw data and a number of assumptions, and so are of uncertain reliability.

The dose estimates indicate strongly that the effects of PAs in human beings are cumulative at very low intake rates. Lower rates of intake of PAs may lead to chronic forms of intoxication, though, at present, there is no evidence on which the degree of risk in these circumstances can be evaluated. The information available on dose-response relationships is very limited, but the data support the conclusion that even low rates of intake of PAs over a period of time may present a health risk and that exposure should be minimized wherever possible.

There has not been any systematic monitoring of PAs in cereal grains, food products, and herbal medicines. Analytical surveys of these materials are feasible, but it would be difficult to design surveys that would give direct estimates of the dietary intake of PAs.

9.2 Acute Effects of Exposure

9.2.1 Acute liver disease

All cases of human intoxication in reported accounts have been in the acute phase of the disease, the dominant symptom being rapidly filling ascites. The disease can affect large subpopulations and, in one study, up to 22.6% of the population was affected.

Children appear to be the most vulnerable group and mortality can be high at the extremes of age. The liver is the principal target organ. In the acute stage of the disease, the liver shows a characteristic centrilobular haemorrhagic necrosis, which in man is accompanied by occlusion of the hepatic veins. However, characteristic veno-occlusive lesions, seen in the central veins of hepatic lobules, may not always be evident in the needle biopsy examination of the liver, but are always apparent on examination of the autopsy material.

9.3 Chronic Effects of Exposure

9.3.1 Cirrhosis of the liver

There is evidence that the administration of a single dose of PA to experimental animals or a single acute episode of illness in man, following brief consumption of PA-containing herbs or PA-contaminated food, may lead to progressive chronic liver disease resulting in cirrhosis. Cirrhosis may also be a consequence of long-term low-dose administration of PAs to experimental animals and possibly also of long-term low intake of PAs by human beings, though there is no proof of the latter. Cirrhosis resulting from the toxic effects of PAs in the advanced stage, may not be distinguishable from that resulting from other causes (sections 6.4.1.5 and 7.4). The Task Group did not find any evidence suggesting that PAs are a causative factor of the specific disease, Indian Childhood Cirrhosis (section 7.5).

9.3.2 Mutagenicity and teratogenicity

Several PAs, PA-derivatives, and related compounds have been shown to produce chromosome aberrations in plants and several cell culture systems, mutagenic effects (Salmonella ("Ame's"), sister chromatid exchanges, and other tests), and teratogenic and fetotoxic effects in experimental animals (sections 6.4.5, 6.4.6, 6.4.7). Chromosomal aberrations have been reported in the blood cells of children suffering from veno-occlusive disease, believed to have been caused by fulvine. The Task Group was not aware of data on the teratogenic/fetotoxic effects of PAs on human beings and was unable to evaluate the potential for these effects in PA exposure.

9.3.3 Cancer of the liver

A relatively large number of people have been exposed, in the past, to PAs and have suffered acute and chronic toxic effects. However, no information is available on the long-term follow-up of these populations, to ascertain whether this type of exposure could have resulted in an increased incidence of liver cancer or other types of cancer. Because of this lack of knowledge, it is not possible, at present, to make an evaluation of the cancer risk due to PAs. However, various PAs have been shown to be carcinogenic for experimental animals, which implies that a potential cancer risk for human beings should be seriously considered.

Of several PAs evaluated for carcinogenicity by IARC (1976, 1983), there is "sufficient or limited evidence" for the carcinogenicity in experimental animals (IARC, 1976) of monocrotaline, retrorsine, isatidine, lasiocarpine, petasitenine, senkirkine, and of extracts of the PA-containing plants <u>Petasites japonicum</u>, <u>Tussilago farara</u>, <u>Symphytum officinale</u>, <u>Senecio longilobus</u>, <u>Senecio numorensis</u>, <u>Farfugium japonicum</u>, and <u>Senecio cannabifolius</u>. These studies were carried out mainly on rats, with few studies on mice or hamsters (section 6.4.8). The carcinogenicity data obtained with other PAs are difficult to evaluate, because of the limited number of treated animals and the lack of adequate numbers as controls. The main target organ is the liver, where liver cell tumours and haemangioendothelial sarcomas were observed. In some instances, tumours in extra-hepatic tissues (lung, pancreas, intestine) were also observed, namely with monocrotaline, retrorsine, and lasiocarpine. Some PAs, for example, retrorsine, have been shown to be carcinogenic after a single dose. The pyrrolic metabolites have also been shown to be carcinogenic for rats.

It may be recalled that several of the PAs involved in human poisoning include the above compounds. It is notable that the dose rates that have been effective in inducing tumours in rats, mostly equivalent to 0.2 - 3 mg/kg body weight per day (Table 17), are roughly similar in magnitude to estimated intake rates (0.49 - 3.3 mg/kg body weight per day) (Table 16) in several episodes of human toxicity. Comparison of the total intakes resulting in human toxicity with the total doses to death observed in the chronic toxicity studies on rats indicates that human beings are more susceptible (section 7.8) and suggests that human beings may survive for sufficient time to develop cancer after only a brief exposure at this level or a longer exposure at a markedly lower level. A more quantitative assessment is not possible on the basis of the available information, and the Task Group stressed the need for appropriate epidemiological studies.

9.3.4 Effects on other organs

Substantiated reports of PA-induced extra-hepatic injury in man are limited to <u>Trichodesma</u> intoxication, in which symptoms and signs were predominantly neurological. The range of organs affected by other PAs in experimental and farm animals suggests that exposure of human beings to other PAs may also carry the potential for extra-hepatic injury.

There are extensive reports of experimental studies in which PAs have been demonstrated to produce the characteristic vascular changes of primary pulmonary hypertension and consequent right ventricular hypertrophy of the heart in rats and non-human primates (section 6.4.2). Susceptibility is age dependent, weanling rats being more vulnerable than older animals. There is only circumstantial evidence of PA-induced pulmonary vascular disease in one patient (section 7.6), but judging by the experimental evidence available, it is possible that human beings may be susceptible to PA-induced cardiopulmonary changes.

In the opinion of the Task Group, the neurological involvement which is a dominant feature in PA-intoxicated horses and is also seen in cows and sheep, cannot be explained solely as a consequence of liver damage. Central nervous system lesions have been demonstrated in sheep, pigs, and rats. Distribution studies of the radiolabelled metabolite, ^{3}H-dehydroretronecine, show increasing accumulation of radioactivity in the brain with time.

Trichodesma alkaloids, structurally related to monocrotaline, are neurotoxic agents. Trichodesma toxicosis in man has been reported only from the USSR, together with several studies on experimental animals. Detailed reports on the pathological findings were not available to the Task Group, but the information available indicated that the central nervous system was the primary target organ (sections 6.4.3 and 7.7).

Stomach and intestinal lesions have been shown in PA-exposed sheep, mice, cows, and rats. Distribution studies with radiolabelled pyrroles showed a high retention of radioactivity in the stomach, consistent with the acid-sensitive nature of the pyrroles. In rats, pyrrolic metabolites are secreted in high concentrations in the bile.

Kidney changes following to PA administration have been shown in mice, pigs, horses, sheep, and monkeys. Pyrrolizidine metabolites have been found covalently bound to kidney DNA in rats. Urinary excretion is a major route of excretion of metabolic products of PAs in rats.

There is no evidence of involvement of organs other than the liver and central nervous system ascribed primarily to PA toxicity in any of the published human case reports. It is possible that under some circumstances, other major organ systems may also be at risk. As bioactivation of PAs has been demonstrated only in the liver, the risk of damage should be expected to be lower in the organs.

9.4 Effects on the Environment

9.4.1 Agriculture

In some countries, PA-containing weeds densely cover areas of up to thousands of square kilometres. Their adverse effects include the covering of pastures, additional costs in agricultural production, and the poisoning of farm animals.

The toxicity of PAs for farm animals, including sheep, cattle, horses, pigs, goats, and poultry, which has been the inspiration for much of the investigation of PA toxicity. In Australia, for example, Heliotropium europaeum and Echium plantagineum cause the death of thousands of animals annually (section 6.2).

9.4.2 Wild-life

By contrast, little is known about the consumption of PA-containing plants by wild-life, or of their individual sensitivities. The death of deer in Louisiana has been ascribed to eating Heliotropium or Crotalaria species, and an experimental study has shown that the rainbow trout (Salmo gairdneri) is sensitive to Senecio jacobaea alkaloids (sections 6.5.1 and 6.5.2).

There is no information on the effects of the alkaloids on field rodents or other seed-eating mammals and birds that might be expected to consume seeds of PA-containing plants and to suffer toxic effects.

9.4.3 Insects

Many species of insects, such as some moths of the family Arctidae and butterflies of the sub-families Danainae and Ithominae, have become dependent on PA-containing plants, using the alkaloids as defensive chemicals and derivatives of them as pheromones and other signalling chemicals. Thus, complete elimination of PA-containing plants in a region might lead to a marked reduction in the local population of insects of this type (section 6.5.3).

9.4.4 Soil and water

There have not been any studies on the fate of PAs when the plants in which they occur wilt and age. If alkaloid is leached into soil or water, it is probably readily degraded by microorganisms since, as a base and ester, it is subject to oxidative and hydrolytic reactions.

REFERENCES

AFZELIUS, B.A. & SCHOENTAL, R. (1967) The ultrastructure of enlarged hepatocytes induced in rats with a single oral dose of retrorsine, pyrrolizidine (Senecio) alkaloids. J. ultrastruct. Res., 20: 328-345.

AIKAT, B.K., BHUSNURMATH, S.R., DATTA, D.V., & CHHUTTANI, P.N. (1978) Veno-occlusive disease in north-west India. Indian J. Pathol. Microbiol., 21: 203-211.

AL-HASANY, M. & MOHAMED, A.S. (1970) Veno-occlusive disease of the liver in Iraq. Arch. dis. Child., 45: 722-724.

ALLEN, J.R. & CARSTENS, L.A. (1968) Veno-occlusive disease in rhesus monkeys. Am. J. vet. Res., 29: 1681-1694.

ALLEN, J.R. & CARSTENS, L.A. (1970) Pulmonary vascular occlusions initiated by endothelial lysis in monocrotaline-intoxicated rats. Exp. mol. Pathol.., 13: 159-171.

ALLEN, J.R. & CARSTENS, L.A. (1971) Monocrotaline induced Budd-Chiari syndrome in monkeys. Am. J. dig. Dis., 16: 111-121.

ALLEN, J.R. & CHESNEY, C.F. (1972) Effect of age on development of cor pulmonale in non-human primates following pyrrolizidine alkaloid intoxication. Exp. mol. Pathol., 17: 220-232.

ALLEN, J.R., CHILDS, G.R., & CRAVENS, W.W. (1960) Crotalaria spectabilis toxicity in chickens. Proc. Soc. Exp. Biol. Med., 104: 434-436.

ALLEN, J.R., LALICH, J.J., & SCHMUTTLE, S.M. (1963) Crotalaria spectabilis induced cirrhosis in turkeys. Lab. Invest., 12: 512-517.

ALLEN, J.R., CARSTENS, L.A., & OLSON, B.E. (1967) Veno-occlusive disease in Macaca speciosa monkeys. Am. J. Pathol., 50: 653-667.

ALLEN, J.R., CARSTENS, L.A., & KATAGIRI, G.J. (1969) Hepatic veins of monkeys with veno-occlusive disease-sequential ultrastructural changes. Arch. Pathol., 87: 279-289.

ALLEN, J.R., CHESNEY, C.F., & FRAZEE, W.J. (1972) Modifications of pyrrolizidine alkaloids intoxication resulting from altered microsomal enzymes. Toxicol. appl. Pharmacol., 23: 470-479.

ALLEN, J.R., HSU, I.-C., & CARSTENS, L.A. (1975) Dehydroretronecine induced rhabdomyosarcomas in rats. Cancer Res., 35: 997-1002.

AMES, M.M. & POWIS, G. (1978) Determination of indicine N-oxide and indicine in plasma and urine by electron-capture gas-liquid chromatography. J. Chromatogr., 166: 519-526.

ARMSTRONG, S.J. & ZUCKERMAN, A.J. (1970) Production of pyrroles from pyrrolizidine alkaloids by human embryo tissue. Nature (Lond.), 228: 569-570.

ARMSTRONG, S.J., ZUCKERMAN, A.J., & BIRD, R.G. (1972) Induction of morphological changes in human embryo liver cells by the pyrrolizidine alkaloid lasiocarpine. Br. J. exp. Pathol., 53: 145-149.

ARORA, R.R., PYARELAL, GHOSH, T.K., MATHUR, K.K., & TANDON, B.N. (1981) Epidemiology of veno-occlusive disease in tribal population of Madhya Pradesh and Bihar. J. commun. Dis. (India), 13: 147-151.

ARSECULERATNE, S.N., GUNATILAKA, A.A.L., & PANABOKKE, R.G. (1981) Studies on medicinal plants of Sri Lanka: occurrence of pyrrolizidine alkaloids and hepatotoxic properties in some traditional medicinal herbs. J. Ethnopharmacol., 4: 159-177.

ASADA, Y., FURUYA, T., & MURAKAMI, N. (1981) Pyrrolizidine alkaloids from Ligularia japonica. Planta Med., 42: 202-203.

ASADA, Y., SHIRAISHI, M., TAKEUCHI, T., OSAWA, Y., & FURUYA, T. (1985) Pyrrolizidine alkaloids from Crassocephalum crepidioides. Planta Med., 51: 539-540.

BARNES, J.M., MAGEE, P.N., & SCHOENTAL, R. (1964) Lesions in the lungs and livers of rats poisoned with pyrrolizidine alkaloids fulvine and its N-oxide. J. Pathol. Bacteriol., 88: 521-531.

BERRY, D.M. & BRAS, G. (1957) Venous occlusion of the liver in Crotalaria and Senecio poisoning. North Am. Vet., 38: 323-327.

BHATTACHARYA, K.J. (1965) Foetal and neonatal responses to hepatotoxic agents. J. Pathol. Bacteriol., 90: 151-161.

BICCHI, C., D'AMATO, A., & CAPPELLETTI, E. (1985) Determination of pyrrolizidine alkaloids in Senecio inaequidens D.C. by capillary gas chromatography. J. Chromatogr., 349: 23-29.

BICK, Y.A.E. (1970) Comparison of the effects of LSD, heliotrine, and X-irradiation on chromosome breakage and the effects of LSD on the rats of cell division. Nature (Lond.), 226: 1165-1167.

BICK, Y.A.E. & CULVENOR, C.C.J. (1971) Effects of dehydroheliotridine, a metabolite of pyrrolizidine alkaloids on chromosome structure and cell division in cultures of animal cells. Cytobios, 3: 245-255.

BICK, Y.A.E. & JACKSON, W.D. (1968) Effects of the pyrrolizidine alkaloid heliotrine on cell division and chromosome breakage in cultures of leucocytes from the marsupial Potorus tridactylus. Aust. J. biol. Sci., 21: 469-481.

BICK, Y.A.E., CULVENOR, C.C.J., & JAGO, M.V. (1975) Comparative effects of pyrrolizidine alkaloids and related compounds on leukocyte cultures from Potorus tridactylus. Cytobios, 14: 151-160.

BIRECKA, H., FROHLICH, M.W., HULL, L., & CHASKES, M.J. (1980) Pyrrolizidine alkaloids of Heliotropium from Mexico and adjacent USA. Phytochemistry, 19: 421-426.

BIRECKA, H., CATALFAMO, J.L., & EISEN, R.N. (1981) A sensitive method for detection and quantitative determination of pyrrolizidine alkaloids. Phytochemistry, 20: 343-344.

BLACK, D.N. & JAGO, M.V. (1970) Interaction of dehydroheliotridine, a metabolite of heliotridine based pyrrolizidine alkaloids, with natural and heat denatured RNA. Biochem. J., 118: 347-353.

BOHLMANN, F., KLOSE, W., & NIKISCH, K. (1979) [Synthesis of dehydroheliotridine and 3-oxy-dehydroheliotridine.] Tetrahedr. Lett., 1979: 3699-3702 (in German).

BOHLMANN, F., ZDERO, C., JAKUPOVIC, J., GRENZ, M., CASTRO, V., KING, R.M., ROBINSON, H., & VINCENT, L.P.D. (1986) Further pyrrolizidine alkaloids and furoeremophilanes from Senecio species. Phytochemistry, 15: 1151-1159.

BOPPRE, M. (1986) Insects pharmacophagously utilizing defensive plant chemicals (pyrrolizidine alkaloids). Naturwissenschaften, 73: 17-26.

BRAGINSKII, B.M. & BOBOKHADZAEV, I. (1965) [Hepatosplenomegaly against the background of heliotropic toxicosis.] Sov. Med., 28: 57-60 (in Russian).

BRAS, G. (1973) Aspects of hepatic vascular diseases. In: Gall, E.A. & Mostofi, F.K., ed. The liver: International Academy of Pathology monograph, Baltimore, Maryland, Williams and Wilkins, pp. 406-530.

BRAS, G. & HILL, K.R. (1956) Veno-occlusive disease of the liver - essential pathology. Lancet, 2: 161-163.

BRAS, G. & WATLER, D.C. (1955) Further observations on the morphology of veno-occlusive disease of the liver in Jamaica. West Indian med. J., 4: 201-211.

BRAS, G., JELLIFFE, D.B., & STUART, K.L. (1954) Veno-occlusive disease of the liver with nonportal type of cirrhosis occurring in Jamaica. Arch. Pathol., 57: 285-300.

BRAS, G., BERRY, D.M., & GYORGI, P. (1957) Plants as etiological factor in veno-occlusive disease of liver. Lancet, 1: 960-962.

BRAS, G., BROOKS, S.E.H., & WATLER, D.C. (1961) Cirrhosis of liver in Jamaica. J. Pathol. Bacteriol., 82: 503-512.

BRAUCHLI, J., LUTHY, J., ZWEIFEL, U., & SCHLATTER, C. (1982) Pyrrolizidine alkaloids from Symphytum officinale L. and their percutaneous absorption in rats. Experientia (Basel), 38: 1085-1087.

BRIGGS, L.H., CAMBIE, R.C., CANDY, B.J., O'DONOVAN, G.M., RUSSELL, R.A., & SEELYE, R.N. (1965) Alkaloids of New Zealand Senecio species. Part II: senkirkine. J. Chem. Soc., C1965: 2492-2498.

BRINK, N.G. (1982) Somatic and teratogenic effects induced by heliotrine in Drosophilia. Mutat. Res., 104: 105-111.

BROOKS, W.E.H., MILLER, C.G., MCKENZIE, K., AUDRETSCH, J.J., & BRAS, G. (1970) Acute veno-occlusive disease of the liver. Fine structure in Jamaican children. Arch. Pathol., 89: 507-529.

BROWN, K.S. (1984) Chemical ecology of dehydropyrrolizidine alkaloids in adult Ithomiinae (Lepidoptera: Nymphalidae). Rev. Bras. Biol., 44: 435-440.

BRUGGEMAN, I.M. & VAN DER HOEVEN, J.C.M. (1985) Induction of SCEs by some pyrrolizidine alkaloids in V79 Chinese hamster cells co-cultured with chick embryo hepatocytes. Mutat. Res., 142: 209-212.

BRUNER, L.H., CARPENTER, L.J., HAMLOW, P., & ROTH, R.A. (1986) Effect of a mixed-function oxidase inducer and inhibitor on monocrotaline pyrrole pneumotoxicity. Toxicol. appl. Pharmacol., 85: 416-427.

BUCKMASTER, G.W., CHEEKE, P.R., ARSCOTT, G.H., DICKINSON, E.D., PIERSON, M.L., & SHULL, L.R. (1977) Response of Japanese quail to dietary and injected pyrrolizidine (Senecio) alkaloid. J. anim. Sci., 45(6): 1322-1325.

BULL, L.B. (1955) The histological evidence of liver damage from pyrrolizidine alkaloids. Aust. vet. J., 31: 33-40.

BULL, L.B. & DICK, A.T. (1959) The chronic pathological effects on the liver of the rat of the pyrrolizidine alkaloids heliotrine, lasiocarpine, and their N-oxides. J. Pathol. Bacteriol., 78: 483-502.

BULL, L.B. & DICK, A.T. (1960) The function of total dose in th production of chronic lethal disease in rats by periodic injections of the pyrrolizidine alkaloid heliotrine. Aust. J. exp. Biol. med. Sci., 38: 515.

BULL, L.B., DICK, A.T., KEAST, J.C., & EDGAR, G. (1956) An experimental investigation of the hepatotoxic and other effects on sheep of consumption of Heliotropium europaeum L. Heliotropium poisoning of sheep. Aust. J. agric. Res., 7: 281-332.

BULL, L.B., DICK, A.T., & MCKENZIE, J.S. (1958) The acute effects of heliotrine and lasiocarpine and their N-oxides on the rat. J. Pathol. Bacteriol., 75: 17-25.

BULL, L.B., CULVENOR, C.C.J., & DICK, A.T. (1968) The pyrrolizidine alkaloids, Amsterdam, North Holland Publishing Co.

BURGUERA, J.A., EDDS, G.T., & OSUNA, O. (1983) Influence of selenium on aflatoxin B_1 or Crotalaria toxicity in turkey poults. Am. J. vet. Res., 44: 1714-1717.

BURNS, J. (1972) The heart and pulmonary arteries in rat fed on Senecio jacobaea. J. Pathol., 106: 187-194.

BUTLER, W.H., MATTOCKS, A.R., & BARNES, J.M. (1970) Lesions in the liver and lungs of rats given pyrrole derivatives of pyrrolizidine alkaloids. J. Pathol., 100: 169-175.

CAMPBELL, J.G. (1956) An investigation of the hepatotoxic effects in the fowl of ragwort (Senecio jacobaea Linn) with special reference to the induction of liver tumours with seneciphylline. Proc. R. Soc. Edinburgh, B66: 111-130.

CAMPBELL, J.G. (1957a) Studies on the influence of sex hormones on the avian liver. II. Acute liver damage in the male fowl and the protective effect of oestrogen, as determined by a liver function test. J. Endocrinol., 15: 346-350.

CAMPBELL, J.G. (1957b) Studies on the influence of sex hormones on avian liver. III. Oestrogen-induced regeneration of the chronically damaged liver. J. Endocrinol., 15: 351-354.

CANDRIAN, U., LUTHY, J., GRAF, U., & SCHLATTER, CH. (1984a) Mutagenic activity of the pyrrolizidine alkaloids seneciphylline and senkirkine in Drosophila and their transfer into rat milk. Food chem. Toxicol., 22: 223-225.

CANDRIAN, U., LUTHY, J., SCHMID, P., SCHLATTER, CH., & GALLASZ, E. (1984b) Stability of pyrrolizidine alkaloids in hay and silage. J. agric. food Chem., 32: 935-937.

CANDRIAN, U., LUTHY, J., & SCHLATTER, CH. (1985) In vivo binding of retronecine-labelled (^{3}H), seneciphylline, and ^{3}H-senecionine to DNA of rat liver, lung and kidney. Chem.-biol. Interact., 54: 57-69.

CARILLO, L. & AVIADO, D.M. (1969) Monocrotaline induced pulmonary hypertension and p-chlorophenylalanine. Lab. Invest., 20: 213-218.

CARSTENS, L.A. & ALLEN, J.R. (1970) Arterial degeneration and glomerular hyalinization in the kidney of monocrotaline intoxicated rats. Am. J. Pathol., 60: 75-92.

CHALMERS, A.H., CULVENOR, C.C.J., & SMITH, L.W. (1965) Characterization of pyrrolizidine alkaloids by gas, thin layer, and paper chromatography. J. Chromatogr., 20: 270-277.

CHEEKE, P.R. & GORMAN, G.R. (1974) Influence of dietary protein and sulphur amino-acid levels on the toxicity of Senecio jacobaea (tansy ragwort) to rats. Nutr. Rep. Int., 9: 197-207.

CHEEKE, P.R. & PIERSON-GOEGER, M.L. (1983) Toxicity of Senecio jacobaea and pyrrolizidine alkaloids in various laboratory animals and avian species. Toxicol. Lett., 18: 343-349.

CHEN, K.K. (1945) Pharmacology. Hepatotoxic alkaloids. Ann. Rev. Physiol., 7: 695-697.

CHEN, K.K., HARRIS, P.N., & ROSE, C.L. (1940) The action and toxicity of platyphylline and seneciphylline. J. Pharmacol. exp. Ther., 68: 130-140.

CHESNEY, C.F. & ALLEN, J.R. (1973a) Resistance of the guinea pig to pyrrolizidine alkaloid intoxication. Toxicol. appl. Pharmacol., 26: 385-392.

CHESNEY, C.F. & ALLEN, J.R. (1973b) Endocardial fibrosis associated with monocrotaline induced pulmonary hypertension in non-human primates (Macaca arctoides). Am. J. vet. Res., 34: 1577-1581.

CHESNEY, C.F., HSU, I.C., & ALLEN, J.R. (1974) Modifications of the in vitro metabolism of the hepatotoxic pyrrolizidine alkaloid, monocrotaline. Res. Commun. chem. Pathol. Pharmacol., 8: 567-570.

CHOPRA, R.N., ed. (1933) Indigenous drugs of India, Calcutta, The Art Press.

CLARK, A.M. (1959) Mutagenic activity of the alkaloid heliotrine in Drosophila. Nature (Lond.), 183: 731-732.

CLARK, A.M. (1976) Naturally occurring mutagens. Mutat. Res., 32: 361-174.

COOK, B.A., SINNHUBER, J.R., THOMAS, P.J., OLSON, T.A., SILVERMAN, T.A., JONES, R., WHITEHEAD, V.M., & ROYMANN, F.B. (1983) Hepatic failure secondary to indicine N-oxide toxicity. A pediatric oncology group study. Cancer, 52: 61-63.

COOK, J.W., DUFFY, E., & SCHOENTAL, R. (1950) Primary liver tumours in rats following feeding with alkaloids of Senecio jacobaea. Br. J. Cancer, 4: 405-410.

CRAIG, A.M., SHEGGEBY, G., & WICKS, C.E. (1984) Large scale extraction of pyrrolizidine alkaloids from tansy ragwort (Senecio jacobaea). Vet. hum. Toxicol., 26: 108-111.

CULVENOR, C.C.J. (1968) Tumour inhibitory activity of pyrrolizidine alkaloids. J. pharm. Sci., 57: 1112-1117.

CULVENOR, C.C.J. (1978)) Pyrrolizidine alkaloids: occurrence and systematic importance in angiosperms. Bot. Notiser, 131: 473-486.

CULVENOR, C.C.J. (1980) Alkaloids and human disease. In: Smith, R.L. & Bababunmi, E.A., ed. Toxicology in the tropics, London, Taylor & Francis Ltd., pp. 124-141.

CULVENOR, C.C.J. (1983) Estimated intakes of pyrrolizidine alkaloids by humans. A comparison with dose rates causing tumours in rats. J. Toxicol. environ. Health, 11: 625-635.

CULVENOR, C.C.J. (1985) Pyrrolizidine alkaloids: some aspects of the Australian involvement. Trends pharmacol. Sci., 6: 18-22.

CULVENOR, C.C.J. & JAGO, M.V. (1979) Carcinogenic plant products and DNA. In: Grover, P.L., ed. Chemical carcinogens and DNA, Boca Raton, Florida, CRC Press, Vol. 1, 161 pp.

CULVENOR, C.C.J., DOWNING, D.T., EDGAR, J.A., & JAGO, M.V. (1969) Pyrrolizidine alkaloids as alkylating and antimitotic agents. N.Y. Acad. Sci., 163: 837-847.

CULVENOR, C.C.J., EDGAR, J.A., SMITH, L.W., & TWEEDDALE, H.J. (1970a) Dihydropyrrolizines. III. Preparation and reactions of derivatives related to pyrrolizidine alkaloids. Aust. J. Chem., 23: 1853-1867.

CULVENOR, C.C.J., EDGAR, J.A., SMITH, L.W., & TWEEDDALE, H.J. (1970b) Dehydrolizines. IV. Manganese dioxide oxidation of 1,2-dihydropyrrolizidines. Aust. J. Chem., 23: 1869-1879.

CULVENOR, C.C.J., EDGAR, J.A., JAGO, M.V., OUTTERIDGE, A., PETERSON J.E., & SMITH, L.W. (1976a) Hepato- and pneumotoxicity of pyrrolizidine alkaloids and derivatives in relation to molecular structure. Chem.-biol. Interact., 12: 299-324.

CULVENOR, C.C.J., EDGAR, J.A., SMITH, L.W., & HIRONO, I. (1976b) The occurrence of senkirkine in Tussilago farfara. Aust. J. Chem., 29: 229-233.

CULVENOR, C.C.J., CLARKE, M., EDGAR, J.A., FRAHN, J.L., JAGO, M.V., PETERSON, J.E., & SMITH, L.W. (1980a) Structure and toxicity of the alkaloids of Russian comfrey (Symphytum x uplandicum Nyman), a medicinal herb and item of human diet. Experientia (Basel), 36: 377-389.

CULVENOR, C.C.J., EDGAR, J.A., FRAHN, J.L., & SMITH, L.W. (1980b) The alkaloids of Symphytum x uplandicum (Russian comfrey). Aust. J. Chem., 33: 1105-1113.

CULVENOR, C.C.J., EDGAR, J.A., & SMITH, L.W. (1981) Pyrrolizidine alkaloids in honey from Echium plantagineum L. J. agric. food Chem., 29: 958-960.

CULVENOR, C.C.J., JAGO, M.V., PETERSON, J.E., SMITH, L.W., PAYNE, A.L, CAMPBELL, D.G., EDGAR, J.A., & FRAHN, J.L. (1984) Toxicity of Echium plantagineum (Paterson's curse). I. Marginal toxic effects of Merino wethers from long-term feeding. Aust. J. agric. Res., 35: 293-304.

CULVENOR, C.C.J., EDGAR, J.A., SMITH, L.W., KUMANA, C.R., & LIN, H.J. (1986) Heliotropium lasiocarpum Fisch and Mey identified as cause of veno-occlusive disease due to a herbal tea. Lancet, 1: 978.

DANN, A.T. (1960) Detection of N-oxides of the pyrrolizidine alkaloids. Nature, 4730: 1051.

DANNINGER, T., HAGEMANN, U., SCHMIDT, V., & SCHOENHOEFER, P.S. (1983) Toxicity of pyrrolizidine alkaloid-containing medicinal plants. Pharm. Ztg, 128: 289-303.

DATTA, D.V., KHUROO, M.S., MATTOCKS, A.R., AIKAT, B.K., & CHHUTTANI, P.N. (1978a) Veno-occlusive disease of liver due to Heliotropium plant used as medicinal herb (report of six cases with review of literature). J. Assoc. Phys. India, 26(5): 383-393.

DATTA, D.V., KHUROO, M.S., MATTOCKS, A.R., AIKAT, B.K., & CHHUTTANNI, P.N. (1978b) Herbal medicines and veno-occlusive disease in India. Postgrad. med. J., 54: 511-515.

DAVIDSON, C.S. (1963) Plants and fungi as etiologic agents of cirrhosis. New Engl. J. Med., 268: 1072-1073.

DAVIDSON, J. (1935) The action of retrorsine on rat's liver. J. Pathol. Bacteriol., 40: 285-295.

DEAGEN, J.T. & DEINZER, M.L. (1977) Improvement in the extraction of pyrrolizidine alkaloids. Lloydia, 40: 395-397.

DEAN, R.E. & WINWARD, A.H. (1974) An investigation into the possibility of tansy ragwort poisoning of black-tailed deer. J. wildl. Dis., 10: 166-169

DEINZER, M.L., THOMSON, P.A., BURGETT, D.M., & ISAACSON, D.L. (1977) Pyrrolizidine alkaloids: their occurrence in honey from tansy ragwort (S. jacobaea). Science, 195: 497-499.

DEINZER, M.L., THOMSON, P.A., GRIFFIN, D., & DICKINSON, E. (1978) Sensitive analytical method for pyrrolizidine alkaloids. The mass spectra of retronecine derivatives. Biomed. mass Spectrom., 5: 175-179.

DELFOSSE, E.S. (1985) Re-evaluation of the biological control program for Heliotropium europaeum in Australia. In: Delfosse, E.S., ed. Proceedings of the VI International Symposium on Biological Control of Weeds, 19-25 August 1984, Vancouver, Canada, Ottawa Agriculture Canada, pp. 735-42.

DELFOSSE, E.S. & CULLEN, J.M. (1985a) CSIRO division of Entomology Submission to the enquiries into biological control of Echium plantagineum L., Paterson's curse/Salvation Jane. Plant Prot. Q., 1: 24-40.

DELFOSSE, E.S. & CULLEN, J.M. (1985b) Echium plantagineum: catalyst for conflict and change in Australia. In: Delfosse, E.S., ed. Proceedings of the VI International Symposium on Biological Control of Weeds, 19-25 August, 1984, Vancouver, Canada, Ottawa Agriculture Canada pp. 249-292.

DELORME, P., JAY, M., & FERRY, S. (1977) Phytochemical inventory in indigenous Boraginaceae: study of the alkaloids and polyphenolic compounds (anthocyanic and flavonoid compounds). Plant. Méd. Phytothér., 11: 5-11. (in french)

DE WAAL, H.L. & VAN TWISK, P. (1964) The chemical composition of four Senecio species from Kruger National Park. Koedoe (South Africa), 7: 40-42.

DICK, A.T., DANN, A.T., BULL, L.B., & CULVENOR, C.C.J. (1963) Vitamin B_{12} and the detoxication of hepatotoxic pyrrolizidine alkaloids in rumen liquor. Nature (Lond.), 197: 207-208.

DICKINSON, J.O. (1980) Release of pyrrolizidine alkaloids into milk. Proc. West. Pharmacol. Soc., 23: 377-379.

DICKINSON, J.O., COOKE, M.P., KING, R.R., & MOHAMED, P.A. (1976) Milk transfer of pyrrolizidine alkaloids in cattle. J. Am. Vet. Med. Assoc., 169: 1192-1196.

DIMENNA, G.P., KRICK, T.P., & SEGALL, H.J. (1980) Rapid high performance liquid chromatography isolation of monoesters, diesters, and macrocyclic diesters and pyrrolizidine alkaloids from Senecio jacobaea and Amsinckia intermedia. J. Chromatogr., 192: 474-478.

DOLINSKAYA, K.N. (1952) [Pathomorphology of heliotropic hepatic dystrophy in children.] In: Milenkov, S.M. & Kizhaikin, Y., ed. [Collection of scientific papers on Toxic Hepatitis with Ascites,] Tashkent, Publishing House of the University of Central Asia, pp. 165-172 (in Russian).

DREIFUSS, P.A. (1984) A study of the mass spectra of pyrrolizidine alkaloids by negative ion chemical ionization and ms/ms: the identification of a monoester pyrrolizidine alkaloid in Eupatorium rugosum. Diss. Abstr. Int. B., 45: '1183-1184.

DRIVER, H.E. & MATTOCKS, A.R. (1984) The toxic effects in rats of some synthanecine carbamate and phosphate esters analogous to hepatotoxic pyrrolizidine alkaloids. Chem.-biol. Interact., 51: 201-218.

DUBROVINSKII, S.B. (1946) [About the alimentary toxicosis caused by heliotrope.] J. Sov. Prot. Health, 6: 17-21 (in Russian).

DUBROVINSKII, S.B. (1952) [The etiology of toxic hepatitis with ascites.] In: Millenkov, S.M. & Kizhaikin, Y., ed. [Collection of scientific papers on Toxic Hepatitis with Ascites,] Tashkent, USSR,] Tashkent, Publishing House of the University of Central Asia, pp. 9-25 (in Russian).

EASTMAN, D.F. & SEGALL, H.J. (1982) A new pyrrolizidine alkaloid metabolite, 19-hydroxysenecionine, isolated from in vitro mouse hepatic microsomes using high performance liquid chromatography. Drug Metab. Disp., 10: 696-699.

EASTMAN, D.F., DIMENNA, G.P., & SEGALL, H.J. (1982) Covalent binding of two pyrrolizidine alkaloids, senecionine, and seneciphylline to hepatic macromolecules and their distribution, excretion and transfer into milk of lactating mice. Drug Metab. Disp., 10: 236-240.

EDGAR, J.A. (1982) Pyrrolizidine alkaloids sequestered by Solomon Island Danaine butterflies. The feeding preferences of the Danainae and Ithomiinae. J. Zool. (Lond.), 196: 385-399.

EDGAR, J.A. (1985) Gas chromatography of pyrrolizidine alkaloids. In: Seawright, A.A., Hegarty, M.P., James, L.F., & Keeler, R.F., ed. Plant toxicology. Proceedings of the Australia - USA Poisonous Plants Symposium, Brisbane, Australia, 14-18 May 1984, Brisbane, Queensland Poisonous Plants Committee, pp. 227-234.

EISENSTEIN, D. & HUXTABLE, R.J. (1979) Approaches to the treatment of pyrrolizidine toxicosis. In: Cheeke, P.R., ed. Proceedings of the Symposium on Pyrrolizidine (Senecio) Alkaloids: Toxicity, Metabolism, and Poisonous Plant Control Measures, Corvallis, Oregon, USA, 23-24 February 1979, Oregon, Nutrition Research Institute, pp. 109-113.

EL DAREER, S.M., TILLERY, K.F., LLOYD, H.H., & HILL, D.L. (1982) Disposition of indicine N-oxide in mice and monkeys. Cancer Treat. Rep., 66: 183-186.

EMMEL, M.W. (1948) Crotalaria poisoning in cattle. J. Am. Vet. Med. Assoc., 113: 164.

EMMEL, M.W., SANDERS, D.A., & HENLEY, W.W. (1935) Crotalaria spectabilis seed poisoning in swine. J. Am. Vet. Med. Assoc., 86: 43-49.

EVANS, J.V., PENG, A., & NIELSEN, C.J. (1979) The gas chromatographic mass spectrometric analysis of the new antitumour drug indicine-N-oxide utilizing a novel reaction accompanying trimethysilylation. Biomed. mass Spectrom., 6: 38-43.

EVANS, J.V., DALEY, S.K., MCCLUSKY, G.A., & NIELSEN, C.J. (1980) Direct quantitative analysis of indicine N-oxide in cancer patient samples by gas chromatography using the internal standard heliotrine N-oxide including a mass spectral comparison of their trimethylsilyl derivatives. Biomed. mass Spectrom., 7: 65-73.

FARRINGTON, K.J. & GALLAGHER, C.H. (1960) Complexes of copper with some pyrrolizidine alkaloids and with some of their esterifying acids. Aust. J. biol. Sci., 13: 600-603.

FISH, M.S., SWEELEY, C.C., JOHNSON, N.M., LAWRENCE, E.P., & HORNING, E.C. (1956) Chemical and enzymic rearrangements of N,N-dimethylaminoacid oxides. Biochem. Biophys. Acta, 21: 196-197.

FISHMAN, A.P. (1974) Dietary pulmonary hypertension. Circ. Res., 35: 657-660.

FORSYTH, A.A. (1968) British poisonous plants, London, Her Majesty's Stationery Office.

FOX, D.W., HART, M.C., BERGESON, P.S., JARRETT, P.B., STILLMAN, A.E., & HUXTABLE, R.J. (1978) Pyrrolizidine (Senecio) intoxication mimicking Reye's syndrome. J. Paediatr., 93: 980-982.

FRAHN, J.L., CULVENOR, C.C.J., & MILLS, J.A. (1980) Preparative separation of the pyrrolizidine alkaloids, intermedine and lycopsamine, as their borate complexes. J. Chromatogr., 195: 379-383.

FURMANOWA, M., GUZEWSKA, J., & BELDOWSKA, B. (1983) Mutagenic effects of aqueous extracts of Symphytum officinale L. and of its alkaloidal fractions. J. appl. Toxicol., 3: 127-130.

FURUYA, T. & HIKICHI, M. (1971) Alkaloids and triterpenoids of Symphytum officinale. Phytochemistry, 10: 2217-2220.

FUSHIMI, K., KATO, K., KATO, T., MATSUBARA, N., & HIRONO, I. (1978) Carcinogenicity of flower stalks of Petasites japonicus Maxim in mice and Syrian golden hamsters. Toxicol. Lett., 1: 291-294.

GADELLA, T.W.J., KLIPHUIS, E., & HUIZING, H.J. (1983) Cyto- and chemotaxonomical studies on the sections officinalia and coerulea of the genus Symphytum. Bot. Helv., 93: 169-192.

GANEY, P.E., FINK, G.D., & ROTH, R.A. (1985) The effect of dietary restriction and altered sodium intake on the cardiopulmonary toxicity of monocrotaline pyrrole. Toxicol. appl. Pharmacol., 78: 55-62.

GARDINER, M.R., ROYCE, R., & BOKOR, A. (1965) Studies on Crotalaria crispata, a newly recognized cause of Kimberley horse disease. J. Pathol. Bacteriol., 89: 43-55.

GHANEM, J. & HERSHKO, C. (1981) Veno-occlusive disease and primary hepatic vein thrombosis in Israeli Arabs. Isr. J. med. Sci., 17: 339-347.

GHODSI, F. & WILL, J.A. (1981) Changes in pulmonary structure and function induced by monocrotaline intoxication. Am. J. Physiol., 240: H149-H155.

GILRUTH, J.A. (1903) Hepatic cirrhosis affecting horses and cattle (so-called "Winton disease"), Wellington, New Zealand Department of Agriculture, pp. 228-279 (11th annual report).

GOEGER, D.E., CHEEKE, P.R., SCHMITZ, J.A., & BUHLER, D.R. (1982) Toxicity of tansy ragwort (Senecio jacobaea) to goats. Am. J. vet. Res., 43: 252-254.

GOEGER, D.E., CHEEKE, P.R., RAMSDELL, H.S., NICHOLSON, S.S., & BUHLER, D.R. (1983) Comparison of the toxicities of Senecio jacobaea, Senecio vulgaris and Senecio glabellus in rats. Toxicol. Lett., 15: 19-23.

GONZALEZ, A.G., DE LA FUENTE, G., REINA, M., & LOYOLA, L.A. (1986a) Pyrrolizidine alkaloids from Senecio phillipicus and Senecio illinitus. Planta Med., 52: 160.

GONZALEZ, E., GARCIA, R., LEMUS, I., & ERAZO, S. (1986b) Pharmacological studies on senecionine, a pyrrolizidine alkaloid from Senecio fistulosus Poepp. Ex less. (hualtata). An. R. Acad. Farm., 52(1): 123-132.

GRASES, P.J. & BEKER, S.G. (1972) Veno-occlusive disease of the liver. A case from Venezuela. Am. J. Med., 53: 511-516.

GRAY, A.I., NIC, A.N., TSAOIR, E., & WATERMAN, P.G. (1983) Hepatotoxic alkaloids and allantoin in Symphytum tuberosum. J. Pharm. Pharmacol., 35(Suppl): 13.

GREEN, C.R. & CHRISTIE, G.S. (1961) Malformations in foetal rats induced by the pyrrolizidine alkaloid, heliotrine. Br. J. exp. Pathol., 42: 369-378.

GREEN, M.H.L. & MURIEL, W.J. (1975) Use of repair-deficient strains of Escherichia coli and liver microsomes to detect and characterise DNA damage caused by the pyrrolizidine alkaloids heliotrine and monocrotaline. Mutat. Res., 28: 331-336.

GREEN, C.E., SEGALL, H.J., & BYARD, J.L. (1981) Metabolism, cytotoxicity, and genotoxicity of the pyrrolizidine alkaloid senecionine in primary cultures of rat hepatocytes. Toxicol. appl. Pharmacol., 60: 176-185.

GUENGERICH, F.P. (1977) Separation and purification of multiple forms of microsomal cytochrome P-450. J. biol. Chem., 252: 3970-3979.

GUENGERICH, F.P. & MITCHELL, M.B. (1980) Metabolic activation of model pyrroles by cytochrome P-450. Drug Metab. Disp., 8: 34-38.

GUIDUGLI, F.H., PESTCHANKER, M.J., DESALMERON, M.S.A., & GIORDANO, O.S. (1986) 1-hydroxyplatyphyllide, a norsequiterpene lactone from Senecio gilliesiano. Phytochemistry, 25: 1923-1926.

GUNER, N. (1986) Alkaloids of Heliotropium suaveolens. J. nat. Prod., 49: 369.

GUPTA, P.S., GUPTA, G.D., & SHARMA, M.L. (1963) Venoocclusive disease of the liver. Br. med. J., 1: 1184-1186.

HABS, H. (1982) Senecio numorensis fuchsii. Carcinogenic and mutagenic activity of an alkaloidal extract from a phytotherapeutic drug. Dtsch. Apoth. Ztg, 122: 799-804.

HABS, H., HABS, M., MARQUARDT, H., RODER, E., SCHMAHL, D., & WIEDENFELD, H. (1982) Carcinogenic and mutagenic activity of an alkaloidal extract of Senecio numorensis spp. fuchsii. Arzneimittelforsch, 32: 144-148.

HAGGLUND, K.M., L'EMPEREUR, K.M., ROBY, M.R., & STERMITZ, F.R. (1985) Latifoline and latifoline N-oxide from Hackelia floribunda, the western false forget-me-not. J. nat. Prod., 48: 638-639.

HAMMOUDA, F.M., RIZK, A.M., ISMAIL, S.I., ATTEYA, S.Z., GHALEB, H.A., MADKOUR, M.K., POWAND, A.E., & WOOD, G. (1984) Poisonous plants contaminating edible ones and toxic substances in plant foods. Part 3: pyrrolizidine alkaloids from Heliotropium digynum Forssk (= H. luteum, Poir). Pharmazie, 39: 703-705.

HARDING, J.D.J., LEWIS, G., DONE, J.T., & ALLCROFT, R. (1964) Experimental poisoning by Senecio jacobaea in pigs. Pathol. Vet., 1: 204-220.

HARRIS, C., REDDY, J., CHIGA, M., & SVOBODA, D. (1969) Polyribosomal disaggregation by lasiocarpine. Biophys. Acta 180: 587-589.

HARRIS, P.N. & CHEN, K.K. (1970) Development of hepatic tumours in rats following ingestion of Senecio longilobus. Cancer Res., 30: 2881-2886.

HARRIS, P.N., ROSE, C.L., & CHEN, K.K. (1957) Hepatotoxic and pharmacological properties of heliotrine. Arch. Pathol., 64: 152-157.

HARTMANN, T. & ZIMMER, M. (1986) Organ-specific distribution and accumulation of pyrrolizidine alkaloids during the life history of 2 annual Senecio species. J. plant Physiol., 122: 67-80.

HASHEM, M. (1939) Etiology and pathology of types of liver cirrhosis in Egyptian children. J. Egypt. Med. Assoc., 22: 319-354.

HAYASHI, Y. (1966) Excretion and alteration of monocrotaline in rats after a subcutaneous injection (abstract). Fed. Proc., 25: 688.

HAYASHI, Y. & LALICH, J.J. (1967) Renal and pulmonary alterations induced by a single injection of monocrotaline. Proc. Soc. Exp. Biol. Med., 124: 392-396.

HAYASHI, Y. & LALICH, J.J. (1968) Protective effect of mercaptoethylamine and cysteine against monocrotaline intoxication in rats. Toxicol. appl. Pharmacol., 12: 36-43.

HAYASHI, Y., HUSSA, J.F., & LALICH, J.J. (1967) Cor pulmonale in rats. Lab. Invest., 16: 875-881.

HAYASHI, Y., KATO, M., OTSUKA, H. (1979) Inhibitory effects of diet reduction on Monocrotaline intoxication in rats. Toxicol. Lett., 3: 151-155.

HAYASHI, Y., SHIMADA, M., & KATAYAMA, H. (1977) Experimental insulinoma in rats after a single administration of monocrotaline. Toxicol. Lett., 1: 41-44.

HAYASHI, Y., KOKUBO, T., TAKAHASHI, M., FURUKAWA, F., OTSUKA, H., & HASHIMOTO, K. (1984) Correlative morphological and biochemical studies on monocrotaline-induced pulmonary alterations in rats. Toxicol. Lett., 21: 65-77.

HAYES, M.A., ROBERTS, E., & FARBER, E. (1985) Initiation and selection of resistant hepatocyte nodules in rats given the pyrrolizidine alkaloids lasiocarpine and senecionine. Cancer Res., 45: 3726-3734.

HEATH, D., SHABA, J., WILLIAMS, A., SMITH, P., & KOMBE, A. (1975) A pulmonary hypertension-producing plant from Tanzania. Thorax, 30: 399-404.

HENDRICKS, J.D., SINNHUBER, R.O., HENDERSON, M.D., & BUHLER, D.R. (1981) Liver and kidney pathology in rainbow trout (Salmo gairdneri) exposed to dietary pyrrolizidine (Senecio) alkaloids. Exp. mol. Pathol., 35: 170.183.

HILL, K.R. (1960) Worldwide distribution of seneciosis in man and animals. Proc. R. Soc. Med., 53: 281-282.

HILLIKER, K.S. & ROTH, R.A. (1984) Alteration of monocrotaline pyrrole-induced cardiopulmonary effects in rats by hydrallazine, dexamethasone or sulphinpyrazone. Br. J. Pharmacol., 82: 375-380.

HILLIKER, K.S., BELL, T.G., & ROTH, R.A. (1983) Monocrotaline pyrrole-induced pulmonary hypertension in fawn-hooded rats with platelet storage pool deficiency: 5 hydroxytryptamine uptake by isolated, perfused lungs. Thromb. Haemostasis (Stuttgart), 50: 844-847.

HILLIKER, K.S., BELL, T.G., LORIMER, D., & ROTH, R.A. (1984) Effects of thrombocytopaenia in monocrotaline pyrrole-induced pulmonary hypertension. Am. J. Physiol., 246: H747-H753.

HIRONO, I., SHIMIZU, M., FUSHIMI, K., MORI, H., & KATO, K. (1973) Carcinogenic activity of Petasites japonicus Maxim, a kind of coltsfoot. Gann Monogr. Cancer Res., 64: 527-528.

HIRONO, I., MORI, H., & CULVENOR, C.C.J. (1976) Carcinogenic activity of coltsfoot, Tussilago farfara L. Gann Monogr. Cancer Res., 67: 125-129.

HIRONO, I., MORI, H., YAMADA, K., HIRATA, Y., HAGA, M., TATEMATSU, H., & KANIE, S. (1977) Carcinogenic activity of petasitenine, a new pyrrolizidine alkaloid isolated from Petasites japonicus Maxim. J. Natl Cancer Inst., 58: 1155-1157.

HIRONO, I., MORI, H., & HAGA, M. (1978) Carcinogenic activity of Symphytum officinale. J. Natl Cancer Inst., 61: 865-869.

HIRONO, I., HAGA, M., FUJII, M., MATSUURA, S., MATSUBARA, N., NAKAYAMA, M., FURUYA, T., HIKICHI, M., TAKANASHI, H., UCHIDA, E., HOSAKA, S., & UENO, I. (1979a) Induction of hepatic tumours in rats by senkirkine and symphytine. J. Natl Cancer Inst., 63: 469-472.

HIRONO, I., MORI, H., HAGA, M., FUJII, M., YAMADA, K., TAKANASHI, H., UCHIDA, E., HOSAKA, S., UENO, I., MATSUSHIMA, I., UMEZAVA, K., & SHIRAI, A. (1979b) Edible plants containing carcinogenic pyrrolizidine alkaloids in Japan. In: Miller, B.C. ET AL., ed. Naturally occurring carcinogens, mutagens, and modulators of carcinogenesis, Baltimore, Maryland, University Park Press, pp. 79-87.

HIRONO, I., UENO, I., AISO, S., YAMAJI, T., & HAGA, M. (1983) Carcinogenic activity of Farfagium japonicum and Senecio cannabifolius. Cancer Lett., 20: 191-198.

HOET, P., ASTIGUETA, M., & FRISQUE, A.M. (1981) Phytochemical study of Crotalaria nitens. Rev. Latinoam. Quim., 12: 34-35.

HOOPER, P.T. (1972) Spongy degeneration in the brain in relation to hepatic disease and ammonia toxicity in domestic animals. Vet. Res., 90: 37-38.

HOOPER, P.T. (1974) The pathology of Senecio jacobaea poisoning of mice. J. Pathol., 113: 227-230.

HOOPER, P.T. (1975a) Spongy degeneration in the central nervous system of domestic animals. Part I: morphology. Acta neuropathol. (Berlin), 31: 325-334.

HOOPER, P.T. (1975b) Spongy degeneration in the central nervous system of domestic animals. Part III: occurrence and pathogenesis - hepatocerebral disease caused by hyperammonaemia. Acta neuropathol. (Berlin), 31: 343-351.

HOOPER, P.T. (1975c) Experimental acute gastro-intestinal disease caused by the pyrrolizidine alkaloid, lasiocarpine. J. comp. Pathol., 85: 341-349.

HOOPER, P.T. (1978) Pyrrolizidine alkaloid poisoning-pathology with particular reference to differences in animal and plant species. In: Keeler, R.F., Van Kampen, K.R., & James, L.F., ed. Effects of poisonous plants on livestock, New York, Academic Press, pp. 161-176.

HOOPER, P.T. & SCANLAN, W.A. (1977) Crotalaria retusa poisoning in pigs and poultry. Aust. vet. J., 53: 109-114.

HOOPER, P.T., BEST, S.M., & MURRAY, D.R. (1974) Hyperammonaemia and spongy degeneration of the brain in sheep affected with hepatic necrosis. Res. vet. Sci., 16: 216-222.

HOOSON, J. & GRASSO, P. (1976) Cytotoxic and carcinogenic response to monocrotaline pyrrole. J. Pathol., 118: 121-127.

HOU JINQUI, XIA YUDIN, YU ZHANSHI, AN YAN, & TANG YIXAI (1980) [Veno-occlusive disease of the liver with report of 2 cases.] Chung Hua Nei ko Tsa Chih, 19: 187-191 (in Chinese).

HSU, I.C., ALLEN, J.R., & CHESNEY, C.F. (1973) Identification and toxicological effects of dehydroretronecine, a metabolite of monocrotaline. Proc. Soc. Exp. Biol. Med., 144: 834-838.

HSU, I.C., SHUMAKER, R.C., & ALLEN, J.R. (1974) Tissue distribution of tritium labelled dehydroretronecine. Chem.-biol. Interact., 8: 163.

HUIZING, H.J., DE BOER, R., & MALINGRE, T.M. (1981) Preparative ion-pair high performance liquid chromatography and gas chromatography of pyrrolizidine alkaloids from comfrey. J. Chromatogr., 214: 257-262.

HURLEY, J.V. & JAGO, M.V. (1975) Pulmonary oedema in rats given dehydromonocrotaline: a topographic and electron-microscopic study. J. Pathol., 117: 23-32.

HUXTABLE, R.J. (1980) Herbal teas and toxins: novel aspects of pyrrolizidine poisoning in the United States. Perspect. Biol. Med., 24: 1-14.

HUXTABLE, R., PAPLANUS, S., & LAUGHARN, J. (1977) The prevention of monocrotaline induced right ventricular hypertrophy. Chest, 71: 308-310.

HUXTABLE, R.J., CIARAMITARO, D., & EISENSTEIN, D. (1978) The effect of a pyrrolizidine alkaloid, monocrotaline and a pyrrole, dehydroretronecine, on the biochemical functions of the pulmonary endothelium. Mol. Pharmacol., 14: 1189-1203.

HUXTABLE, R.J., LUTHY, J., & ZWEIFEL, V. (1986) Toxicity of comfrey-pepsin preparations: letters to the editor. New Engl. J. Med., 315: 1095.

IARC (1976) Pyrrolizidine alkaloids. In: Some naturally occurring substances, Lyons, International Agency for Research on Cancer, pp. 265-342 (IARC Monograph on the Evaluation of Carcinogenic Risk of Chemicals to Man, Vol. 10).

IARC (1983) Some food additives, feed additives and naturally occurring substances, Lyons, International Agency for Research on Cancer, pp. 207-245 (IARC Monograph on the Evaluation of Carcinogenic Risk of Chemicals to Humans, Vol. 31).

INDIAN COUNCIL OF MEDICAL RESEARCH (1955) Infantile cirrhosis of the liver in India. Indian J. med. Res., 43: 723-747.

ISMAILOV, N.I. (1948a) [On clinical features, etiology and pathogenesis of heliotropic dystrophy (toxic hepatitis with ascites).] Clin. Med., 8: 28 (in Russian).

ISMAILOV, N.I. (1948b) [Heliotropic toxicosis (toxic hepatitis with ascites),] Tashkent, Academy of Sciences of Uzbekistan, pp. 1-120 (in Russian).

ISMAILOV, N.I., MADZHIDOV, N.M., MAGRUPOV, A.I., MAKHKAMOV, G.M., & MUKMINOVA, S.G. (1970) [Clinical signs, diagnosis and treatment of Trichodesma toxicosis (alimentary toxic encephalopathy),] Tashkent, Meditsina, p. 85 (in Russian).

JAGO, M.V. (1969) The development of the hepatic megalocytosis of chronic pyrrolizidine alkaloid poisoning. Am. J. Pathol., 56: 405-422.

JAGO, M.V. (1970) A method for the assessment of the chronic hepatotoxicity of pyrrolizidine alkaloids. Aust. J. exp. Biol. med. Sci., 48: 93-103.

JAGO, M.V. (1971) Factors affecting the chronic hepatotoxicity of pyrrolizidine alkaloids. J. Pathol., 105: 1-11.

JAGO, M.V., LANIGAN, G.W., BINGLEY, J.B., PIEREY, D.W.T., WHITTEN, J.H., & TITCHEN, D.A. (1969) Excretion of the pyrrolizidine alkaloid heliotrine in the urine and bile of sheep. J. Pathol., 98: 115-128.

JAGO, M.V., EDGAR, J.A., SMITH, L.W., & CULVENOR, C.C.J. (1970) Metabolic conversion of heliotrine based pyrrolizidine alkaloids to dehydroheliotridine. Mol. Pharmacol., 6: 402-406.

JELLIFFE, D.B., BRAS, G., & STUART, K.L. (1954a) Venoocclusive disease of the liver. Paediatrics, 14: 334-339.

JELLIFFE, D.B., BRAS, G., & STUART, K.L. (1954b) The clinical picture of veno-occlusive disease of the liver in Jamaican children. Ann. trop. Med. Parasitol., 48: 386-396.

JELLIFFE, D.B., BRAS, G., & MUKHERJEE, K.L. (1957) Veno-occlusive disease of the liver and Indian childhood cirrhosis. Arch. dis. Child., 32: 369-385.

JOHNSON, A.E. (1976) Changes in calves and rats consuming milk from cows fed chronic lethal doses of Senecio jacobaea (tansy ragwort). Am. J. vet. Res., 37: 107-110.

JOHNSON, A.E. (1979) Toxicity of tansy ragwort to cattle. In: Cheeke, P.R., ed. Proceedings of the Symposium on Pyrrolizidine (Senecio) Alkaloids: Toxicity, Metabolism, and Poisonous Plant Control Measures, Corvallis, Oregon, USA, 23-24 February 1979, Oregon, Nutrition Research Institute, pp. 129-134.

JOHNSON, A.E. (1982) Failure of mineral-vitamin supplements to prevent tansy ragwort (Senecio jacobaea) toxicosis in cattle. Am. J. vet. Res., 43: 718-723.

JOHNSON, A.E. & MOLYNEUX, R.J. (1984) Toxicity of thread leaf groundsel (Senecio douglasii var. longilobus) to cattle. Am. J. vet. Res., 45: 26-31.

JOHNSON, A.E., MOLYNEUX, R.J., & STUART, L.D. (1985) Toxicity of Riddel's groundsel (Senecio riddelli) to cattle. Am. J. vet. Res., 46: 577-582.

JOHNSON, W.D. (1981) Mechanism of in vitro acute toxicity of dehydromonocrotaline a metabolite of pyrrolizidine alkaloid monocrotaline. Toxicologist, 1: 107-108.

JOHNSON, W.D., ROBERTSON, K.A., POUNDS, J.G., & ALLEN, J.R. (1978) Dehydroretronecine-induced skin tumours in mice. J. Natl Cancer Inst., 61: 85-89.

JULIEN, M.H., ed. (1982) Biological control of weeds. A world catalog of agents and their target weeds, Curepe, Trinidad and Toabago, Commonwealth Institute of Biological Control.

JUNEJA, T.R., GUPTA, R.L., & SAMANTA, S. (1984) Activation of monocrotaline, fulvine and their derivatives to toxic pyrroles by some thiols. Toxicol. Lett., 21: 185-189.

KAMPANZEV, N.N. (1952) [Experimental heliotropic hepatitis.] In: Milenkov, S.M. & Kizhaikin, Y., ed. [Collection of scientific papers on Toxic Hepatitis with Ascites,] Tashkent, Publishing House of the University of Central Asia, pp. 165-172 (in Russian).

KAY, J.M. & HEATH, D. (1966) Observation on the pulmonary arteries and heart weight of rats fed on Crotalaria spectabilis seeds. J. Pathol., 92: 385-394.

KAY, J.M. & HEATH, D. (1969) Crotalaria spectabilis: the pulmonary hypertension plant, Springfield, Illinois, Charles C. Thomas, p. 38.

KAY, J.M., HARRIS, P., & HEATH, D. (1967a) Pulmonary hypertension produced in rats by ingestion of Crotalaria spectabilis seeds. Thorax, 22: 176-179.

KAY, J.M., GILLUND, T.D., & HEATH, D. (1967b) Mast cells in the lungs of rats fed on Crotalaria spectabilis seeds. Am. J. Pathol., 51: 1031-1044.

KAY, J.M., HEATH, D., SMITH, P., BRAS, G., & SUMMERELL, J. (1971a) Fulvine and pulmonary circulation. Thorax, 26: 249-261.

KAY, J.M., SMITH, P., & HEATH, D. (1971b) Aminorex and the pulmonary circulation. Thorax, 26: 262-270.

KAY, J.M., KEANE, P.M., & SUYAMA, K.L. (1985) Pulmonary hypertension induced in rats by monocrotaline and chronic hypoxia is reduced by p-chlorophenylalanine. Respiration, 47: 48-56.

KEDZIERSKI, B. & BUHLER, D.R. (1985) Configuration of necine pyrroles - toxic metabolites of pyrrolizidine alkaloids. Toxicol. Lett., 25: 115-119.

KEDZIERSKI, B. & BUHLER, D.R. (1986) The formation of 6,7-dihydro-7-hydroxy-1-hydroxymethyl-5H-pyrrolizine, a metabolite of pyrrolizidine alkaloids. Chem.-biol. Interact., 57: 217-222.

KHANIN, M.N. (1948) [The etiology of toxic hepatitis with ascites.] Arch. Pathol. USSR, 1: 42-47 (in Russian).

KIM, H.L. & JONES, L.P. (1982) Protective effects of butylated anisole, ethoxyquin and disulfiran on acute pyrrolizidine alkaloid poisoning in mice. Res. Comm. chem. Pathol. Pharmacol., 36: 341-344.

KIRFEL, A., WILL, G., WIEDENFELD, H. & RODER, E. (1980) (1aR,6bR,10R,11R)-9,15-dioxo-10-hydroxy-10,11,13-trimethyl-1a,2,3,6b-tetrabydro-5H-pyrrolizino-[1a,6b,6a,c],8-dioxa-15-cis-tridecene,$C_{18}H_{25}NO_5$. Cryst. Styruct. Comm., 9: 353-361.

KNIGHT, A.P., KIMBERLING, C.V., STERMITZ, F.R., & ROBY, M.R. (1984) Cynoglossum officinale (hound's tongue) - a cause of pyrrolizidine alkaloid poisoning in horses. J. Am. Vet. Med. Assoc., 185: 647-650.

KOEKEMOER, M.J. & WARREN, F.L. (1951) The occurrence and preparation of the N-oxides. An improved method of extraction of the Senecio alkaloids. J. Chem. Soc., C1951: pp. 66-68.

KOLETSKY, A., OYASU, R., & REDDY, J.K. (1978) Mutagenicity of the pyrrolizidine (Senecio) alkaloid, lasiocarpine in the salmonella/microsome test. Lab. Invest., 38: 352 (Abstract).

KOVACH, J.S., MOERTEL, C.G., & HAHN, R.G. (1979a) A phase 1 study of indicine-N-oxide. Proc. Am. Assoc. Cancer Res., 20: 357.

KOVACH, J.S., AMES, M.M., POWIS, G., MOERTEL, C.G., HAHN, R.G., & CREAGAN E.T. (1979b) Toxicity and pharmacokinetics of a pyrrolizidine alkaloid, indicine-N-oxide, in humans. Cancer Res., 39: 4540-4544.

KRAUS, C., ABEL, G., & SCHIMMER, O. (1985) [Studies on the chromosome damaging effect of some pyrrolizidine alkaloids in human lymphocytes in vitro.] Planta Med., 51: 89-91 (in German).

KRISHNAMACHARI, K.A.V.R., BHAT, R.V., KRISHNAMURTHY, D., KRISHNASWAMY, K., & NAGARAJAN (1977) Aetiopathogenesis of endemic ascites in Sarguja district of Madhya Pradesh. Indian J. med. Res., 65: 672-678.

KUHARA, K., TAKANASHI, H., HIRONO, I., FURUYA, T., & ASADA, Y. (1980) Carcinogenic activity of clivorine, a pyrrolizidine alkaloid isolated from Ligularia dentata. Cancer Lett., 10: 117-122.

KUMANA, C.R., NG, M., LIN, H.J., KO, W., WU, P.C., & TODD, D. (1983) Hepatic veno-occlusive disease due to toxic alkaloid herbal tea - letter to the editor. Lancet, 10 December: 1360-1361.

KUMANA, C.R., NG, M., LIN, H.J., KO, W., WU, P.C., & TODD, D. (1985) Herbal tea induced hepatic veno-occlusive disease: quantification of toxic alkaloid exposure in adults. Gut, 26: 101-104.

KUROZUMI, T., TANAKA, K., KIDO, M., & SHOYAMA, Y. (1983) Monocrotaline-induced renal lesions. Exp. mol. Pathol., 39: 377-386.

LAFRANCONI, W.M. & HUXTABLE, R.J. (1983) Changes in angiotensin-converting enzyme activity in lungs damaged by the pyrrolizidine alkaloid, monocrotaline. Thorax, 38: 307-309.

LAFRANCONI, W.M. & HUXTABLE, R.J. (1984) Hepatic metabolism and pulmonary toxicity of monocrotaline using isolated perfused liver and lung. Biochem. Pharmacol., 33: 2479-2484.

LAFRANCONI, W.M., DUHAMEL, R.C., BRENDEL, K., & HUXTABLE,R.J. (1984) Differentiation of the cardiac and pulmonary toxicity of monocrotaline, a pyrrolizidine alkaloid. Biochem. Pharmacol., 33: 191-197.

LAFRANCONI, W.M., OHNUMA, S., & HUXTABLE, R.S. (1985) Biliary excretion of novel pneumotoxic metabolites of the pyrrolizidine alkaloid monocrotaline. Toxicon, 23: 903-992.

LALICH, J.J. & EHRHART, L.A. (1962) Monocrotaline induced pulmonary arteritis in rats. J. atheroscler. Res., 2: 482-492.

LALICH, J.J. & MERKOW, L. (1961) Pulmonary arteritis produced in rats by feeding Crotalaria spectabilis. Lab. Invest., 10: 744-750.

LANCET (1984) Pyrrolizidine alkaloids (editorial). Lancet, 1: 201-202.

LANGLEBEN, D. & REID, L.M. (1985) Effect of methylprednisolone on monocrotaline-induced pulmonary vascular disease and right ventricular hypertrophy. Lab. Invest., 52: 298-303.

LANIGAN, G.W. (1971) Metabolism of pyrrolizidine alkaloids in the ovine rumen. III. The competitive relationship between heliotrine metabolism and methanogenesis in rumen fluid in vitro. Aust. J. agric. Res., 22: 123-130.

LANIGAN, G.W. (1972) Metabolism of pyrrolizidine alkaloids in ovine rumen. IV. Effects of chloral hydrate and halogenated methanes on rumen methanogenesis and alkaloid metabolism in fistulated sheep. Aust. J. agric. Res., 23: 1085-1091.

LANIGAN, G.W. & SMITH, L.W. (1970a) Metabolism of pyrrolizidine alkaloids in the ovine rumen. I. Formation of 7 α-hydroxy-1α-methyl-8α-pyrrolizidine from heliotrine and lasiocarpine. Aust. J. agric. Res., 21: 493-500.

LANIGAN, G.W. & SMITH, L.W. (1970b) Metabolism of pyrrolizidine alkaloids in the ovine rumen. II. Formation of 7 α-hydroxy-1α-methyl-8α-pyrrolizidine from heliotrine and lasiocarpine. Aust. J. agric. Res., 22: 123-130.

LANIGAN, G.W. & WHITTEM, J.H. (1970) Cobalt pellets and Heliotropium europaeum poisoning in penned sheep. Aust. vet. J., 46: 17-21.

LANIGAN, G.W., PAYNE, A.L., & PETERSON, J.E. (1978) Antimethanogenic drugs and Heliotropium europaeum poisoning in penned sheep. Aust. J. Agric. Res., 29: 1281-1292.

LAWS, L. (1968) Toxicity of Crotalaria mucronata to sheep. Aust. vet. J., 44: 453-455.

LETENDRE, L., SMITHSON, W.A., GILCHRIST, G.S., BURGERT, E.O., Jr, HOGLAND, C.H., AMES, M.M., POWIS, G., & KOVACH, J.S. (1981) Activity of indicine-N-oxide in refractory acute leukemia. Cancer, 47: 437-441.

LETENDRE, L., LUDWIG, J., PERRAULT, J., SMITHSON, W.A., & KOVACH, J.S. (1984) Hepatocellular toxicity during the treatment of refractory acute leukemia with indicine-N-oxide. Cancer, 54: 1256-1259.

LEVINE, O.R., HARRIS, R.C., BLANCE, W.A., & MELLINS, R.B. (1973) Progressive pulmonary hypertension in children with portal hypertension. J. Pediatr., 83: 964-972.

LIN, J.J., LIU, C., & SVOBODA, D.J. (1974) Long-term effects of aflatoxin B_1 and vocal hepatitis on marmoset liver: a preliminary report. Lab. Invest., 30: 267-278.

LUTHY, J., ZWEIFEL, U., SCHLATTER, C., & BENN, M.H. (1980) [Pyrrolizidine alkaloids in coltsfoot (Tussilago farfara L.) of various sources.] Mitt. Geb. Lebensm. Hyg., 71: 73-80 (in German with English summary).

LUTHY, J., ZWEIFEL, U., KARLHUBER, B., & SCHLATTER, C. (1981) Pyrrolizidine alkaloids of Senecio alpinus L. and their detection in feedstuffs. J. agric. food Chem., 29: 302-305.

LUTHY, J., HEIM, TH., & SCHLATTER, CH. (1983) Transfer of [^{3}H] pyrrolizidine alkaloids from Senecio vulgaris L. and metabolites into rat milk and tissues. Toxicol. Lett., 17: 283-288.

LUTHY, J., BRAUCHLI, J., ZWEIFEL, U., SCHMID, P., & SCHLATTER, CH. (1984) [Pyrrolizidine alkaloid in arzneipflanzen der Boraginaceen: Borago officinalis L. and Pulmonaria officinalis L.] Pharm. Acta Helv., 59: 242-246 (in German).

LYFORD, C.L., VERGAVA, G.G., & MOELLER, D.D. (1976) Hepatic veno-occlusive disease originating in Ecuador. Gastroenterology, 70: 105-108.

MCCOMISH, M., BODEK, I., & BRAUFMAN, A.R. (1980) Quantitation of the antineoplastic agent indicine-N-oxide in human plasma by differential pulse polarography. J. pharmacol. Sci., 69: 727-729.

MCCOY, J.W., ROBY, M.R., & STERMITZ, F.R. (1983) Analysis of plant alkaloid mixtures by ammonia chemical ionization mass spectroscopy. J. nat. Prod., 46: 894-900.

MCGEE, J.O'D., PATRICK, R.S., WOOD, C.B., & BLUMGART, L.H. (1976) A case of veno-occlusive disease of the liver in Britain associated with herbal tea consumption. J. clin. Pathol., 29: 788-794.

MCGRATH, J.P.M., DUNCAN, J.R., & MUNNELL, J.F. (1975) Crotalaria spectabilis toxicity in swine: characterization of the renal glomerular lesion. J. comp. Pathol., 85(2): 185-194.

MCLEAN, E.K. (1969) The early sinusoidal lesion in experimental veno-occlusive disease. Br. J. exp. Pathol., 50: 223-229.

MCLEAN, E.K. (1970) The toxic actions of pyrrolizidine (Senecio) alkaloids. Pharmacol. Rev., 22: 429-483.

MCLEAN, E.K. (1974) Senecio and other plants as liver poisons. Isr. J. med. Sci., 10: 436-440.

MCLEAN, E.K. & HILL, K.R. (1969) Portal hypertension in acute experimental veno-occulsive disease of the liver in rats. Br. J. exp. Pathol., 50: 37-41.

MCLEAN, E.K., BRAS, G., & GYORGY, P. (1964) Veno-occlusive lesions in livers of rats fed Crotalaria fulva. Br. J. exp. Pathol., 45: 242-247.

MAKSUDOV, A.M. (1952) [Heliotropic toxic hepatitis with ascites.] In: Milenkov, S.M. & Kizhaikin, Y., ed. [Collection of scientific papers on Toxic Hepatitis with Ascites,] Tashkent, Publishing House of the University of Central Asia, pp. 103-116 (in Russian).

MARTIN, P.A., THORBURNE, M.K., HUTCHINSON, S., BRAS, G., & MILLER, C.G. (1972) Preliminary findings of chromosome studies in rats and humans with veno-occlusive disease. Br. J. exp. Pathol., 53, 374-380.

MATTOCKS, A.R. (1961) Extraction of heat-labile alkaloids from plants. Nature (Lond.), 191: 1281-1282.

MATTOCKS, A.R. (1967a) Spectrophotometric determination of unsaturated pyrrolizidine alkaloids. Anal. Chem., 39: 443-447.

MATTOCKS, A.R. (1967b) Detection of pyrrolizidine alkaloids on thin-layer chromatograms. J. Chromatogr., 27: 505-508.

MATTOCKS, A.R. (1968a) Toxicity of pyrrolizidine alkaloids. Nature (Lond.), 217: 723-728.

MATTOCKS, A.R. (1968b) Spectrophotometric determination of pyrrolizidine alkaloids: some improvements. Anal. Chem., 40: 1749.

MATTOCKS, A.R. (1969) Dehydropyrrolizine derivatives from unsaturated pyrrolizidine alkaloids. J. Chem. Soc., C1969: 1155-1162.

MATTOCKS, A.R. (1971a) Synthetic compounds with toxic properties similar to those of pyrrolizidine alkaloids and their pyrrolic metabolites. Nature (Lond.), 232: 476.

MATTOCKS, A.R. (1971b) The occurrence and analysis of pyrrolizidine alkaloid N-oxides. Xenobiotica, 1: 451-453.

MATTOCKS, A.R. (1971c) Hepatotoxic effects due to pyrrolizidine alkaloid N-oxides. Xenobiotica, 1: 563-565.

MATTOCKS, A.R. (1971d) A field test for N-oxides of unsaturated pyrrolizidine alkaloids. Trop. Sci., 13: 65-70.

MATTOCKS, A.R. (1972a) Toxicity and metabolism of Senecio alkaloids. In: Harborne, J.B., ed. Phytochemical ecology, London, New York, Academic Press, pp. 179-200.

MATTOCKS, A.R. (1972b) Acute hepatotoxicity and pyrrolic metabolites in rats dosed with pyrrolizidine alkaloids. Chem.-biol. Interact., 5: 227-242.

MATTOCKS, A.R. (1973) Mechanisms of pyrrolizidine alkaloid toxicity. Pharmacology and the future of man. In: Proceedings of the 5th International Congress on Pharmacology, San Francisco, 1972, Basel, Karger, Vol. 2, pp. 114-123.

MATTOCKS, A.R. (1977) Tissue distribution of radioactivity in rats given tritiated analogues of hepatotoxic pyrrolizidine alkaloids. Xenobiotica, 7: 665-670.

MATTOCKS, A.R. (1980) Toxic pyrrolizidine alkaloids in comfrey. Lancet, 22 November: 1136-1137.

MATTOCKS, A.R. (1981a) Liver cell enlargement in rats given hydroxymethyl pyrroles analogous to pyrrolizidine alkaloid metabolites, followed later by the hepatotoxin dimethylnitrosamine. Toxicol. Lett., 8: 201-205.

MATTOCKS, A.R. (1981b) Relation of structural features to pyrrolic metabolites in livers of rats given pyrrolizidine alkaloids and derivatives. Chem.-biol. Interact., 35: 301-310.

MATTOCKS, A.R. (1981c) A simple preparation of dehydroretronecine using potassium nitrosodisulphonate. Chem. Ind., 7: 251.

MATTOCKS, A.R. (1982) Hydrolysis and hepatotoxicity of retronecine diesters. Toxicol. Lett., 14: 111-116.

MATTOCKS, A.R. (1986) Chemistry and toxicology of pyrrolizidine alkaloids, London, New York, Academic Press.

MATTOCKS, A.R. & BIRD, I. (1983) Pyrrolic and N-oxide metabolites formed from pyrrolizidine alkaloids by hepatic microsomes in vitro: relevance to in vivo hepatotoxicity. Chem.-biol. Interact., 43: 209-222.

MATTOCKS, A.R. & CABRAL, J.R.P. (1979) Effects of some pyrrolic and dehydropyrrolizine esters on mouse skin: a preliminary study. Tumori, 65: 289-293.

MATTOCKS, A.R. & CABRAL, J.R.P. (1982) Carcinogenicity of some pyrrolizidine alkaloid metabolites and analogues. Cancer Lett., 17: 61-66.

MATTOCKS, A.R. & DRIVER, H.E. (1983) A comparison of the pneumotoxicity of some pyrrolic esters and similar compounds analogous to pyrrolizidine alkaloid metabolites given intravenously to rats. Toxicology, 27: 159-177.

MATTOCKS, A.R. & DRIVER, H.E. (1987) Toxic actions of senaetnine, a new pyrrolizidine alkaloid, in rats. Toxicol. Lett., 38: 315-319.

MATTOCKS, A.R. & JUKES, R. (1987) New improved field test for toxic pyrrolizidine alkaloids. J. nat. Prod., 50: 161-166.

MATTOCKS, A.R. & LEGG, R.F. (1980) Antimitotic activity of dehydroretronecine, a pyrrolizidine alkaloid metabolite, and some analogous compounds in a rat liver parenchymal cell line. Chem.-biol. Interact., 30: 325-336.

MATTOCKS, A.R. & WHITE, I.N.H. (1970) Estimation of metabolites of pyrrolizidine alkaloids in animal tissues. Anal. Biochem., 38: 529-535.

MATTOCKS, A.R. & WHITE, I.N.H. (1971a) The conversion of pyrrolizidine alkaloids to dihydropyrrolizine derivatives by rat-liver microsomes in vitro. Chem.-biol. Interact., 3: 383-396.

MATTOCKS, A.R. & WHITE, I.N.H. (1971b) Pyrrolic metabolites from non-toxic pyrrolizidine alkaloids. Nat. new Biol., 231: 114-115.

MATTOCKS, A.R. & WHITE, I.N.H. (1973) Toxic effects and pyrrolic metabolites in the liver of young rats given the pyrrolizidine alkaloid retrorsine. Chem.-biol. Interact., 6: 297-306.

MATTOCKS, A.R. & WHITE, I.N.H. (1976) The distribution of [^{3}H]-synthanecine A bis-n-ethylcarbamate and its metabolites in the rat. Chem.-biol. Interact., 15: 173-184.

MEYRICK, B.O. & REID, L.M. (1979) Development of pulmonary arterial changes in rats fed Crotalaria spectabilis. Am. J. Pathol., 94: 37-50.

MEYRICK, B.O. & REID, L.M. (1982) Crotalaria-induced pulmonary hypertension: uptake of ^{3}H-thymidine by the cells of the pulmonary circulation and alveolar walls. Am. J. Pathol., 106: 84-94.

MILKOWSKY, A.S. (1985) The synthesis and anti-tumour activity of monocyclic analogs of indicine-N-oxide. Diss. Abstr. Int., 45(3): 844.

MILLER, W.C., RICE, D.L., KREUSEL, R.G., & BEDROSSIAN, C.W.M. (1978) Monocrotaline model of non-cardiogenic pulmonary edema in dogs. J. appl. Physiol., 45: 962-965.

MIRANDA, C.L., BUHLER, D.R., & CHEEKE, P.R. (1979) The effect of tansy ragwort consumption on mineral metabolism in the rat. In: Cheeke, P.R., ed. Proceedings of the Symposium on Pyrrolizidine (Senecio) Alkaloids: Toxicity, Metabolism, and Poisonous Plant Control Measures, Corvallis, Oregon, USA, 23-24 February 1979, Oregon, Nutrition Research Institute, pp. 61-64.

MIRANDA, C.L., CARPENTER, H.M., CHEEKE, P.R., & BUHLER, D.R. (1981a) Effect of ethoxyquin on the toxicity of pyrrolizidine alkaloid monocrotaline and on hepatic long metabolism in mice. Chem.-biol. Interact., 37: 95-107.

MIRANDA, C.L., HENDERSON, M.C., & BUHLER, D.R. (1981b) Dietary copper enhances the hepatotoxicity of Senecio jacobaea in rats. Toxicol. appl. Pharmacol., 60(3): 418-423.

MIRANDA, C.L., REED, R.L., CHEEKE, P.R., & BUHLER, D.R. (1981c) Protective effects of butylated hydroxyanisole against the acute toxicity of monocrotaline in mice. Toxicol. appl. Pharmacol., 59: 424-430.

MIRANDA, C.L., BUHLER, D.R., RAMSDELL, H.S., CHEEKE, P.R., & SCHMITZ, J.A. (1982a) Modifications of chronic hepatotoxicity of pyrrolizidine (Senecio) alkaloids by butylated hydroxyanisole and cysteine. Toxicol. Lett., 10: 177-182.

MIRANDA, C.L., HENDERSON, M.C., BUHLER, D.R., & SCHMITZ, J.A. (1982b) Comparative effects of antioxidants on the toxicity of mixed pyrrolizidine alkaloids from Senecio jacobaea in mice. J. Toxicol. environ. Health, 9(5-6): 933-939.

MIRANDA, C.L., HENDERSON, M.C., REED, R.L., SCHMITZ, J.A., & BUHLER, D.R. (1982c) Protective action of zinc against pyrrolizidine alkaloid-induced hepatotoxicity in rats. J. Toxicol. environ. Health, 9: 359-366.

MIROCHNIK, M.F. (1938) [Clinical course and etiopathogenesis of toxic hepatites with ascites.] Medical Institute, scientific papers of the second therapeutic clinic, Vol I, Tashkent, Medical Literature Publishers.

MISER, J.S., MISER, A.W., SMITHSON, W.A., COCCIA, P.F., AMES, M.M., DAVIS, D.M., HUGHES, C.S., & GILCHRIST, G.S. (1982) A phase I trial of indicine-N-oxide in childhood malignancy. Proc. Am. Soc. Clin. Oncol., 1: 137.

MNUSHKIN, A.S. (1949) [Some materials to clinical features and pathogenesis of toxic hepatitis with ascites.] Sov. Med., 1: 8-9 (in Russian).

MNUSHKIN, A.S. (1952) [The clinical features, pathogenesis and treatment of toxic hepatitis with ascites.] In: Milenkov, S.M. & Kizhaikin, Y., ed. [Collection of scientific papers on Toxic Hepatitis with Ascites,] Tashkent, Publishing House of the University of Central Asia, pp. 91-98 (in Russian).

MOHABBAT, O., SRIVASTAVA, R.N., YOUNOS, M.S., SEDIQ, G.G., MENZAD, A.A., & ARAM, G.N. (1976) An outbreak of hepatic veno-occlusive disease in north-western Afghanistan. Lancet, 7 August: 269-271.

MOLTENI, A., WARD, W.F., TS'AO, C.-S., PORT, C.D., & SOLLIDAY, N.H. (1984) Monocrotaline-induced pulmonary endothelial dysfunction in rats. Proc. Soc. Exp. Biol. Med., 176: 88-94.

MOLYNEUX, R.J. & ROITMAN, J.N. (1980) Specific detection of pyrrolizidine alkaloids on thin-layer chromatograms. J. Chromatogr., 195: 412-415.

MOLYNEUX, R.J., JOHNSON, A.E., ROITMAN, J.N., & BENSON, M.E. (1979) Chemistry of toxic range plants. Determination of pyrrolizidine alkaloid content and composition in Senecio species by nuclear magnetic resonance spectroscopy. J. agric. food Chem., 27: 494-499.

MORI, H., SUGIE, S., YOSHIMI, N., ASADA, Y., FURUYA, T., & WILLIAMS, G.M. (1985) Genotoxicity of a variety of pyrrolizidine alkaloids in the hepatocyte primary culture-DNA repair test using rat, mouse, and hamster hepatocytes. Cancer Res., 45: 3125-3129.

MUNIER, R. (1953) Separation of alkaloids from their N-oxides by paper chromatography. Bull. Soc. Chim. Biol., 35: 1225.

NARDI, N.B., GIMMLER, M.C., & LEITE, B.G. (1980) Antimitotic action of integerrimine in rats. Rev. Bras. Genet., 3: 387-392.

NATIONAL CANCER INSTITUTE (1978) Bioassay of lasiocarpine for possible carcinogenecity, Bethesda, Maryland, National Cancer Institute, US Department of Health, Education and Welfare (Report of Carcinogenesis Testing Programme, Division of Cancer Cause and Prevention).

NAYAK, N.C., SAGREIYA, K., & RAMALINGASWAMI, V. (1969) Indian childhood cirrhosis - the nature and significance of cytoplasmic hyaline of hepatocytes. Arch. Pathol., 88: 631-637.

NEWBERNE, P.M. (1968) The influence of a low isotrope diet on response of maternal and fetal rats to lasiocarpine. Cancer Res., 28: 2327-2337.

NEWBERNE, P.M. & ROGERS, A.E. (1973) Nutrition, monocrotaline and aflatoxin B_1 in liver carcinogenesis. Plant Food Man, 1: 23-31.

NEWBERNE, P.M., WILSON, R., & ROGERS, A.E. (1971) Effects of a low-lipotrope diet on the response of young male rats to the pyrrolizidine alkaloid monocrotaline. Toxicol. appl. Pharmacol., 18: 387-397.

NEWBERNE, P.M., CHAN, W.C., & ROGERS, A.E. (1974) Influence of light, riboflavin and carotene on the response of rats to the acute toxicity of aflatoxin and monocrotaline. Toxicol. appl. Pharmacol., 28: 200-208.

NICHOLS, W.C., MOERTL, C.G., RUBIN, J., SCHULT, A.J., & BRITTEL, J.C. (1981) Phase II trial of indicine-N-oxide (NSC132319) in patients with advanced colorectal carcinoma. Cancer Treat. Rep., 65: 337-339.

NIWA, H., ISHIWATA, H., & YAMADA, K. (1985) Isolation of petasitenine, a carcinogenic pyrrolizidine alkaloid from Farfugium japonicum. J. nat. Prod., 48: 10083-1004.

NOLAN, J.P., SCHEIG, R.L., & KLATSKIN, G. (1966) Delayed hepatitis and cirrhosis in weanling rats following a single small dose of the Senecio alkaloid, lasiocarpine. Am. J. Pathol., 49: 129-151.

OHNUMA, T., SRIDHAR, K.S., RATNER, L.H., & HOLLAND, J.F. (1982) Phase I study of indicine-N-oxide in patients with advanced cancer. Cancer Treat. Rep., 66(7): 1509-1515.

OHTSUBO, K., ITO, Y., SAITO, M., FURUYA, T., & HIKICHI, M. (1977) Hypertrophy of pulmonary arteries and arterioles with cor pulmonale in rats induced by seneciphylline, a pyrrolizidine alkaloid. Experientia (Basel), 33: 498-499.

OLSON, J.W., HACKER, A.D., ALTIERE, R.J., & GILLESPIE, M.N. (1984) Polyamines and the development of monocrotaline-induced pulmonary hypertension. Am. J. Physiol., 247: H682-H685.

OLSON, J.W., ATKINSON, J.E., HACKER, A.D., ALTIERE, R.J., & GILLESPIE, M.N. (1985) Suppression of polyamine biosynthesis prevents monocrotaline induced pulmonary oedema and arterial medial thickening. Toxicol. appl. Pharmacol., 81: 91-99.

ORD, M.J., HERBERT, A., & MATTOCKS, A.R. (1985) The ability of bifunctional and monofunctional pyrrole compounds to induce sister chromatid exchange (SCE) in human lymphocytes and mutations in Salmonella typhimurium. Mutat. Res., 149: 485-493.

PANTONE, D.J., BROWN, S.M., & WOMERSLEY, C. (1985) Biological control of fiddleneck. California Agric., 39: 4-9.

PECKHAM, J.C., SANGSTER, L.P., & JONES, O.H. (1974) Crotalaria spectabilis poisoning in swine. J. Am. Vet. Med. Assoc., 165: 633-638.

PEDERSEN, E. (1975) Pyrrolizidine alkaloids in Danish species of the family Boraginaceae. Arch. Pharm. Chem. Sci. Ed., 3: 55-64.

PERCY, J.J. & PIERCE, A.E. (1971) Immunosuppressive activity of the pyrrolizidine alkaloid metabolite dehydroheliotridine. Immunology, 21: 273-280.

PERSAUD, J.V. & HOYTE, D.A. (1974) Pregnancy and progeny in rats treated with the pyrrolizidine alkaloid fulvine. Exp. Pathol., 9: 59-63.

PESTCHANKER, M.J. & GIORDANO, O.S. (1986) Pyrrolizidine alkaloids from five Senecio species. J. nat. Prod., 49: 722-723.

PESTCHANKER, M.J., ASCHERI, M.S., & GIORDANO, O.S. (1985a) Uspallatine, a pyrrolizidine alkaloid from Senecio uspallatensis. Phytochemistry, 24: 1622-1624.

PESTCHANKER, M.J., ASCHERI, M.S., & GIORDANO, O.S. (1985b) Pyrrolizidine alkaloids from Senecio subulatus and S. glandulosus. Planta Med., 51: 165-166.

PETERSON, J.E. (1965) Effects of pyrrolizidine alkaloid lasiocarpine N-oxide on nuclear and cell division in the liver of rats. J. Pathol. Bacteriol., 89: 153-171.

PETERSON, J.E. & CULVENOR, C.C.J. (1983) Plant and fungal toxins. In: Keeler, R.F. & Tu, A.T., ed. Handbook of natural toxins, New York, Marcel Dekker, Vol. 1, pp. 637-681.

PETERSON, J.E. & JAGO, M.V. (1980) Comparison of the toxic effects of dehydroheliotridine and heliotrine in pregnant rats and their embryos. J. Pathol., 131: 339-355.

PETERSON, J.E. & JAGO, M.V. (1984) Toxicity of Echium plantagineum (Paterson's curse): pyrrolizidine alkaloid poisoning in rats. Aust. J. agric. Res., 35: 305-316.

PETERSON, J.E., SAMUEL, A., & JAGO, M.V. (1972) Pathological effects of dehydroheliotridine, a metabolite of heliotridine-based pyrrolizidine alkaloids in the young rat. J. Pathol., 107: 107-189.

PETERSON, J.E., JAGO, M.V., REDDY, J.K., & JARRETT, R.G. (1983) Neoplasia and chronic disease associated with the prolonged administration of dehydroheliotridine to rats. J. Natl Cancer Inst., 70: 381-386.

PETRY, T.W., BOWDEN, G.P., HUXTABLE, R.J., & SIPES, I.G. (1984) Characterization of hepatic DNA damage induced in rats by the pyrrolizidine alkaloid monocrotaline. Cancer Res., 44: 1505-1509.

PETRY, T.W., BOWDEN, G.P., BUHLER, D.R., & SIPES, I.G. (1986) Genotoxicity of the pyrrolizidine alkaloid jacobine in rats. Toxicol. Lett., 32: 275-281.

PIERCY, P.L. & RUSOFF, L.L. (1946) Crotalaria spectabilis poisoning in Louisiana livestock. J. Am. Vet. Med. Assoc., 108: 69-73.

PIERSON, M.L., CHEEKE, P.R., & DICKINSON, E.O. (1977) Resistance of the rabbit to dietary pyrrolizidine (Senecio) alkaloid. Res. Commun. chem. Pathol. Pharmacol., 16: 561-564.

PIETERS, L.A.C. & VLIETINCK, A.J. (1985) Quantitative 'H Fourier transform nuclear magnetic resonance spectroscopic analysis of mixtures of pyrrolizidine alkaloids from Senecio vulgaris. Fresenius' Z. Anal. Chem., 321: 355-358.

PIETERS, L.A.C. & VLIETINCK, A.J. (1986) Comparison of high-performance liquid chromatography with ^{1}H nuclear magnetic resonance spectrometry for the quantitative analysis of pyrrolizidine alkaloids from Senecio vulgaris. J. liq. Chromatogr., 9: 745-755.

PLESTINA, R. & STONER, H.B. (1972) Pulmonary oedema in rats given monocrotaline pyrrole. J. Pathol., 106: 235-249.

POOL, B.L. (1982) Genotoxic activity of an alkaloidal extract of Senecio nemorensis spp. fuchsii in Salmonella typhimurium and Escherichia coli systems (letter to the editor). Toxicology, 24: 351-355.

POWIS, G., AMES, M.M., & KOVACH, J.S. (1979) Metabolic conversion of indicine-N-oxide to indicine in rabbits and humans. Cancer Res., 39: 3564-3570.

PRABHU, M.B. (1940) Infantile cirrhosis of liver. Indian J. Pediatr., 7: 121-134.

RADHAKRISHNA RAO, M.V. (1935) Histopathology of the liver in infantile biliary cirrhosis. Indian J. med. Res., 23: 69-90.

RAJAGOPALAN, T.R. & NEGI, R.K.S. (1985) Alkaloids from Doronicum pardalianches Linn. Indian J. Chem., 24B, 882.

RAKIETEN, N., GORDON, B.S., BEATTY, A., COONEY, D.A., DAVIS, R.D., & SCHIEN, P.S. (1971) Pancreatic islet cell tumours produced by the combined action of streptozotocin and nicotinamide. Proc. Soc. Exp. Biol. Med., 137: 280-283.

RAMALINGASWAMI, V. & NAYAK, N.C. (1969) Liver disease in India. In: Popper, H. & Schaffner, F., ed. Progress in liver diseases, New York, London, Grune and Stratton, pp. 222-235.

RAMSDELL, A.S. & BUHLER, D.R. (1981) High performance liquid chromatographic analysis of pyrrolizidine (Senecio) alkaloids using a reversed phase styrene-divinylbenzene resin column. J. Chromatogr., 210: 154-158.

RAO, M.S. & REDDY, J.K. (1978) Malignant neoplasms in rats fed lasiocarpine. Br. J. Cancer, 37: 289-293.

RAPPAPORT, A.M., KNOBLAUCH, M., SUMMERSKILL, J., & BRAS, G. (1967) Experimental veno-occlusive disease. Gastroenterology, 54: 164.

RATNOFF, O.D. & MIRICK, G.S. (1949) Influence of sex upon the lethal effects of an hepatotoxic alkaloid, monocrotaline. Bull. Johns Hopkins Hosp., 84: 507-25.

RECHICIGL, M., Jr, ed. (1983) CRC Handbook on naturally occurring food toxicants, Boca Raton, Florida, CRC Press, 266 pp.

RESCH, J.F. & MEINWALD, J. (1982) A revised structure for acetylheliosupine. Phytochemistry, 21: 2340-2431.

RHODES, K. (1957) Two types of liver disease in Jamaican children. West Indian med. J., 6: 1-29.

RIDKER, P.M., OHKUMA, S., MCDERMOTT, W.V., TREY, C., & HUXTABLE, R.J. (1985) Hepatic veno-occlusive disease associated with consumption of pyrrolizidine alkaloid containing dietary supplements. Gastroenterology, 88: 1050-1054.

RIZK, A.M., HAMMOUDA, F.M., ISMAIL, S.I., GHALEB, H.A., MADKOUR, M.K., POHLAND, A.E., & WOOD, G. (1983) Poisonous plants contaminating edible ones and toxic substances in plant foods. I. Alkaloids from Senecio desfontainei. Fitoterapia, 54: 115-121.

ROBERTSON, K.A. (1982) Alkylation of N^2 in deoxyguanosine by dehydroretronecine, a carcinogenic metabolite of the pyrrolizidine alkaloid monocrotaline. Cancer Res., 42: 8-14.

ROBINS, D.J. (1982) The pyrrolizidine alkaloids. In: Herz, W., Grisebach, H., & Kirby, G.W., ed. Progress in the chemistry of organic natural products, Vienna, New York, Springer-Verlag, pp. 115-203.

RODER, E., WEIDENFELD, H., & STENGL, P. (1981) Pyrrolizidine alkaloids senecionine and retrorsine from Senecio inaequidens. Planta Med., 41: 412-413.

RODER, E., WIEDENFELD, H., & KNOZINGER-FISCHER (1984a) Pyrrolizidine alkaloid from Senecio abrotanifolius spp. abroatanifolius, spp. abrotanifolius, and var. tiroliensis. Planta Med., 41: 412-413.

RODER, E., WIEDENFELD, H., & BRITZ-KIRSTGEN, R. (1984b) Pyrrolizidine alkaloids from Senecio cacaliaster. Phytochemistry, 23: 1761-1763.

ROGERS, A.E. & NEWBERNE, P.M. (1971) Lasiocarpine factors influencing its toxicity and effects in liver cell division. Toxicol. appl. Pharmacol., 18: 356-366.

ROITMAN, J.N. (1981) Comfrey and liver damage. Lancet, 25 April: 944.

ROITMAN, J.N. (1983) Ingestion of pyrrolizidine alkaloids: a health hazard of global proportions. In: Finlay, J.W. & Schwass, D.E., ed. Xenobiotics in foods and feeds, Washington DC, American Chemical Society, pp. 345-378 (ACS Series).

ROSE, A.L., GARDNER, C.A., MCDONNELL, J.D., & BULL, L.B. (1957a) Field and experimental investigation of "walk-about" disease of horses (Kimberley horse disease) in northern Australia: Crotalaria poisoning in horses. Part II. Aust. vet. J., 33: 49-62.

ROSE, A.L., GARDNER, C.A., MCDONNELL, J.D., & BULL, L.B. (1957b) Field and experimental investigation of "walk about" disease of horses (Kimberley horse disease) in northern Australia. Part I. Aust. vet. J., 33: 25-33.

ROSE, C.L., HARRIS, P.N., & CHEN, K.K. (1959) Some pharmacological action of supinine and lasiocarpine. J. Pharmacol. exp. Ther., 126: 179-184.

ROSE, E.F. (1972) Senecio species: toxic plants used as food and medicine in the Transkei. S. Afr. Med. J., 46: 1039-1043.

ROSENFIELD, I. & BEATH, O.A. (1945) Tissue changes induced by Senecio riddellii. Am. J. clin. Pathol., 15: 407-412.

ROTH, R.A., DOTZLAF, L.A., BARANYI, B., & HOOK, J.B. (1981) Effect of monocrotaline ingestion on liver, kidney and lung of rat. Toxicol. appl. Pharmacol., 60: 193-203.

ROYES, K. (1948) Infantile hepatic cirrhosis in Jamaica. Caribb. med. J., 10: 16-48.

SAFOUH, M. & SHEHATA, A.H. (1965) Hepatic vein occlusion disease of Egyptian children. J. Pediatr., 67: 415-432.

ST. GEORGE-GRAMBAUER, T.D. & RAC, R. (1962) Hepatogenous chronic copper poisoning in sheep in south Australia due to consumption of Echium plantagineum L. (Salvation Jane). Aust. vet. J., 38: 288-293.

SALASPURO, M. & SIPPONEN, P. (1976) Demonstration of an intracellular copper-binding protein in orcein staining in long standing cholestatic liver diseases. Gut, 17: 787-790.

SAMUEL, A. & JAGO, M.V. (1975) Localization in the cell cycle of the antimitotic action of the pyrrolizidine alkaloid, lasiocarpine, and its metabolite, dehydroheliotridine. Chem.-biol. Interact., 10: 185-197.

SANDERS, D.A., SHEALY, A.L., & EMMEL, M.W. (1936) The pathology of Crotalaria spectabilis Roth poisoning in cattle. J. Am. Vet. Med. Assoc., 89: 150-156.

SAVIN, I.G., (1983) [Research of the influence of heliotrin on the microsomal mono oxygenase systems of the rat liver.] N.E. Pyrogov 2nd Moscow State Medical Institute. (Thesis in biological chemistry) (in Russian).

SAVVINA, K.I. (1952) Pathological anatomy of atrophic hepatic cirrhosis. Arch. Pathol., 14: 65-70.

SCHOENTAL, R. (1959) Liver lesions in young rats suckled by mothers treated with the pyrrolizidine (Senecio) alkaloids, lasiocarpine, and retrorsine. J. Pathol. Bacteriol., 77: 485-504.

SCHOENTAL, R. (1961) Herbal medicines and liver disease. J. exp. med. Sci., 4: 126-135.

SCHOENTAL, R. (1963) Liver disease and 'natural' hepatotoxins. Bull. World Health Organ., 29: 823-833.

SCHOENTAL, R. (1968) Toxicology and carcinogenic action of pyrrolizidine alkaloids. Cancer Res., 28: 2237-2246.

SCHOENTAL, R. (1970) Hepatotoxic activity of retrorsine, senkirkine and hydroxysenkirkine in newborn rats, and the role of epoxides in carcinogenesis by pyrrolizidine alkaloids and aflatoxins. Nature (Lond.), 227: 401-402.

SCHOENTAL, R. (1975) Pancreatic islet-cell and other tumours in rats given heliotrine, a monoester pyrrolizidine alkaloid, and nicotinamide. Cancer Res., 35: 2020-2024.

SCHOENTAL, R. & BENSTED, J.P.M. (1963) Effects of whole body irradiation and of partial hepatectomy on the liver lesions induced in rats by a single dose of retrorsine, a pyrrolizidine (Senecio) alkaloid. Br. J. Cancer, 17: 242-251.

SCHOENTAL, R. & CAVANAGH, J.B. (1972) Brain and spinal cord tumours in rats treated with pyrrolizidine alkaloids. J. Natl Cancer Inst., 49: 665-671.

SCHOENTAL, R. & COADY, A. (1968) The hepatotoxicity of some Ethiopian and east African plants including some used as traditional medicines. East Afr. med. J., 45: 577-579.

SCHOENTAL, R. & HEAD, M.A. (1955) Pathological changes in rats as result of treatment with monocrotaline. Br. J. Cancer, 9: 229-237.

SCHOENTAL, R. & HEAD, M.A. (1957) Progression of liver lesions produced in rats by temporary treatment with pyrrolizidine (Senecio) alkaloids, and the effects of betaine and high casein diet. Br. J. Cancer, 11: 535-544.

SCHOENTAL, R. & MAGEE, P.N. (1957) Chronic liver changes in rats after a single dose of lasiocarpine, a pyrrolizidine (Senecio) alkaloid. J. Pathol. Bacteriol., 74: 305-319.

SCHOENTAL, R. & MAGEE, P.N. (1959) Further observation on the subacute and chronic liver changes rats after a single dose of various pyrrolizidine (Senecio) alkaloids. J. Pathol. Bacteriol., 78: 471-482.

SCHOENTAL, R. & MATTOCKS, A.R. (1960) Hepatotoxic activity of semisynthetic analogues of pyrrolizidine alkaloids. Nature (Lond.), 185: 842-843.

SCHOENTAL, R., HEAD, M.A., & PEACOCK, P.R. (1954) Senecio alkaloids: primary liver tumours in rats as a result of treatment with (1) mixture of alkaloids from S. jacobaea Lin, (2) retrorsine, (3) isatidine. Br. J. Cancer, 8: 458-465.

SCHOENTAL, R., FOWLER, M.E., & COADY, A. (1970) Islet cell tumours of the pancreas found in rats given pyrrolizidine alkaloids from Amsinckia intermedia Fisch and Mey and from Heliotropium supinum. Cancer Res., 30: 2127-2131.

SEGALL, H.J. (1979a) Reversed phase isolation of pyrrolizidine alkaloids. Liq. Chromatogr., 2: 429-436.

SEGALL, H.J. (1979b) Preparative isolation of pyrrolizidine alkaloids derived from Senecio vulgaris. Liq. Chromatogr., 2: 1319-1323.

SEGALL, H.J., DALLAS, J.L., & HADDON, W.F. (1984) Two dihydropyrrolizine alkaloid metabolites isolated from mouse hepatic microsomes in vitro. Drug Metab. Disp., 12: 68-71.

SEGALL, H.J., WILSON, D.W., DALLAS, J.L., & HADDON, W.F. (1985) Trans-4-hydroxy-2-hexenol: a reactive metabolite from the macrocyclic pyrrolizidine alkaloid senecionine. Science, 229: 472-475.

SEGER, C.L., NEWSON, J.D., ROTH, E.E., & HUTCHINSON, W.R. (1969) Chronic toxic hepatitis in deer from a Louisiana coastal marsh. Bull. Wildlife Disease Assoc. Proc. Ann. Conf., pp. 295-6.

SELZER, G. & PARKER, R.G.F. (1951) Senecio poisoning exhibiting as Chiari's syndrome: a report of 12 cases. Am. J. Pathol., 27: 885-907.

SENER, B., TEMIZER, H., TEMIZER, A., & KARAKAYA, A.E. (1986) High performance liquid chromatographic determination of alkaloids in Senecio vernalis. J. Pharm. Belg., 41: 115-117.

SHARMA, R.K., KHAJURIA, G.S., & ATAL, C.K. (1965) Thin layer chromatography of pyrrolizidine alkaloids. J. Chromatogr., 19: 433-434.

SHERLOCK, S. (1968) Diseases of liver and biliary system, 4th ed., Oxford, Edinburgh, Blackwell Scientific Publishers, 250 pp.

SHTENBERG, A.I. & ORLOVA, N.V. (1955) [The question of the etiology of the so-called "Ozhalangar Encephalitis".] Vopr. Pitan., 14: 27-31 (in Russian).

SHULL, L.R., BUCKMASTER, G.W., & CHEEKE, P.R. (1976) Factors influencing pyrrolizidine (Senecio) alkaloid metabolism: species, liver sulphydryls and rumen fermentation. J. anim. Sci., 43: 1247-1253.

SHUMAKER, R.C., ROBERTSON, K.A., HSU, I.C., & ALLEN, J.R. (1976) Neoplastic transformation in tissues of rats exposed to monocrotaline or dehydroretronecine. J. Natl Cancer Inst., 56: 787-789.

SIDDIQI, M.A., SURI, K.A., SURI, M.P., & ATAL, C.K. (1978a) Novel pyrrolizidine alkaloid from Crotalaria nana. Phytochemistry, 17: 2143-2144.

SIDDIQI, M.A., SURI, K.A., SURI, O.P., & ATAL, C.K. (1978b) Genus Crotalaria. Part 34. Cronaburmine, a new pyrrolizidine alkaloid from Crotalaria nana Burm. Indian J. Chem., 16B: 1132-1133.

SMITH, L.W. & CULVENOR, C.C.J. (1981) Plant sources of hepatotoxic pyrrolizidine alkaloids. J. nat. Prod., 44: 129-152.

SMITH, T.E., WEISBACH, H., & UDENFRIEND S. (1962) Studies on the mechanism of monoaminooxidase: metabolism of $\underline{N},\underline{N}$-dimethyltryptamine and $\underline{N},\underline{N}$-dimethyltryptamine-$\underline{N}$-oxide. Biochemistry, 1: 137.

SOBIN, L.H., FETRAT, M.E., & ANWER, M.A. (1969) Hepatic lesions in Afghanistan. Trop. geogr. Med., 21: 27-29.

SRIVASTAVA, R.N., MOHABBAT, O., GHANI, A.R., & ARAM, G.N. (1978) Veno-occlusive disease of the liver. Indian. Paediatr., 15: 143-146.

SRUNGBOONMEE, S. & MASKASAME, C. (1981) A preliminary study on the toxicity of Crotalaria juncea to cattle. Sattawaphaet San, 32: 91-107.

STEIN, H. (1957) Veno-occlusive disease of liver in African children. Br. med. J., 29 June: 1496-1499.

STEIN, H. & ISAACSON, C. (1962) Veno-occlusive disease of the liver. Br. med. J., 10 February: 372-374.

STENMARK, K.R., MORGANROTH, M.L., REMIGIO, L.K., VOELKEL, N.F., MURPHY, R.C., HENSON, P.M., MATHIAS, M.M., & REEVES, J. (1985) Alveolar inflammation and arachidonate metabolism in monocrotaline-induced pulmonary hypertension. Am. J. Physiol., 248: H859-H866.

STILLMAN, A.E., HUXTABLE, R.J., CONSROE, P., KOHNEN, P., & SMITH, S. (1977) Hepatic veno-occlusive disease due to pyrrolizidine poisoning in Arizona. Gastroenterology, 73: 349-352.

STIRLING, G.A., BRAS, G., & URQUHART, A.E. (1962) The early lesions in veno-occlusive disease of the liver. Arch. dis. Child., 37: 535-538.

STOYEL, C. & CLARK, A.M. (1980) The transplacental micronucleus test. Mutat. Res., 74: 393-398.

STUART, K.L. & BRAS, G. (1955) Clinical observations on veno-occlusive disease of the liver in Jamaican adults. Br. med. J., 2: 348-352.

STUART, K.L. & BRAS, G. (1956) Veno-occlusive disease of the liver in Barbados. West Indian med. J., 5: 33-36.

STUART, K.L. & BRAS, G. (1957) Veno-occlusive disease of the liver. Q. J. Med., 26: 291-315.

STYLES, J., ASHBY, J., & MATTOCKS, A.R. (1980) Evaluation in vitro of several pyrrolizidine alkaloid carcinogens: observation on the essential pyrrolic nucleus. Carcinogenesis, 1: 161-164.

SUFFNESS, M. & CORDELL, G.A. (1985) Antitumour alkaloids. In: Brossi, A., ed. The alkaloids, New York, Academic Press, pp. 1-347.

SUGITA, T., HYERS, T.M., DAUBER, I.M., WAGNER, W.W., MCMURTRY, I.F., & REEVES, J.T. (1983a) Lung leak precedes right ventricular hypertrophy in monocrotaline treated rats. J. appl. Physiol., 54: 371-374.

SUGITA, T., STENMARK, K.R., WAGNER, W.W., HENSON, P.M., HENSON, J.E., HYERS, T.M., & REEVES, J.T. (1983b) Abnormal alveolar cells in monocrotaline induced pulmonary hypertension. Exp. lung Res., 5: 201-215.

SUNDARESON, A.E. (1942) An experimental study of placental permeability to cirrhogenic poisons. J. Pathol. Bacteriol., 54: 289-298.

SVÓBODA, D. & REDDY, J.K. (1972) Malignant tumours in rats given lasiocarpine. Cancer Res., 32: 908-912.

SVOBODA, D. & REDDY, J.K. (1974) Lasiocarpine induced, transplantable squamous cell carcinoma of rat skin. J. Natl Cancer Inst., 53: 1415-1418.

SVOBODA, D. & SOGA, J. (1966) Early effects of pyrrolizidine alkaloids on the fine structure of rat liver cells. Am. J. Pathol., 48: 347-373.

SWICK, R.A., CHEEKE, P.R., & BUHLER, D.R. (1979) Factors affecting the toxicity of dietary tansy ragwort to rats. In: Cheeke, P.R., ed. Proceedings of the Symposium on Pyrrolizidine (Senecio) Alkaloids: Toxicity, Metabolism, and Poisonous Plant Control Measures, Corvallis, Oregon, USA, 23-24 February 1979, Oregon, Nutrition Research Institute.

SWICK, R.A., CHEEKE, P.R., GOEGER, D.E., & BUHLER, D.R. (1982a) Effect of dietary Senecio jacobaea and injected Senecio alkaloids and monocrotaline on guinea pigs. J. anim. Sci., 55: 1411-1416.

SWICK, R.A., CHEEKE, P.R., & BUHLER, D.R. (1982b) Subcellular distribution of hepatic copper, zinc and iron and serum ceruloplasmin in rats intoxicated by oral pyrrolizidine (Senecio) alkaloids. J. anim. Sci., 55(6): 1425-1430.

SWICK, R.A., CHEEKE, P.R., PATTON, N.M., & BUHLER, D.R. (1982c) Absorption and excretion of pyrrolizidine (Senecio) alkaloids and their effects on mineral metabolism in rabbits. J. anim. Sci., 55: 1417-1424.

SWICK, R.A., CHEEKE, P.R., MIRANDA, C.L., & BUHLER, D.R. (1982d) The effect of consumption of the pyrrolizidine alkaloid containing plant Senecio jacobaea on iron and copper metabolism in the rat. J. Toxicol. environ. Health, 10: 757-768.

SWICK, R.A., CHEEKE, P.R., RAMSDELL, H.S., & BUHLER, D.R. (1983) Effect of sheep rumen fermentation and methane inhibition on the toxicity of Senecio jacobaea. J. anim. Sci., 56: 645-651.

TAKANASHI, H., UMEDA, M., & HIRONO, I. (1980) Chromosomal aberrations and mutation in cultured mammalian cells induced by pyrrolizidine alkaloids. Mutat. Res., 78: 67-77.

TAKEOKA, O., ANGEVINE, M., & LALICH, J. (1962) Stimulation of mast cells in rat fed various chemicals. Am. J. Pathol., 40: 545-554.

TANDON, B.N., TANDON, H.D., TANDON, R.K., NARENDRANATHAN, M., & JOSHI, Y.K. (1976) Epidemic of veno-occlusive disease in central India. Lancet, 7 August: 271-272.

TANDON, B.N., TANDON, H.D., KOSHY, A., NARENDRANATHAN, M., JOSHI, Y.K., TANDON, R.K., BHARGAVA, S., RAJANI, M., BHATIA, M.L., MANCHANDA, S.C., & KASTURI, T.E. (1977) Epidemiological, clinical, biochemical and haemodynamic study of veno-occlusive disease of the liver due to Crotalaria alkaloids in India. J. All Ind. Inst. Med. Sci., 3: 165-175.

TANDON, B.N., TANDON, H.D., & MATTOCKS, A.R. (1978) Study of an epidemic of veno-occlusive disease in Afghanistan. Indian J. med. Res., 68: 84-90.

TANDON, H.D. & TANDON, B.N. (1975) Epidemic of liver disease - Gulran District, Herat Province, Afghanistan, Alexandria, World Health Organization, Regional Office for the Eastern Mediterranean (Assignment report No. EM/AFG/OCD/001/RB).

TANDON, H.D., TANDON, B.N., TANDON, R., & NAYAK, N.C. (1977) A pathological study of the liver in an epidemic outbreak of veno-occlusive disease. Indian J. med. Res., 65: 679-684.

TANDON, H.D., TANDON, B.N., & MATTOCKS, A.R. (1978) An epidemic of veno-occlusive disease of the liver in Afghanistan. Am. J. Gastroenterol., 72: 607-613.

TANDON, R.K., TANDON, B.N., TANDON, H.D., BHATIA, M.L., BHARGAVA, S., LAL, P., & ARORA, R.R. (1976) Study of an epidemic of veno-occlusive disease in India. Gut, 17: 849-855.

TANNER, M.S. & PORTMANN, B. (1981) Indian childhood cirrhosis. Arch. dis. Child., 56: 4-6.

TAYLOR, S., BELT, R.J., HAAS, C.D., & HOOGSTRATEN, B. (1983) Phase I trial of indicine-N-oxide on two dose schedules. Cancer, 51: 1988-1991.

TEILUM, G. (1949) Endophlebitis hepatica obliterans. Acta pathol. microbiol., 26: 147-166.

TEMIZER, A., ONAR, A.N., SENER, B., TEMIZER, H., & KARAKAYA, A.E. (1985) Determination of alkaloids by differential pulse polarography. I. Senecio alkaloids. J. Pharm. Belg., 40(2): 75-78.

TEREKHOV, G.N. (1939) [Pathomorphology of toxic hepatitis with ascites.] In: Mirochnik, M.F., ed. [Functional, diagnostic and pathological changes of toxic hepatitis with ascites,] Tashkent, State Publishing House of Science Technology and Socio-economic Literature of Uzbekistan, pp. 145-200 (in Russian).

TEREKHOV, G.N. (1952) [The pathomorphology of alimentary toxic dystrophy with ascites.] In: Milenkov, S.M. & Kizhaikin, Y., ed. [Collection of scientific papers on Toxic Hepatitis with Ascites,] Tashkent, Publishing House of the University of Central Asia, pp. 148-164 (in Russian).

THORPE, E. & FORD, E.J.H. (1968) Development of hepatic lesions in calves fed with ragwort (Senecio jacobaea). J. comp. Pathol., 78: 195-205.

TITTEL, G., HINZ, H., & WAGNER, H. (1979) Quantitative determination of pyrrolizidine alkaloids in Symphyti radix by HPLC. Planta Med., 37: 1-8.

TUCHWEBER, B., KOVAKS, K., JAGO, M.V., & BEAULIEU, T. (1974) Effect of steroidal and non-steroidal microsomal enzyme inducers on the hepatotoxicity of pyrrolizidine alkaloids in rats. Res. Commun chem. Pathol. Pharmacol., 7: 459-480.

TUCKER, A., BRYANT, S.E., FROST, H.H., & MIGALLY, N. (1983) Chemical sympathectomy and serotonin inhibition reduce monocrotaline-induced right ventricular hypertrophy in rats. Can. J. Physiol. Pharmacol., 61: 356-362.

TURNER, J.H. & LALICH, J.J. (1965) Experimental cor pulmonale in the rat. Arch. Pathol., 79: 409-418.

VALDIVIA, E., SONNAD, J., HAYASHI, Y., & LALICH, J.J. (1967a) Experimental interstitial pulmonary oedema. Angiology, 18: 378-383.

VALDIVIA, E., LALICH, J., HAYASHI, Y., & SONNAD, J. (1967b) Alterations in pulmonary alveoli after a single injection of monocrotaline. Arch. Pathol., 84: 64-76.

VAN DER WATT, J.J., PURCHASE, I.F.H., & TUSTIN, R.C. (1972) The chronic toxicity of retrorsine, a pyrrolizidine alkaloid, in vervet monkeys. J. Pathol., 107: 279-287.

VILLARROEL, L.H., TORRES, R.G., NAVARRO, J.M., & FAJARDO, V.M. (1985) Senecionine and seneciphylline from Senecio patagonicus. Fitoterapia, 56: 250-251.

VISCONTINI, M. & GILHOF-SCHAUFELBERGER, H. (1971) Synthesis of (±) dehydroheliotridine. Helv. chim. Acta, 54: 449-456.

WAGENVOORT, C.A., DINGEMANS, K.P., & LOTGERING, G.G. (1974a) Electron microscopy of pulmonary vasculature after application of fulvine. Thorax, 29: 511-521.

WAGENVOORT, C.A., WAGENVOORT, N., & DIJK, H.J. (1974b) Effect of fulvine on pulmonary arteries and veins of the rat. Thorax, 29: 522-529.

WAGNER, H., NEIDHARDT, U., & TITTEL, G. (1981) TLC and HPLC analysis of pyrrolizidine N-oxide alkaloids of Symphyti radix. Planta Med., 41: 232-239.

WAKIM, K.G., HARRIS, P.N., & CHEN, K.K. (1946) The effects of senecionine on the monkey. J. Pharmacol. exp. Ther., 87: 38-45.

WALKER, K.H. & KIRKLAND, P.D. (1981) Senecio lautus toxicity in cattle. Aust. vet. J., 57: 1-7.

WATT, J.M. & BREYER-BRANDWIJK, M.G. (1962) Medicinal and poisonous plants of southern and eastern Africa, Edinburgh, London, E. and S. Livingstone, 584 p.

WEHNER, F.C., THIEL, P.G., & VAN RENSBURG, S.J. (1979) Mutagenicity of alkaloids in the Salmonella/microsome system. Mutat. Res., 66: 187-190.

WESTON, C.F.M., COOPER, B.T., DAVIES, J.D., & LEVINE, D.F. (1987) Veno-occlusive disease of the liver secondary to ingestion of comfrey. Br. med. J., 295: 183.

WHITE, I.N.H. (1976) The role of liver glutathion in the acute toxicity of retrorsine to rats. Chem.-biol. Interact., 13: 333-342.

WHITE, I.N.H. (1977) Excretion of pyrrolic metabolites in bile of rats given retrorsine or the bis-n-ethylcarbamate of synthanecine A. Chem.-biol. Interact., 16: 169-180.

WHITE, I.N.H. & MATTOCKS, A.R. (1971) Some factors affecting the conversion of pyrrolizidine alkaloids to N-oxides and to pyrrolic derivatives in vitro. Xenobiotica, 1: 503-505.

WHITE, I.N.H. & MATTOCKS, A.R. (1972) Reactions of dihydropyrrolizidine with deoxyribonucleic acid in vitro. Biochem. J., 128: 291-297.

WHITE, I.N.H., MATTOCKS, A.R., & BUTLER, W.H. (1973) The conversion of the pyrrolizidine alkaloid retrorsine to pyrrolic derivatives in vivo and in vitro and its acute toxicity to various animal species. Chem.-biol. Interact., 6: 207-218.

WHITE, R.D., KRUMPERMAN, P.H., CHEEKE, P.R., & BUHLER, D.R. (1983) An evaluation of acetone extracts from six plants in the Ames mutagenicity test. Toxicol. Lett., 15: 25-31.

WHITE, R.D., SWICK, R.A., CHEEKE, P.R. (1984) Effect of dietary copper and molybdenum on tansy ragwort (Senecio Jacobaea) toxicity in sheep. Am. J. vet. Res., 45: 159-161.

WHO (1980) Inventory of medicinal plants used in the different countries, Geneva, World Health Organization (Document No. DPM/80.3).

WICKRAMANAYAKE, P.P., ARBOGAST, B.L., BUHLER, D.R., DEINZER, M.L., & BURLINGAME, A.L. (1985) Alkylation of nucleosides and nucleotides by dehydroretronecine: characterization of covalent adducts by liquid secondary ion mass spectrometry. J. Am. Chem. Soc., 107: 2485-2488.

WIEDENFELD, H. (1982) Two pyrrolizidine alkaloids from Gynura scandens. Phytochemistry, 21: 2767-2768.

WIEDENFELD, H., PASTEWKA, U., STENGL, P., & ROEDER, E. (1981) On the gas chromatographical determination of the pyrrolizidine alkaloids of some Senecio species. Planta Med., 41: 124-128.

WIEDENFELD, H., RODER, E., & ANDERS, E. (1985) Pyrrolizidine alkaloids from seeds of Crotalaria scassellatii. Phytochemistry, 24: 376-378.

WILLIAMS, A.O., EDINGTON, G.M., & OBAKPONOVWE, P.C. (1967) Hepatocellular carcinoma in infancy and childhood in Ibadan, western Nigeria. Br. J. Cancer, 21(3): 474-482.

WILMOT, F.C. & ROBERTSON, G.W. (1920) Senecio disease or cirrhosis of the liver due to Senecio poisoning. Lancet, 23 October: 848-849.

WITTIG, M. & STEPHEN, C. (1964) Modification of microsomal lipid peroxidation and drug metabolism by cytoplasmic copper. Res. Commun. Chem. Pathol. Pharmacol., 44: 477-493.

WONG, R.Y. & ROITMAN, J.N. (1984) Structure and absolute configuration of (+)-doronine-benzene (1:1). Acta Crystallogr., Sect. C: Cryst. Struc. Commun., C40: 163-166.

WURM, H. (1939) [A cluster of endophlebilis hepatica obliterans in the age group of suckling infants.] Klin. Wochenschr., 18: 1527-1531 (in German).

YAMANAKA, H., NAGAO, M., SUGIMURA, T., FURUYA, T., SHIRAI, A., & MATSUSHIMA, T. (1979) Mutagenicity of pyrrolizidine alkaloids in the Salmonella/mammalian microsome test. Mutat. Res., 68: 211-216.

YULDASHEVA, L.N. & SULTANOVA, R.G. (1983) [Oxidative reactions in rat liver tissue under conditions of chronic heliotrine hepatitis.] Vopr. Med. Khim., 29: 81-85 (in Russian).

YUNUSOV, S.YU. & PLEKHANOVA, N.V. (1959) [The alkaloids of Trichodesma incanum. The structure of incanine and trichesmine.] Zh. Obshch. Khim., 29: 677-684 (in Russian).

ZHELTOVA, L.I. (1952) [The clinical course of toxic hepatitis.] In: Milenkov, S.M. & Kizhaikin, Y., ed. [Collection of scientific papers on Toxic Hepatitis with Ascites,] Tashkent, Publishing House of the University of Central Asia, pp. 76-90 (in Russian).

APPENDIX I

Pyrrolizidine Alkaloids and Their Plant Sources

ISOLATIONS OF TOXIC PYRROLIZIDINE ALKALOIDS - (8701081148)

Alkaloid	Plant Sources	Reference	Reference Location
19-Acetoxysenkirkine	Senecio laricifolius H.B.K.	Bohlmann et al. (1986)	A
6-Acetylanacrotine	Crotalaria agatiflora Schweinf.	Culvenor & Smith (1972)	B281
6-Acetyl-trans-anacrotine	Crotalaria agatiflora Schweinf.	Culvenor & Smith (1972)	B281
Acetylcrotaverrine	Crotalaria vernucosa L. C. walkeri Arnott	O.P. Suri et al. (1976) K.A. Suri et al. (1976)	B339 B340
7-Acetylechinatine	Cynoglossum amabile Stapf. and Drummond Lindelofia spectabilis Lehm. Symphytum asperum Lepech. S. officinale Linn.	Culvenor & Smith, unpubl. Rao et al. (1974) Pedersen (1975b) Pedersen (1975b)	B62 A A
Acetylgynuramine	Gynura scandens O. Hoffm.	Wiedenfeld (1982)	C
Acetylheliosupine	Cynoglossum officinale L. Myosotis sylvatica Hoffm. Symphytum asperum Lepech. S. officinale Linn.	Pedersen (1970); Resch & Meinwald (1982) Resch & Meinwald (1982) Pedersen (1975) Pedersen (1975)	B43, A A A A
Acetylindicine	Heliotropium indicum L.	Mattocks (1967a)	B71
7-Acetylintermedine	Borago officinale L. Symphytum aspera S. officinale Linn. S. x uplandicum Nyman	Luthy et al. (1984) Roitman (1981) Roder et al. (1982) Culvenor et al. (1980a), (1980b)	A A C A, A
Acetyllasiocarpine	Heliotropium europaeum L.	Culvenor et al. (1975)	B69

- 276 -

Compound	Species	Reference	Code
7-Acetyllycopsamine	Amsinckia menziesii (Lehm.) Nels. & Macbr.	Roitman, (1983a)	C
	Anchusa officinalis L.	Pedersen (1975); Broch-Due & Aasen (1980)	A, B32
	Borago officinale L.	Luthy et al. (1984)	A
	Symphytum aspera	Roitman (1981)	A
	S. officinale Linn.	Huizing & Malingre (1981)	C
	S. x uplandicum Nyman	Culvenor et al. (1980a, 1980b)	A, A
3a'-Acetyllycopsamine	Amsinckia menziesii (Lehm.) Nels. & Macbr.	Roitman (1983a	C
7-Acetylmadurensine	Crotalaria agatiflora Schweinf.	Culvenor & Smith (1972)	B81
7-Acetyl-cis-madurensine	Crotalaria agatiflora Schweinf.	Culvenor & Smith (1972)	B81
7-Acetylscorpioidine	Myosotis scorpioides L.	Resch et al. (1982)	C
Acetylseneciphylline	Senecio pterophorus DC.	Edgar et al. (1976)	B246
18-Acetylsenkirkine	Senecio illinitus Phill.	Gonzalez et al. (1986a)	A
	S. kirkii Hook. f. ex Kirk	Briggs et al. (1965)	A
	S. tenuifolius Burm.	Bhakuni & Gupta (1982)	C
Acetylsyneilesine	Syneilesis palmata Maxim.	Hikichi & Furuya (1976)	B79
Amabiline	Borago officinale L.	Luthy et al. (1984)	A
	Cynoglossum amabile Stapf. et Drummond	Culvenor & Smith (1967)	B33
	C. glochidiatum Wall. ex Lindl.	K.A. Suri et al. (1975a)	B36
	Eupatorium cannabinum L.	Luthy et al. (1984)	A
	Lindelofia angustifolia (Schrenk) Brand.	K.A. Suri et al. (1975a)	B36
Anacrotine	Crotalaria agatifolia Schweinf.	Culvenor & Smith (1972)	B281
	C. incana L.	Mattocks (1968)	B302
	C. laburnifolia L.	Snehelata et al. (1966); Sawhney et al. (1967)	B309, 291
	C. laburnifolia L. subsp. eldomae	Crout (1972)	B312
	C. micans Link.	Atal et al. (1966a)	B280
	C. verrucosa L.	Subramanian & Nagarajan (1967)	B338

- 277 -

Name	Species	Reference	Code
Anadoline	Symphytum orientale	Ulubelen & Doganca (1971); Culvenor et al. (1975)	B103, 104
	S. tuberosum L.	Ulubelen & Ocal (1977)	B105
Angelylechimidine (or isomer)	Symphytum asperum Lepech.	Gadella et al. (1983)	A
	S. x uplandicum Nyman	Gadella et al. (1983)	A
7-Angelylheliotridine trachelanthate	Heliotropium supinum L.	Crowley & Culvenor (1959)	B83
7-Angelylheliotridine viridiflorate	Heliotropium supinum L.	Crowley & Culvenor (1959)	B83
7-Angelylheliotrine	Heliotropium digynum Forssk.	Hammouda et al. (1984)	A
	H. eichwaldii Steud ex DC.	O.P. Suri et al. (1975)	B63
7-Angelyl-9-sarracinyl-retronecine	Senecio triangularis Hook.	Rueger & Benn (1983)	C
6-Angelyl-trans-anacrotine	Crotalaria agatiflora Schweinf.	Culvenor & Smith (1972)	B281
Asperumine	Echium vulgare L.	Karimov et al. (1975)	B50
	Symphytum asperum Lepech.	Man'ko et al. (1969); Man'ko & Kotowskii (1970); Man'ko et al. 1970	B95, 96 B97
	S. caucasicum Bieb.	Man'ko et al. (1972); Mel'kumova et al. 1974	B98, 99
Axillaridine	Crotalaria axillaris Ait.	Crout (1968b, 1969)	B287, 288
	C. scassellatii Chiov	Wiedenfeld et al. (1985)	A
Axillarine	Crotalaria axillaris Ait.	Crout (1968b, 1969)	B287, 288
	C. scassellatii Chiov	Wiedenfeld et al. (1985)	A
Bisline	Senecio othonniformis Fourcade	Coucourakis & Gordon-Gray (1970)	B224
	S. petasitis DC.	Gonzalez et al. (1973)	B225
Brachyglottine	Brachyglottis repanda Forst. et Forst.	White, pers. commun.	

Carategine	Lindelofia tschimganica	Akramov et al. (1965)	B87
	Rindera oblongifolia M. Pop.	Akramov et al. (1965)	B87
	Solenanthus karateginus Lipsky	Akramov et al. (1964)	B93
Chlorodeoxysceleratine	Senecio latifolius DC. (S. sceleratus)	Gordon-Gray (1967)	B261
Clivorine	Ligularia brachyphylla Hand. Mazz.	Klasek et al. (1971)	B134
	L. clivorum	Klasek et al. (1967, 1969, 1970); Birnbaum et al. (1971)	B135, 136, 137, 138
	L. dentata (A. Gray) Hara	Klasek et al. (1971)	B134
	L. elegans (Cass.)	Klasek et al. (1971)	B134
Crispatine	Crotalaria candicans W. & A.	Suri et al. (1982)	C
	C. crispata F. Muell. ex Benth.	Culvenor & Smith (1963)	B294
	C. lunata Beddome ex Polhill	Rothschild et al. (1979)	C
	C. madurensis R. Wight	Habib et al. (1971)	B315
Crobarbatine	Crotalaria barbata R. Graham ex R. Wight Walk.-Arn.	Puri et al. (1973)	B289
Cromadurine	Crotalaria madurensis Wight	Rao et al. (1974, 1975b)	B62, 316
Cronaburmine	Crotalaria nana Burm.	Siddiqi et al. (1978b)	A
Crosemperine	Crotalaria aegyptiaca Benth.	Zalkow et al. (1979)	B419
	C. semperflorens Vent.	Atal et al. (1967)	B331
Crotaflorine	Crotalaria agatiflora Schweinf.	Culvenor & Smith (1972)	B281
Crotafoline	Crotalaria laburnifolia L. subsp. eldomae	Crout (1972)	B312
Crotalarine	Crotalaria burhia Buch-Ham.	Ali & Adil (1973); Rao et al. (1975a)	B292, 293
Crotaleschenine	Crotalaria leschenaultii	Suri & Atal (1967); Smith et al. (1988)	B313

Crotananine	Crotalaria nana Burm.	Siddiqi et al. (1978a)	A
Crotastriatine	Crotalaria pallida Ait. (syn. C. mucronata Desv., C. striata DC.)	Gandhi et al. (1968); Batra et al. (1975)	B323, 324
Crotaverrine	Crotalaria verrucosa L.	O.P. Suri et al. (1976)	B339
	C. walkeri Arnott	K.A. Suri et al. (1976)	B340
Cruentine A	Senecio cruentus DC.	Chu & Chu (1964)	B172
Cruentine B	Senecio cruentus DC.	Chu & Chu (1964)	B172
Curassavinine	Heliotropium curassavicum Linn.	Mohanraj et al. (1982a)	C
Cynaustine	Borago officinale L. (or amabiline ?)	Larson et al. (1984)	C
	Cynoglossum australe R. Br.	Culvenor & Smith, 1967	B33
	C. lanceolatum Forsk.	Suri et al. 1975a	B36
Deoxyaxillarine	Crotalaria scassellatii Chiov	Wiedenfeld et al. 1985	A
Diacetyllycopsamine	Amsinckia menziesii (Lehm.) Nels. & Macbr.	Roitman, 1983a	C
Dibenzoylretronecine	Caccinea glauca Savi.	Siddiqi et al. 1978a	B346
Dicrotaline	Crotalaria dura J.M. Wood et Evans	Marais, 1944; Adams & Van Duuren, 1953a	B295, 296
	C. globifera E. Mey.	Marais, 1944; Adams & Van Duuren, 1953a	B295, 296
N-(Dihydropyrrolizino-methyl)heliotrine chloride	Heliotropium europaeum L.	Culvenor & Smith, 1969	B68
15,20-Dihydroxyeruci-foline	Senecio dolichodoryius Cuatr.	Bohlmann et al. 1986	A
Dihydroxytriangularine	Alkanna tinctoria Tausch	Roder et al. 1984b	C

Doronenine	Senecio abrotanifolius ssp. abrotanifolius	Roder et al. 1984a	A
	S. abrotanifolius ssp. abrotanifolius var tiroliensis	Roder et al. 1984a	A
	S. doronicum L.	Roder et al. 1979a, 1980; Kirfel et al. 1980	B176, C, A
Doronine	Doronicum macrophyllum	Alieva et al. 1976	B122
	Senecio abrotanifolius ssp. abrotanifolius	Roder et al. 1984a	A
	S. abrotanifolius ssp. abrotanifolius var tiroliensis	Roder et al. 1984a	A
	S. clevelandii E.L. Greene	Wong & Roitman, 1984	A
	S. othonnae Bieb.	Khalilov et al. 1977	B223
Echimidine	Amsinckia intermedia Fisch & Mey.	Frahn et al. 1980	A
	Echium italicum L.	Culvenor & Smith, unpubl.	B48
	E. plantagineum L.	Culvenor, 1956	A
	Symphytum asperum Lepech.	Roitman, 1981	B98
	S. caucasicum Bieb.	Man'ko et al. 1972	B15, 100
	S. officinale Linn.	Furuya & Araki, 1968; Huizing & Malingre, 1979	B102, 103
	S. orientale L.	Ulubelen & Boganca, 1970, 1971	A
	S. peregrinum Ledeb.	Gadella et al. 1983	B105
	S. tuberosum L.	Ulubelen & Ocal, 1977	A, A
	S. x uplandicum Nyman	Culvenor et al. 1980a, 1980b	
Echinatine	Anchusa arvensis (L.) Bieb.	Pedersen, 1975	A
	Cynoglossum amabile Stapf. and Drummond	Culvenor and Smith, 1967	B33
	C. creticum	Zalkow et al. 1979	B419
	C. officinale L.	Sykulska, 1962; Jerzmanowska & Sykulska, 1964; Knight et al. 1984	B41, 42
	C. pictum Ait.	Man'ko & Marchenko, 1971b	C
	Eupatorium maculatum L.	Tsuda & Marion, 1963	B44
	E. cannabinum	L.Pedersen, 1975a	B129
	E. purpureum	Mills, 1967	B128
	E. rugosum	Dreifuss, 1984	B130
	Heliotropium indicum L.	Hoque et al. 1976	A, B72

- 281 -

	H. suaveolens Bieb.	Guner, 1986	A
	H. supinum L.	Crowley & Culvenor, 1959	B83
	Lappula glochidiata	Suri et al. 1978	B84
	Lindelofia angustifolia (Schrenk) Brand.	Rao et al. 1974; Suri et al. 1975a	B62, 36
	L. spectabilis Lehm.	Rao et al. 1974; Suri et al. 1975a	B62, 36
	L. stylosa (Kar. et Kir.) Brand.	Kiyamitdinova et al. 1967	B75
	L. tschimganica	Akramov et al. 1965	B87
	Paracynoglossum imeritinum (Kusn.) M. Pop.	Man'ko & Marchenko, 1971b	B88
	Prestonia sp.	Edgar, 1985	A
	Rindera austroechinata M. Pop.	Akramov et al. 1965	B87
	R. baldshuanica Kusnezov	Akramov et al. 1965	B87
	R. cyclodonta	Akramov et al. 1967a	B59
	R. echinata Regel	Men'shikov & Denisova, 1953	B92
	R. oblongifolia M. Pop.	Akramov et al. 1965	B87
	Solenanthus circinnatus Ledeb.	Akramov et al. 1964	B93
	S. coronatus	Kiyamitdinova et al. 1967	B75
	S. karateginus Lipsky	Akramov et al. 1964	B93
	Symphytum asperum Lepech.	Man'ko et al. 1970b	B97
	S. caucasicum Bieb.	Man'ko et al. 1972	B98
	S. officinale Linn.	Man'ko et al. 1970a; Huizing & Malingre, 1979	B101, C
Echiumine	Amsinckia hispida (Ruiz et Pav.) I.M. Johnston	Culvenor & Smith, 1966a	B30
	A. intermedia Fisch et C. Mey.	Culvenor & Smith, 1966a	B30
	A. lycopsoides Lehm.	Culvenor & Smith, 1966a	B30
	Echium plantagineum L.	Culvenor, 1956	B48
Emiline	Emilia flammea Cass.	Kohlmunzer & Tomczyk, 1969; Tomczyk & Kohlmunzer, 1971; Kohlmunzer et al. 1971	B123, 124 B125
13,19-Epoxyseneciphylline	Senecio megaphyllus Green.	Bohlmann et al. 1986	A
	S. usgorensis Cuatr.	Bohlmann et al. 1986	A
13,19-Epoxyspartiodine	Senecio megaphyllus Green.	Bohlmann et al. 1986	A

Erucifoline	Senecio aegypticus L.	Klasek et al. 1968a	B145
	S. erraticus Berthol. subsp. barbaraeifolius Krock.	Schroter & Santavy, 1960; Sedmera et al. 1972	B179, 180
	S. erucifolius L.	Kompis & Santavy, 1962; Sedmera et al. 1972	B183, 180
Europine	Heliotropium arbainense	Zalkow et al. 1979	B419
	H. digynum Forssk.	Hammouda et al. 1984	A
	H. europaeum L.	Culvenor, 1954	B66
	H. lasiocarpum Fisch and Mey.	Culvenor et al. 1986	A
	H. maris-mortui	Zalkow et al. 1978	B74
	H. rotundifolium	Zalkow et al. 1978	B74
	H. suaveolens Bieb.	Guner, 1986	A
	Trichodesma africana	Zalkow et al. 1979	B419
Floricaline	Cacalia floridana	Cava et al. 1968	B119
Floridanine	Cacalia floridana	Cava et al. 1968	B119
	Doronicum macrophyllum	Alieva et al. 1976	B122
	Senecio aureus L.	Roder et al. 1983	C
	S. erraticus Berthol.	Gaiduk et al. 1974	B177
	S. othonnae Bieb.	Khalilov & Telezhenetskaya, 1973b	B222
Florosenine	Cacalia floridana	Cava et al. 1968	B119
	Senecio aureus L.	Roder et al. 1983	C
	S. fluviatilis Wallr.	Klasek et al. 1973b	B185
	S. quebradensis Greenm.	Bohlmann et al. 1986	A
Fulvine	Crotalaria berteroana DC. (C. fulva Roxb.)	Schoental, 1963	B297
	C. crispata F. Muell. ex Benth.	Culvenor & Smith, 1963	B294
	C. madurensis R. Wight	Atal et al. 1966a; Habib et al. 1971	B280, 315
	C. paniculata Willd.	Subramanium et al. 1968	B325
Globiferine	Crotalaria globifera E. Mey.	Brown et al. 1984	C
Grahamine	Crotalaria grahamiana R. Wight et Walk.-Arn.	Atal et al. 1969	B299

Grantaline	Crotalaria globifera E. Mey C. virgulata subsp. grantiana (Harv.) Polhill (C. grantiana Harvey)	Brown et al., 1984 Smith & Culvenor, 1984	C C
Grantianine	Crotalaria globifera E. Mey. C. virgulata subsp. grantiana (Harv.) Polhill (C. grantiana Harvey)	Brown et al. 1984 Adams et al. 1942b; Adams & Gianturco, 1956b; Smith & Culvenor, 1984	C B301 B259, C
Gynuramine	Gynura scandens O. Hoffm.	Wiedenfeld, 1982	A
Heleurine	Heliotropium europaeum L. H. indicum L. H. lasiocarpum Fisch and Mey.	Culvenor, 1954 Hoque et al. 1976 Culvenor et al. 1986	B66 B72 A
Heliosupine	Cynoglossum creticum C. officinale C. pictum C. viridiflorum Pallas ex Lehm. Echium vulgare L. Heliotropium supinum L.	Zalkow et al. 1979 Man'ko & Borisyuk, 1957; Man'ko, 1959; Sykulska, 1961 Man'ko & Marchenko, 1971b, 1972b Man'ko, 1972 Man'ko, 1964 Denisova et al. 1953; Crowley & Culvenor, 1959	B419 B38, 39 B40 B44, 45 B34 B49 B82, 83
	Myosotis sylvatica Hoffm. Paracynoglossum imeritinum (Kusn.) Symphytum asperum Lepech. S. officinale Linn.	Culvenor & Smith, unpubl. Man'ko & Marchenko, 1971 Man'ko et al. 1970b Man'ko et al. 1970a	B88 B97 B101
Heliotrine	Heliotropium acutiflorum H. arbainense H. arguzioides Kar. et Kir. H. curassavicum Linn. H. dasycarpum Ledeb. H. digynum Forssk. H. eichwaldi Steud. ex DC. H. europaeum L. H. indicum L.	Akramov et al. 1968 Zalkow et al. 1979 Zolotavina, 1963 Rajagopalan & Batra, 1977b Akramov et al. 1961a Hammouda et al. 1984 Gandhi et al. 1966a Trautner & Neufeld, 1949; Culvenor et al. 1954 Hoque et al. 1976	B51 B419 B53 B56 B54 A B60 B64, 65 B72

	H. lasicarpum Fisch et C. Mey.	Men'shikov, 1932	B73
	H. olgae	Kiyamitdinova et al. 1967; Sheveleva et al. 1969	B75, 76
	H. popovii H. Riedl. subsp. gillianum	Mohabbet et al. 1976	B77
	H. ramosissimum	Habib, 1975; Schoental & Cavanagh, 1972	B79, 78
	H. suaveolens Bieb.	Guner, 1986	A
	H. supinum L.	Pandey et al. 1983	C
	H. transoxanum	Akramov et al. 1968	B51
Heliovinine	Heliotropium curassavicum Linn.	Mohanraj et al. 1982c	C
Heterophylline	Parsonsia heterophylla A. Cunn.	Edgar et al. 1980	B418
	P. spiralis	Edgar et al. 1980	B418
18-Hydroxysenkirkine	Crotalaria laburnifolia L. subsp.	Crout, 1972	B312
19-Hydroxysenkirkine	Senecio laricifolius H.B.K.	Bohlmann et al. 1986	A
Incanine	Heliotropium olgae	Sheveleva et al. 1969	B76
	Trichodesma incanum Alph. DC.	Yunusov & Plekhanova, 1953, 1957, 1959; Tashkhodzhaev et al. 1979a	B110, 111 B112, C
Indicine	Heliotropium amplexicaule Vahl.	Ketterer et al. 1987, (in press)	B70
	H. indicum L.	Mattocks et al. 1961	A
	Prestonia sp.	Edgar, 1985	
Integerrimine	Cacalia hastata L. subsp. orientalis Kitamura	Hayashi et al. 1972	B120
	Crotalaria brevideris Benth. var. intermedia (Kotschy) Polhill	Suri et al. 1975b	B304
	C. brevifolia	Sawhney & Atal, 1966	B290
	C. incana L.	Adams & Van Duuren, 1953b	B187
	C. pallida Ait.	Sawhney et al. 1967	B291
	C. tetragona Roxb.	Puri et al. 1974	B314
	C. zanzibarica Benth. (C. usaramoensis)	Culvenor & Smith, 1966b	B337
	Petasites hybridus L.	Luthy et al. 1983	C

Senecio alpinus L.	Luthy et al. 1981	A
S. antiephorbium (L.) Sch. Bip.	Rodriguez & Gonzalez, 1969	B152
S. brasiliensis DC.	Motidome & Ferreira, 1966a	B163
S. brasiliensis Less. var tripartitus	Nardi et al. 1980	A
S. durieui Gay	Panizo & Rodriguez, 1974	B157
S. erraticus Berthol. subsp. barbaraeifolius Krock.	Santavy et al. 1962	B182
S. faberi Hemsl.	Wei et al. 1982	C
S. formosus	Munoz Quevedos, 1976	B186
S. glandulosus Don ex Hook. et Arn.	Pestchanker et al. 1985b	A
S. inaequidens DC.	Bicchi et al. 1985	A
S. incanus L. subsp. carniolicus (Willd.) Br.-Bl.	Klasek et al. 1968b	B147
S. integerrimus	Manske, 1939a; Roitman et al. 1979	B156, 420
S. kleinia Sch. Bip.	Gonzalez & Calero, 1958b	B205
S. leucostachys Baker	Pestchanker & Giordano, 1986	A
S. magnificus F. Muell.	Gellert & Mate, 1964	B216
S. morrisonensis Hayata	Lu, Sheng-Teh, 1972	B217
S. nebrodensis L. var sicula	Plescia et al. 1976	B218
S. ragonesei Cabr.	Pestchanker & Giordano, 1986	A
S. spathulatus A. Rich.	White, 1969	B158
S. squalidus L.	Kropman & Warren, 1950; Gonzalez & Calero, 1958a	B265, 205
S. tenuifolius Burm.	Bhakuni & Gupta, 1982	C
S. triangularis Hook.	Roitman, 1983b	C
S. vernalis Walst. et Kit.	Sener et al. 1986; Hartmann & Zimmer, 1986	A, A
S. viscosus L.	Barger & Blackie, 1936; Santavy et al. 1962	B264, 182
S. vulgaris L.	Pieters & Vlietinck, 1986	A
Intermedine		
Amsinckia hispida (Ruiz et Pav.) I.M. Johnston	Culvenor & Smith, 1966a	B30
A. intermedia Fisch et C. Mey.	Culvenor & Smith, 1966a	B30
A. lycopsoides Lehm.	Culvenor & Smith, 1966a	B30
A. menziesii (Lehm.) Nels. & Macbr.	Roitman, 1983	C
Borago officinale L.	Luthy et al. 1984	A

Group	Species	Reference		
Isocromadurine	Conoclinium coelestinium (L.) DC	Herz et al. 1981	C	
	Eupatorium compositifolium Walt.	Herz et al. 1981	C	
	Symphytum aspera	Roitman, 1981	A	
	S. officinale Linn.	Roder et al. 1982	C	
	S. x uplandicum Nyman	Culvenor et al. 1980a, 1980b	A,	A
	Trichodesma africana	Zalkow et al. 1979		B419
Isoline	Crotalaria candicans W. and A.	Suri et al. 1982	C	
	C. madurensis R. Wight	Rao et al. 1975c		B317
	Senecio othonniformis Fourcade	Coucourakis & Gordon-Gray, 1970;		B224
		Coucourakis et al. 1972		B225
Jacobine	Crassocephalum crepidioides	Asada et al. 1985	A	
	Senecio alpinus L.	Luthy et al. 1981	A	
	S. brasiliensis DC.	Adams & Gianturco, 1956e		B148
	S. cineraria DC.	Barger & Blackie, 1937		B167
	S. jacobaea L.	Manske, 1931; Blackie, 1937;		B194, 153
		Bradbury & Culvenor, 1954		B198
		Blackie, 1937; Dorosh & Alekseev, 1960		B153, 227
	S. paludosus L.	Alekseev, 1961b		B228
Jacoline	Crassocephalum crepidioides	Asada et al. 1985	A	
	Senecio alpinus L.	Luthy et al. 1981	A	
	S. jacobaea L.	Bradbury & Culvenor, 1954		B198
Jaconine	Senecio alpinus L.	Luthy et al. 1981	A	
	S. jacobaea L.	Bradbury & Culvenor, 1954		B198
Jacozine	Senecio alpinus L.	Luthy et al. 1981	A	
	S. cannabifolius Less	Asada et al. 1982	C	
	S. jacobaea L.	Bradbury & Culvenor, 1954; Culvenor, 1964		B198, 202

Junceine	Crotalaria juncea L.	Adams & Gianturco, 1956a, 1956b, 1956c	B305, B306, B307
	C. wightiana Grah. ex Wight & Arn.	Atal et al. 1966b	B329
Lasiocarpine	Heliotropium arbainense	Zalkow et al. 1979	B419
	H. arborescens L.	Carcamo-Marquez, 1961	B52
	H. curassavicum Linn.	Rajagopalan & Batra, 1977b	B56
	H. digynum Forssk.	Hammouda et al. 1984	A
	H. eichwaldii Steud. ex DC.	Rao et al. 1974	B62
	H. europaeum L.	Culvenor et al. 1954	B65
	H. indicum Linn.	Hoque et al. 1976	B72
	H. lasiocarpum Fisch. et C. Mey.	Men'shikov, 1932	B73
	H. maris mortui	Zalkow et al. 1979	B419
	H. suaveolens Bieb.	Gurer, 1986	A
	H. supinum L.	Pandey et al. 1983	C
	Lappula intermedia	Man'ko & Vasil'kov, 1968	B85
	Symphytum caucasicum	Man'ko et al. 1969	B95
	S. officinale Linn.	Man'ko et al. 1969, 1970a	B95, 101
Latifoline	Cynoglossum latifolium R. Br.	Crowley & Culvenor, 1962	B37
	Hackelia floribunda	Hagglund et al. 1985	A
Ligudentine	Ligularia brachyphylla Hand.-Mazz.	Klasek et al. 1971	B134
	L. dentata (A. Gray) Hara.	Klasek et al. 1971	B134
Ligularidine	Ligularia dentata (A. Gray) Hara	Hikichi et al. 1979 , Asada & Furuya, 1984a	B436, C
Ligularine	Ligularia brachyphylla Hand.-Mazz.	Klasek et al. 1971	B134
	L. dentata (A. Gray) Hara.	Klasek et al. 1971	B134
	L. elegans Cass.	Klasek et al. 1971	B134
Ligularizine	Ligularia dentata (A. Gray) Hara.	Asada & Furuya, 1984a	C

Lycopsamine	Amsinckia hispida (Ruiz et Pav.) I.M. Johnston	Culvenor & Smith, 1966a	B30
	A. intermedia Fisch. et C. Mey.	Culvenor & Smith, 1966a	B30
	A. lycopsoides Lehm.	Culvenor & Smith, 1966a	B30
	A. menziesii (Lehm.) Nels. & Macbr.	Roitman, 1983a	C
	Anchusa officinalis L.	Broch-Due & Aasen,1980	B32
	Borago officinale L.	Larson et al. 1984; Luthy et al. 1984	C, A
	Eupatorium compositifolium Walt.	Herz et al. 1981	C
	Heliotropium steudneri Vatke	Schneider et al. 1975	B80
	Messerschmidia sibirica	Hikichi et al. 1980	C
	Parsonsia eucalyptophylla F. Muell.	Edgar & Culvenor, 1975	B28
	P. straminea (R. Br.) F. Muell.	Edgar & Culvenor, 1975	B28
	Prestonia sp.	Edgar, 1985	A
	Symphytum asperum Lepech.	Roitman,1981	A
	Symphytum officinale Linn.	Huizing & Malingre,1981	C
	Symphytum x uplandicum Nyman	Culvenor et al. 1980a, 1980b	A, A
Madurensine	Crotalaria agatiflora Schweinf.	Atal et al. 1966a	B280
	C. laburnifolia L. subsp. eldomae	Crout, 1972	B312
	C. madurensis R. Wight	Atal et al. 1966a; Mahran et al. 1979	B280, 426
Merenskine	Senecio latifolius DC.	Bredenkamp et al. 1985	C
Monocrotaline	Crotalaria aegyptiaca Benth.	Mahran et al. 1979; Zalkow et al. 1979	B426, 419
	C. assamica Benth.	Crotalaria Research Group, 1974	B286
	C. burhia	Rao et al. 1975	B293
	C. cephalotes Steud. ex A. Rich	Pilbeam et al. 1983	C
	C. crispata F. Muell. ex Benth.	Culvenor & Smith, 1963	B294
	C. cunninghamii R. Br.	Pilbeam et al. 1983	C
	C. grahamiana R. Wight ex Walk.-Am.	Gandhi et al. 1966b; Atal et al. 1969	B298, 299
	C. leschenaultii DC.	Suri & Atal, 1967	B313
	C. leiloba Bartl.	Puri et al. 1974	B314
	C. mitchellii Benth.	Culvenor et al. 1967b	B318
	C. mysorensis Roth.	Sawhney & Atal, 1968	B319
	C. nitens Kunth.	Hoet et al. 1981	A

Group	Species	Reference	Code
	C. novae-hollandiae DC. subsp. lasiophylla (Benth.) A. Lee	Culvenor et al. 1967b	B318
	C. paulina Schrank.	Pilbeam et al. 1983	C
	C. quinquefolia L.	Pilbeam et al. 1983	C
	C. recta Steud. ex A. Rich	Crout, 1968a	B326
	C. retusa L.	Adams & Rogers, 1939; Culvenor & Smith, 1957a	B327, 328
	C. sagittalis L.	Willette & Cammarata, 1972	B330
	C. spectabilis Roth.	Neal et al. 1935; Adams & Rogers, 1939	B325, 327
	C. stipularia Desv.	Puri et al. 1974	B314
	Lindelofia spectabilis Lehm.	Rao et al. 1974	B62
Monocrotalinine	Crotalaria grahamiana Wight et Arn.	Rajagopalan & Batra, 1977a	B300
Myoscorpine	Myosotis scorpioides L.	Resch et al. 1982	C
	Symphytum officinale Linn.	Resch et al. 1982C	C
Neoligularidine	Ligularia dentata (A. Gray) Hara.	Asada & Furuya, 1984a	C
Neopetasitenine	Ligularia japonica	Asada et al. 1981	A
	Petasites japonicus Maxim.	Yamada et al. 1976a	B19
Neosenkirkine	Senecio auricola Bourg.	Panizo & Rodriguez, 1974	B157
	S. grandifolius Less.	Bohlmann et al. 1986	A
	S. pierotii	Asada & Furuya, 1982	C
Neotriangularine	Senecio triangularis Hook.	Roitman, 1983b	C
Nilgirine	Crotalaria pallida Ait.	Atal et al. 1968	B322
Onetine	Senecio othonnae Bieb.	Danilova et al. 1962	B221
Otosenine	Cacalia floridana	Cava et al. 1968	B119
	Doronicum macrophyllum	Alieva et al. 1976	B122
	D. pardalianches Linn.	Rajagopalan & Negi, 1985	A
	Emilia flammea Cass.	Kohlmunzer & Tomczyk, 1969	B123

	Senecio aegyptiacus L.	Klasek et al. 1968a	B145
	S. aureus L.	Resch et al. 1983, Roder et al. 1983	C, C
	S. cineraria DC.	Habib, 1974	B170
	S. desfontainei Druce	Haddad et al. 1963; Klasek et al. 1968a	B173, 145
	S. erraticus Berthol.	Gaiduk et al. 1974	B177
	S. erraticus subsp. barbaraeifolius Krock.	Santavy, 1958; Schroter & Santavy, 1960	B178, 179
	S. fluviatilis Wallr.	Klasek et al. 1973b	B185
	S. jacobaea L.	Akramov et al. 1968	B51
	S. othonnae Bieb.	Zhdanovich & Men'shikov, 1941; Danilova et al. 1962	B220 B221
	S. renardii Winkl.	Danilova & Konovalova, 1950	B249
	S. tomentosus	Adams et al. 1956; Schroter & Santavy, 1960	B271, 179
Parsonsine	Parsonsia heterophylla A. Cunn.	Bygers & Gainsford, 1979; Edgar et al. 1980	B417, 418
	P. spiralis	Edgar et al. 1980	B418
Petasinine	Petasites japonicus Maxim.	Yamada et al. 1978b	B141
Petasitenine	Farfugium japonicum Kitam	Niwa et al. 1985	A
	Petasites japonicus Maxim.	Yamada et al. 1976a, 1976b; Furuya et al. 1976	B19,139 B20
Retroisosenine	Senecio nemorensis L. var. bulgaricus (Vel.) Stoj. et Stef.	Nghia et al. 1976	B219
	S. nemorensis L. var subdecurrens	Klasek et al. 1980a	B422
Retrorsine	Crotalaria spartioides DC.	Bruennerhoff & de Waal, 1961	B334
	C. zanzibarica Benth. (C. usaramoensis)	Culvenor & Smith, 1966b	B337
	Senecio ambrosioides	Adams & Gianturco, 1956e	B148
	S. ampullaceus Hook.	Adams & Govindachari, 1949b; Adams & Looker, 1951; Warren et al. 1950	B149, 151 B150
	S. bipinnatisectus Belcher	White, 1969	B158
	S. brasiliensis DC.	Motidome & Ferreira, 1966a	B163

S. bupleuroides DC.	Sapiro, 1949	B164
S. cineraria DC.	Klasek et al. 1975	B171
S. cruentus DC.	Asada et al. 1982	C
S. cymbaroides	Roitman et al. 1979	B420
S. desfontainei Druce	Rizk et al. 1983	A
S. discolor DC.	Schoental, 1960; Hennig, 1961	B174, 175
S. douglasii DC.	Adams & Govindachari, 1949b; Adams & Looker, 1951	B149, 151
S. eremophilus Richards	Adams & Govindachari, 1949b; Adams & Looker, 1951	B149, 151
S. erucifolius L.	Ferry & Brazier, 1976	B184
S. filaginoides (H. et A.) DC.	Pestchanker & Giordano, 1986	A
S. formosus	Munoz Quevedo, 1976	B186
S. gilliesiano	Guidugli et al. 1986	A
S. glaberrimus DC.	Blackie, 1937	B153
S. glandulosus Don ex Hook. et Arn.	Pestchanker et al. 1985b	A
S. graminifolius N.J. Jacq.	de Waal, 1941	B188
S. griesbachii	Motidome & Ferreira, 1966b	B190
S. ilicifolius Thunb.	de Waal, 1940a, 1940b, 1941; Culvenor & Smith, 1954	B191, 192 B188, 127
S. inaequidens DC.	Roder et al. 1981	A
S. isatideus DC.	Blackie, 1937; de Waal, 1939	B153, 193
S. jacobaea L.	Ferry & Brazier, 1976	B184
S. latifolius DC.	Watt, 1909; Barger et al. 1935	B211, 212
S. longilobus Benth.	Adams & Govindachari, 1949b; Adams & Looker, 1951; Warren et al. 1950	B149, 151 B150
S. paucicalyculatus Klatt	Pretorius, 1949	B230
S. phillipicus Roegel et Koern	Gonzalez et al. 1986a	A
S. pterophorus DC.	de Waal, 1940b, 1941; Culvenor & Smith, 1954	B192, 188 127
S. quadridentatus Labill.	Culvenor & Smith, 1955	B247

S. ragonesei Cabr.	Pestchanker & Giordano, 1986	A, B194, 193
S. retrorsus DC.	Manske, 1931; de Waal, 1939	
S. riddellii Torr. et A. Gray	Roitman et al. 1979	B420
S. riddellii Torr. et A. Gray var. parksii Cory.	Adams & Govindachari, 1949b	B149
S. ruderalis Harvey	Leisegarg, 1950	B254
S. seratophiloides Griseb.	Pestchanker & Giordano, 1986	A
S. spartioides	Roitman et al. 1979	B420
S. subulatus Don ex Hook. et Arn var erectus	Pestchanker et al. 1985b	A
S. swaziensis Compton	Gordon-Gray et al. 1972; Gordon-Gray & Wells, 1974	B268, 270
S. triangularis Hook.	Roitman, 1983b	C
S. uspallatensis	Pestchanker et al. 1985a	A
S. venosus Harvey	Blackie, 1937	B153
S. vernalis Waldst. et Kit.	Roder et al. 1979b	C
S. viminalis Bremek.	de Waal & van Twisk, 1964	A
S. vulgaris L.	Tschu Shun et al. 1960	B277
S. werneriaefolius	Roitman et al. 1979	B420

Retusamine

Crotalaria mitchellii Benth.	Culvenor et al. 1967b	B318
C. mitchellii Benth. subsp. laevis A. Lee	Culvenor et al. 1967b	B318
C. novae-hollandiae DC. subsp. lasiophylla Benth. A. Lee	Culvenor et al. 1967b	B318
C. novae-hollandiae DC. subsp. novae-hollandiae	Culvenor et al. 1967b	B318
C. retusa L.	Culvenor & Smith, 1957a; Wunderlich, 1962	B328, C

Riddelliine

Crotalaria juncea L.	Adams & Gianturco, 1956a	B305
Senecio aegypticus L.	Klasek et al. 1968a	B145
S. ambrosioides	Roitman et al. 1979	B420
S. cruentus DC.	Asada et al. 1982	C
S. cymbaroides	Roitman et al. 1979	B420
S. desfontainei Druce	Haddad et al. 1963; Klasek et al. 1968a	B173, 145

- 293 -

	S. douglassii DC.	Adams & Govindachari, 1949b; Adams & Looker, 1951	B149, 151
	S. eremophilus Richards	Adams & Govindachari, 1949b; Adams & Looker, 1951	B149, 151
	S. longilobus Benth.	Adams & Govindachari, 1949b; Adams & Looker, 1951; Warren et al. 1950	B149, 151 B150
	S. riddellii Torr. et A. Gray	Manske, 1939a; Adams et al. 1942c	B156, 252
	S. riddellii Torr. et A. Gray var. parksii (Cory)	Adams & Govindachari, 1949b	B149
	S. spartioides	Roitman et al. 1979	B420
	S. vernalis Walst. et Kit.	Serer et al. 1986	A
	S. vulgaris	Roitman et al. 1979	B420
Rinderine	Eupatorium altissimum L.	Herz et al. 1981	C
	E. serotinum Michx.	Locock et al. 1966	B131
	Prestonia sp.	Edgar, 1985	A
	Rindera baldshuanica Kusnezov	Akramov et al. 1961c	B91
	Solenanthus turkestanicus Regel et Smirnov (Kusnezov)	Akramov et al. 1962	B94
Sceleratine	Senecio latifolius DC. (S. sceleratus)	de Waal & Pretorius, 1941; de Waal et al. 1963	B257, 260
Scorpioidine	Myosotis scorpioides L.	Resch et al. 1982	C
Sencalenine	Senecio cacaliaster (Lam.)	Roder et al. 1984b	A
Senecicannabine	Senecio cannabifolius Less.	Asada et al. 1982	C
Senecionine	Brachyglottis repanda Forst. et Forst.	Mortimer & White, 1967	B117
	Caltha biflora	Stermitz & Adamovics, 1977	B341
	C. leptosepala	Stermitz & Adamovics, 1977	B341
	Castilleja rhexifolia Rydb.	Stermitz & Suess, 1978	B342
	Castilleja "rhexifolia aff. miniata"	Roby & Stermitz, 1984	C

Crotalaria juncea L.	Adams & Gianturco, 1956a	B305
C. micans Link.	Sethi & Atal, 1964	B284
C. zanzibarica Benth. (C. usaramoensis)	Culvenor & Smith, 1966b	B337
Emilia sonchifolia DC.	Culvenor, unpubl.	
Erechtites hieracifolia (L.) Raf. ex DC.	Manske, 1939b; Culvenor & Smith, 1954	B126, 127
Gynura segetum (Lour.) Merr.	Hua et al. 1983; Liang & Roder, 1984	C, C
Ligularia japonica	Asada et al. 1981	A
Petasites hybridus L.	Luthy et al. 1983	C
P. laevigatus (Willd.) Reichenb.	Massagetov & Kuzovkov, 1953	B142
Senecio aegypticus L.	Klasek et al. 1968a; Gharbo & Habib, 1969	B145, 146
S. alpinus (L.) Scop.	Luthy et al. 1981	A
S. ambrosioides	Adams & Gianturco, 1956e	B148
S. ampullaceus Hook.	Adams & Govindachari, 1949b; Adams & Looker, 1951;	B149, 151
	Warren et al. 1950	B150
S. argentino Baker (vira-vira Hieron)	Pestchanker & Giordano, 1986	A
S. aureus L.	Manske, 1936, 1939a	B155, 156
S. brasiliensis DC.	Fonseca, 1951; Novelli & De Varella, 1945; Adams & Gianturco, 1956e	B162, 161
S. carthamoides Greene	Adams & Govindachari, 1949b; Adams & Looker, 1951	B148 B149, 151
S. cineraria DC.	Barger & Blackie, 1937; Adams & Govindachari, 1949a;	B167, 168
	Alekseev et al. 1962a	B169
S. congestus (R.Br.) DC.	Roder et al. 1982	C
S. cruentus DC.	Asada et al. 1982	C
S. cymbaroides	Roitman et al. 1979	B420
S. desfontainei Druce	Klasek et al. 1968a	B145
S. discolor DC.	Schoental, 1960; Hennig, 1961	B174, 175
S. douglasii DC.	Adams & Govindachari, 1949b; Adams & Looker, 1951	B149, 151
S. eremophilus Richards	Adams & Govindachari, 1949b; Adams & Looker, 1951	B149, 151
S. erraticus Berthol.	Gaiduk et al. 1974	B177

- 294 -

s. erraticus Berthol. subsp. barbaraeifolius Krock.	Santavy, 1958; Schroter & Santavy, 1960; Kompis et al. 1960	B178, 179 B181 B183
s. erucifolius L.	Kompis & Santavy, 1962	A
s. filaginoides (H. et A.) DC.	Pestchanker & Giordano, 1986	A
s. fistulosus Poepp. ex Less.	Gonzalez et al. 1986b	B148
s. fremontii Torr. et A. Gray	Adams & Gianturco, 1956e	A
s. gilliesiano	Guidugli et al. 1986	B187
s. glabellus Turcz. DC.	Adams & Van Duuren, 1953b	B191, 192
s. ilicifolius Thunb.	de Waal, 1940a, 1940b, 1941; Culvenor & Smith, 1954	B188, 127
s. illinitus Phill.	Gonzalez et al. 1986a	A
s. inaequidens DC.	Roder et al. 1981	A
s. integerrimus Nutt.	Manske, 1939a	B156
s. jacobaea L.	Bradbury & Mosbauer, 1956	B199
s. laricifolius H.B.K.	Bohlmann et al. 1986	A
s. lautus Forst. f. ex Willd.	Culvenor unpubl.	A
s. leucostachys Baker	Pestchanker & Giordano, 1986	A
s. longiflorus Sch. Bip.	de Waal & van Twisk, 1964	B15
s. magnificus F. Muell.	Culvenor, 1962	A
s. multilobatus	McCoy et al. 1983	A
s. multivenius Benth. in Oerst	Bohlmann et al. 1986	B218
s. nebrodensis L. var. sicula	Plescia et al. 1976	B423
s. nemorensis L. subsp. fuchsii Gmel.	Wiedenfeld & Roder, 1979	B229
s. pampeanus Cabrera	Novelli, 1958	C
s. pancicii Degen var arnautorum (Velen.) Stoj. Stef. et Kit.	Jizba et al. 1982	C
s. pancicii Degen var pancicii	Jizba et al. 1982	C
s. patagonicus Hook. and Arn.	Villarroel et al. 1985	A
s. petasitis DC.	Gharbo & Habib, 1969	B146
s. pimpinellifolius H.B.K.	Bohlmann et al. 1986	A
s. pseudo-arnica Less.	Manske, 1939a	B156
s. pterophorus DC.	de Waal, 1940b, 1941; Culvenor & Smith, 1954	B192, 188 B127
s. quadridentatus Labill.	Culvenor & Smith, 1955	B247

S. sandrasicus	Temizer et al. 1985	A
S. scandens	Batra & Rajagopalan, 1977	B256
S. seratophiloides Griseb.	Pestchanker & Giordano, 1986	A
S. spartioides	Manske, 1939a; Adams & Gianturco, 1957b	B156, 262
S. spathulatus A. Rich	White, 1969	B158
S. squalidus L.	Barger & Blackie, 1936; Kropman & Warren, 1950	B264, 265
S. subalpinus C. Koch	Trivedi & Santavy, 1963	B267
S. subulatus Don ex Hook. et Arn var erectus	Pestchanker et al. 1985b	A
S. temuifolius Burm.	Bhakuni & Gupta, 1982	C
S. tomentosus	Adams et al. 1956	B271
S. triangularis Hook.	Kupchan & Suffness, 1967	B272
S. uintahensis	Roitman et al. 1979	B420
S. vernalis Waldst. et Kit.	Roder et al. 1979b	C
S. viminalis Bremek.	de Waal & van Twisk, 1964	A
S. viscosus L.	Barger & Blackie, 1936; Santavy et al. 1962	B264, 182
S. vulgaris L.	Grandval & Lajoux, 1895; Barger & Blackie, 1936; Konovalova & Orekhov, 1937a; Tschu Shun et al. 1960	B275, 264 B276 B277
S. wernariaefolius	Roitman et al. 1979	B420
Syneilesis palmata Maxim.	Hikichi & Furuya, 1976	B279
Tussilago farfara	Rosberger et al. 1981	C
Senecio caudatus DC.	Bohlmann et al. 1986	A
7-Senecioyl-9-(2-hydroxy-3-acetylbutyrl)retronecine		
Senecio caudatus DC.	Bohlmann et al. 1986	A
7-Senecioyl-9-(2-hydroxymethyl-2,3-dihydroxybutyrylretronecine		
Senecio caudatus DC.	Bohlmann et al. 1986	A
7-Senecioyl-9-(2-methyl-2,3-dihydroxybutyryl)-retronecine		

7-Senecioylretronecine	Senecio cacaliaster (Lam.)	Roder et al. 1984b	C
	S. caudatus DC.	Bohlmann et al. 1986	A
	S. triangularis Hook.	Rueger & Benn, 1983a	C
	S. variabilis Sch. Bip.	Bohlmann et al. 1986	A
9-Senecioylretronecine	Senecio caudatus DC.	Bohlmann et al. 1986	A
	S. variabilis Sch. Bip.	Bohlmann et al. 1986	A
7-Senecioyl-9-sarracinyl-retronecine	Senecio cacaliaster (Lam.)	Roder et al. 1984b	C
	S. caudatus DC.	Bohlmann et al. 1986	A
	S. triangularis Hook.	Rueger & Benn, 1983	C
	S. ungeniensis Thell.	Bohlmann et al. 1986	A
	S. variabilis Sch. Bip.	Bohlmann et al. 1986	A
Seneciphylline	Adenostyles alliarae	Yakhontova et al. 1976	B115
	A. glabra	Wiedenfeld et al. 1984	C
	A. rhombifolius (Willd.) M. Pimen.subsp. platyphylloides	Pimenov et al. 1975	B116
	Crotalaria juncea L.	Adams & Gianturco, 1956a	B305
	Erechtites hieracifolia (L.) Raf. ex DC.	Manske, 1939a; Culvenor & Smith, 1954	B126, 127
	Senecio alpinus (L.) Scop.	Klasek et al. 1968b; Luthy et al. 1981	B147, A
	S. ambrosioides	Adams & Gianturco, 1956e	B148
	S. ampullaceus Hook.	Adams & Govindachari, 1949b; Adams & Looker, 1951; Warren et al. 1950	B149, 151
			B150
	S. aquaticus Hill	Blackie, 1937; Evans & Evans, 1949	B153, 154
	S. borysthenicus	Red'ko, 1956; Alekseev, 1961a	B159, 160
	S. brasiliensis DC.	Fonseca, 1951; Novelli & de Varella, 1945; Adams & Gianturco, 1956e	B162, 161
	S. cannabifolius Less.	Alekseev, 1964; Asada et al. 1982	B148
	S. carthamoides Greene	Adams & Govindachari, 1949b; Adams & Looker, 1951	B165, C
			B149, 151
	S. chrysanthemoides	Wali & Handa, 1964	B166
	S. cineraria DC.	Barger & Blackie, 1937; Adams & Govindachari, 1949a	B167, 168

S. cruentus DC.	Asada et al. 1982	C
S. cymbaroides	Roitman et al. 1979	B420
S. desfontainei Druce	Gharbo & Habib, 1969	B146
S. douglasii DC.	Adams & Govindachari, 1949b; Adams & Looker, 1951	B149, 151
S. eremophilus Richards	Adams & Govindachari, 1949b; Adams & Looker, 1951	B149, 151
S. erraticus Berthol. subsp. barbaraeifolius Krock.	Kompis et al. 1960; Santavy et al. 1962	B181, 182
S. erucifolius L.	Kompis & Santavy, 1962	B183
S. fluviatilis Wallr.	Klasek et al. 1973b	B185
S. fremontii Torr. et A. Gray	Adams & Gianturco, 1956e	B148
S. grandifolia	Glonti, 1958	B189
S. ilicifolius Thunb.	de Waal, 1940a, 1940b, 1941;	B191, 192
S. incanus L. subsp. carniolus (Willd.) Br.-Bl.	Culvenor & Smith, 1954	B188, 127
S. jacobaea L.	Klasek et al. 1968b	B147
S. krylovii	Blackie, 1937; Bradbury & Culvenor, 1954	B153, 198
S. kubensis Grossh.	Sapunova & Ban'kovskii, 1968	B207
S. lampsanoides	Khalilov & Telezhenetskaya, 1973a	B208
S. laricifolius H.B.K.	Khalilov & Damirov, 1974	B210
S. latifolius DC.	Bohlmann et al. 1986	A
S. longiflorus Sch. Bip.	Danilova et al. 1960	B213
S. longilobus Benth.	de Waal & van Twisk, 1964	A
	Adams & Govindachari, 1949b; Adams & Looker, 1951	B149, 151
S. minimus Poir.	White, 1969	B158
S. multivenius Benth. in Oerst	Bohlmann et al. 1986	A
S. othonnae Bieb.	Zhdanovich & Men'shikov, 1941;	B220
S. palmatus Pall.	Danilova et al. 1962	B221
S. paludosus L.	Alekseev, 1960	B226
	Blackie, 1937; Dorosh & Alekseev, 1960; Alekseev, 1961b	B153, 227
S. pancicii Degen var arnautorum (Velen.) Stoj. Stef. et Kit.	Jizba et al. 1982	B228 C

- 299 -

S. panicicii Degen var panicii	Jizba et al. 1982	C
S. patagonicus Hook. and Arn.	Villarroel et al. 1985	A
S. paucifolius S.G. Gmel.	Alekseev & Ban'kovskii, 1965	B231
S. phillipicusKoegel et Koern	Gonzalez et al. 1986a	A
S. platyphylloides Somm. et Lev.	Murav'eva, 1964b, 1965; Dauksha, 1970	B233, 234
		B235
S. platyphyllus (Bieb.) DC.	Orekhov, 1935; Konovalova & Orekhov, 1938;	B236, 237
	Konovalova, 1951	B238
	Chemova & Murav'eva, 1974	B243
S. pojarkovae	Khalilov et al. 1972	B245
S. propinquus Ait.	de Waal, 1940b, 1941; Culvenor &	B192,
S. pterophorus DC.	Smith, 1954	188
		B127
S. quadridentatus Labill.	Culvenor & Smith, 1955	B247
S. racemosus	Khmel, 1961	B248
S. renardii Winkl.	Danilova & Konovalova, 1950	B249
S. rhombifolius (Willd.) Sch. Bip.	Khalilov & Telezhenetskaya, 1973a	B208
S. scandens	Batra & Rajagopalan, 1977	B256
S. spartioides Torr. et A. Gray	Manske, 1939a; Adams & Gianturco,	B156,
	1957b	262
S. spathulatus	Benn et al. 1979	B263
S. stenocephalus Maxim.	Konovalova & Orekhov, 1937b	B266
S. subalpinus C. Koch.	Trivedi & Santavy, 1963; Klasek	B267,
	et al. 1968b	147
S. vernalis Walst. et Kit.	Sener et al. 1986; Hartmann & Zimmer, 1986	A, A
S. vulgaris L.	Barger & Blackie, 1936; Konovalova &	B264,
	Orenkhov, 1937a;	276
	Tschu Shun et al. 1960	B277
Senecivermine		
Senecio inaequidens	Bicchi et al. 1985	A
S. seratophylloides Griseb.	Pestchanker & Giordano, 1986	A
S. vernalis Waldst. et Kit.	Roder et al. 1979b	C

Senkirkine	Brachyglottis repanda Forst. et Forst.	Mortimer & White, 1967	B117
	Crotalaria laburnifolia subsp. eldomae	Crout, 1972	B312
	Farfugium japonicum Kitam.	Furuya et al. 1971	B133
	Petasites albus L.	Luthy et al. 1983	C
	P. hybridus L.	Luthy et al. 1983	C
	P. japonicus Maxim.	Yamada et al. 1978a	B140
	P. laevigatus (Willd.) Reichenb.	Massagetov & Kuzovkov, 1953	B142
	Senecio antieuphorbium (L.) Sch. Bip.	Rodriguez & Gonzalez, 1969	B152
	S. desfontainei Druce	Rizk et al. 1983	A
	S. grandifolius Less.	Bohlmann et al. 1986	A
	S. illinitus Phill.	Gonzalez et al. 1986a	A
	S. jacobaea L.	Akramov et al. 1968	B51
	S. kirkii Hook. f. ex Kirk.	Briggs et al. 1948; Briggs et al. 1965	B203, 204
	S. kleinia Sch. Bip.	Rodriguez et al. 1967	B206
	S. laricifolius H.B.K.	Bohlmann et al. 1986	A
	S. pierotii	Asada & Furuya, 1982	C
	S. procerus L. var. procerus Stoj. Stef. et Kit.	Jovceva et al. 1978	B244
	S. quebradensis Greenm.	Bohlmann et al. 1986	A
	S. renardii Winkl.	Danilova & Konovalova, 1950; Briggs et al. 1965	B249, 204
	S. tenuifolius Burm.	Bhakuni & Gupta, 1982	C
	S. vernalis Waldst. et Kit.	Roder et al. 1979b	C
	S. uintahensis	Roitman et al. 1979	B420
	Tussilago farfara	Culvenor et al. 1976b; Borka & Onshus, 1979; Luthy et al, 1980	A,B425, A
Sincamidine	Amsinckia intermedia Fisch. et C. Mey.	Culvenor & Smith, 1966a	B30
Spartioidine	Senecio spartioides Torr. et A. Gray	Manske, 1939a; Adams & Gianturco, 1957	B156, 262
	S. vulgaris L.	Pieters & Vlietinck, 1986	A
Spectabiline	Crotalaria spectabilis Roth	Culvenor & Smith, 1957b	B336
Spiracine	Parsonsia spiralis Wall.	Edgar et al. 1980	B418
Spiraline	Parsonsia spiralis Wall.	Edgar et al. 1980	B418

Spiranine	Parsonsia spiralis Wall.	Edgar et al. 1980	B418
Supinine	Borago officinale L.	Luthy et al. 1984	A
	Eupatorium cannabinum L.	Pederson, 1975a	B128
	E. serotinum Michx.	Locock et al. 1966	B131
	E. stoechadosmum Hanse	Furuya & Hikichi, 1973	B132
	Heliotropium europeum L.	Culvenor, 1954	B66
	H. indicum L.	Hoque et al. 1976	B72
	H. supinum L.	Men'shikov & Gurevich, 1949; Crowley & Culvenor, 1959	B81, 83
	Tournefortia sarmentosa Lam.	Crowley & Culvenor, 1955	B107
	Trichodesma zeylanicum (Burm. f.) R. Br.	O'Kelly & Sargeant, 1961	B113
Swazine	Senecio barbellatus DC.	Gordon-Gray & Wells, 1974	B270
	S. swaziensis Compton	Gordon-Gray et al. 1972; Laing & Sommerville, 1972	B268, 269
Symlandine	Symphytum asperum Lepech.	Roitman, 1981	A
	S. officinale Linn.	Roder et al. 1982	C
	S. tuberosum L.	Gray et al. 1983	A
	S. x uplandicum Nyman	Culvenor et al. 1980a, 1980b	A, A
Symphytine	Myosotis scorpioides L.	Resch et al. 1982	C
	Symphytum aspera	Roitman, 1981	A
	S. officinale Linn.	Furuya & Araki, 1968; Furuya & Hikichi, 1971	B15, 16
	S. peregrinum Ledeb.	Gadella et al. 1983	A
	S. x uplandicum Nyman	Culvenor et al. 1980a, 1980b	A, A
Syneilesine	Syneilesis palmata Maxim.	Hikichi & Furuya, 1974, 1976	B278, 279
Triangularine	Alkanna tinctoria Tausch	Roder et al. 1984b	C
	Senecio triangularis Hook.	Roitman, 1983b	C
Trichodesmine	Crotalaria globifera E. Mey.	Brown et al. 1984	C
	C. juncea L.	Adams & Gianturco, 1956a	B305
	C. lunata Beddome ex Polhill	Rothschild et al. 1979	C
	C. recta Steud ex A. Rich.	Crout, 1968a	B326
	C. wightiana Grah. ex Wight & Arn.	Atal et al. 1966b	B329
	C. tetragona Roxb.	Puri et al. 1974	B314

	Heliotropium arguzioides Kar. et Kir.	Akramov et al. 1961a	B54
	Trichodesma incanum Alph. DC.	Men'shikov & Rubinstein, 1935;	B108
		Men'shikov, 1936; Yunusov &	B109,
		Plekhanova, 1957, 1959;	B111, 112
		Tashkhodzhaev et al. 1979	C
Uluganine	Ulugbeckia tschimganica (B.Fedtsch.) Zak.	Khasanova et al. 1974	B114
Uplandicine	Symphytum x uplandicum Nyman	Culvenor et al. 1980a, 1980b	A, A
Usaramine	Crotalaria breviderns Benth. var. intermedia (Kotschy) Polhill	Suri et al. 1975b	B304
	C. brevifolia	Sawhney et al. 1967	B291
	C. incana L.	Sawhney & Atal, 1970a	B303
	C. pallida Ait.	Sawhney et al. 1967	B291
	C. zanzibarica Benth. (C. usaramoensis)	Culvenor & Smith, 1966b	B337
	Senecio glandulosus Don ex Hook. et Arn.	Pestchanker et al. 1985b	A
	S. seratophylloides Griseb	Pestchanker & Giordano, 1986	A
	S. vulgaris L.	Pieters & Vlietinck, 1986	A
Uspallatine	Senecio argentino Baker (vira-vira Hieron)	Pestchanker et al. 1985a	A
	S. leucostachys Baker	Pestchanker & Giordano, 1986	A
	S. seratophiloides Griseb	Pestchanker & Giordano, 1986	A
	S. uspallatensis	Pestchanker et al. 1985a	A
Yamataimine	Cacalia yatabei Maxim.	Hikichi et al. 1978	B121

A - References in this publication.
B - References in Smith & Culvenor, J. Nat. Prod., 44, 129-152 (1981), with reference number.
C - References in Mattocks, Chemistry and Toxicology of Pyrrolizidine Alkaloids, 1986.

APPENDIX II

Table 1. Plants containing hepatotoxic pyrrolizidine alkaloids

Plant	Constituent alkaloids	Plant part	Reference[a]
Apocynaceae			
Fernaldia pandurata (syn. Urechites karwinsky Mueller)	loroquin	root	B 29
Parsonsia eucalyptophylla (F. Muell.)	lycopsamine	aeriel	B 28
Parsonsia heterophylla A. Cunn.	parsonsine heterophylline	whole	B 417 B 418
Parsonsia spiralis Wall.	heterophylline parsonsine spiracine spiranine spiraline	leaf	B 418
Parsonsia straminea (R. Br.) F. Muell.	lycopsamine	aerial	B 28
Parsonsia estonia sp.	echinatine		A Edgar (1985)
Boraginaceae			
Alkanna tinctoria Tausch	7-angelylretronecine dihydroxytriangularine triangularine		C Roder et al. (1984b)
Amsinckia hispida (Ruiz et Pav.) M. Johnston	intermedine lycopsamine echiumine	whole	B 30

Appendix II, Table 1 (contd).

Boraginaceae (contd)

Species	Alkaloids	Part	Ref
Amsinckia intermedia Fisch et C. Mey	intermedine lycopsamine echiumine sincamidine echimidine	whole	B 30
Amsinckia lycopsoides Lehm.	intermedine lycopsamine echiumine	whole	B 30
Amsinckia menziesii (Lehm.) Nels and Macbr.	7-acetyllycopsamine 3'-acetyllycopsamine diacetyllycopsamine lycopsamine intermedine	aerial	C Roitman (1983a)
Anchusa arvensis (L.) Bieb.	echinatine (or diastereoisomer)	whole	B 31
Anchusa officinalis L.	7-acetyllycopsamine (or diastereoisomer) lycopsamine	whole	B 31 B 32
Asperugo procumbens L.	supinine (or diastereoisomer) lycopsamine (or diastereoisomer)	whole	B 31
Borago officinalis L.	lycopsamine amabiline supinine intermedine acetylintermedine acetyllycopsamine thesinine	aerial, root seed, flower	C Larson et al. (1984) A Luthy et al. (1984)

Appendix II, Table 1 (contd).

Plant	Constituent alkaloids	Plant part	Reference[a]
Boraginaceae (contd)			
Cynoglossum amabile Stapf & Drummond	amabiline echinatine	whole	B 33, 34
Cynoglossum australe R. Br.	cynaustine cynaustraline heliosupine	whole	B 33
Cynoglossum creticum	heliosupine echinatine	aerial	B 419
Cynoglossum glochidiatum Wall. ex Lindl.	amabiline	whole	B 36
Cynoglossum lanceolatum Forsk	cynaustraline cynaustine	whole	B 36
Cynoglossum latifolium R. Br.	latifoline 7-angelylretronecine	aerial	B 37
Cynoglossum officinale L.	heliosupine echinatine acetylheliosupine 7-angelylheliotridine	aerial root, aerial aerial	B 38, 39, 40 B 41, 42 B 43
Cynoglossum pictum Ait.	heliosupine echinatine pictumine	root, aerial aerial	B 44, 45 B 46
Cynoglossum viridiflorum Pallas ex Lehm.	viridiflorine heliosupine	root	B 47 B 34

Appendix II, Table 1 (contd).

Boraginaceae (contd)			
Echium plantagineum L. (Echium lycopsis L.)	echiumine echimidine	aerial	b 48
Echium vulgare L.	heliosupine asperumine echinatine (or diastereoisomer)	aerial aerial whole	B 49 B 50 B 31
Hackelia floribunda	latifoline		A Hägglund et al. (1985)
Heliotropium acutiflorum	heliotrine	aerial	B 51
Heliotropium arbainense	heliotrine europine lasiocarpine	aerial	B 419
Heliotropium arborescens L. (Heliotropium peruvianum L.)	lasiocarpine	aerial	B 52
Heliotropium arguzioides Kar. et Kir.	heliotrine trichodesmine	aerial aerial, root	B 53 B 54, 55
Heliotropium curassavicum Linn.	heliotrine lasiocarpine angelylheliotridine	whole	B 56
	curassavine heliovicine	aerial	B 57
	trachelanthamidine acetylcurassavine heliocurassavinine heliocurassavine heliocoromandaline heliocurassavicine curassanecine curassavinine coromandalinine heliovinine	aerial	B 58 C Mohanraj et al. (1982)

Appendix II, Table 1 (contd).

Plant	Constituent alkaloids	Plant part	Reference[a]
Boraginaceae (contd)			
Heliotropium dasycarpum Ledeb.	heliotrine	aerial, root	B 54
		seed	B 59
Heliotropium digynum (Heliotropium luteum)	heliotrine lasiocarpine europine angelylheliotrine		A Hammouda et al. (1984)
Heliotropium eichwaldi Steud. ex DC.	heliotrine lasiocarpine 7-angelylheliotrine	whole aerial aerial	B 60, 61 B 62 B 63
Heliotropium europaeum	heliotrine lasiocarpine	whole	B 64, 65
	europine supinine heleurine	whole	B 66, 67
	N-dihydropyrrolizinomethyl- heliotrine chloride	whole	B 68
	acetyllasiocarpine	whole	B 69
Heliotropium indicum L.	indicine acetylindicine	aerial aerial	B 70 B 71
	indicinine echinatine supinine heleurine lasiocarpine	aerial	B 72
Heliotropium lasiocarpum Fisch. et Mey.	heliotrine lasiocarpine europine heleurine	aerial	B 73 A Culvenor et al. (1986)

Appendix II, Table 1 (contd).

Boraginaceae (contd)

Heliotropium maris-mortui	europine	aerial	B 74
	lasiocarpine	aerial	B 419
Heliotropium olgae	heliotrine	aerial, root	B 75
Heliotropium popovii H. Riedl. subsp. gillianum H. Riedl.	heliotrine	seed	B 77
Heliotropium ramosissimum (Lehm.) DC. (syn. Heliotropium persicum L., Heliotropium undulatum)	heliotrine	aerial	B 78, 79
	heleurine		
	supinine		
	lasiocarpine		
Heliotropium rotundifolium	europine	aerial	B 74
Heliotropium steudneri Vatke	lycopsamine	leaf	B 80
Heliotropium suaveolens Bieb.	heliotrine	aerial	A Guner (1986)
	lasiocarpine		
	europine		
	echinatine		
Heliotropium supinum L.	supinine	root	B 81
	heliosupine	root	B 82
	echinatine	whole	B 83
	7-angelylheliotridine		
	7-angelylheliotridine viridiflorate		
	7-angelylheliotridine trachelanthate		
	heliotrine	seed, leaf	C Pandey et al. (1983)
	lasiocarpine		
Heliotropium transoxanum	heliotrine	aerial	B 51
Lappula glochidiata	echinatine	aerial	B 84

Appendix II, Table 1 (contd).

Plant	Constituent alkaloids	Plant part	Reference[a]
Boraginaceae (contd)			
Lappula intermedia	lasiocarpine	aerial	B 85
Lindelofia angustifolia (Schrenk) Brand.	echinatine amabiline	aerial	B 62, 36
Lindelofia spectabilis Lehm.	echinatine 7-acetylechinatine monocrotaline	aerial	B 62, 63
Lindelofia stylosa (Kar. et Kir.) Brand	viridiflorine echinatine lindelofine	aerial seed	B 86 B 75
Lindelofia tschimganica	carategine echinatine viridiflorine	aerial	B 87
Lithospermum officinale L.	acetylechimidinylretronecine (or diastereoisomer)	whole	B 31
Messerschmidia sibirica	lycopsamine angelylretronecine	whole	C Hikichi et al. (1980)
Myosotis scorpioides L. (syn. Myosotis palustris L.)	7-acetylscorpioidine scorpioidine symphytine myoscorpine	aerial	C Resch et al. (1982)
Paracynoglossum imeritinum (Kusn.) M. Pop.	heliosupine echinatine	aerial, root	B 88 B 89, 90
Rindera austroechinata M. Pop.	echinatine	whole, seed	B 87

Appendix II, Table 1 (contd).

Boraginaceae (contd)			
Rindera baldshuanica Kusnezov	rinderine	aerial	B 91
	echinatine	aerial	B 87
	trachelanthamine		
	turkestanine		
Rindera cyclodonta Bge.	echinatine	aerial	B 59
Rindera echinata Regel	echinatine	aerial	B 92
	trachelanthamine	aerial	B 59
Rindera oblongifolia M. Pop.	carategine	aerial	B 87
	echinatine		
	turkestanine		
Solenanthus circinnatus Ledeb.	echinatine	seed, aerial, root	B 93
Solenanthus coronatus	echinatine	aerial	B 75
Solenanthus karateginius Lipsky	carategine	aerial	B 93
	echinatine		
Solenanthus turkestanicus	rinderine	aerial	B 94
	turkestanine		
Symphytum asperum Lepech.	aspermine	aerial, root	B 95, 96
	echinatine	aerial, root	B 97
	acetylheliosupine (or diastereoisomer)	whole	B 31
	7-acetyllycopsamine	leaf, root	C Roitman (1981)
	intermedine		
	symlandine		
	7-acetylintermedine		
	symphytine		
	lycopsamine		
	echimidine		
	angelylechimidine (or diastereoisomer)	root	A Gadella et al. (1983)

Appendix II, Table 1 (contd).

Plant	Constituent alkaloids	Plant part	Reference[a]
Boraginaceae (contd)			
Symphytum caucasicum Bieb.	lasiocarpine	aerial, root	B 95
	aspermine	aerial, root	B 98, 99
	echinatine		
	echimidine		
Symphytum officinale Linn.	symphytine	root	B 15, 16
	echimidine		B 100
	lasiocarpine	aerial, root	B 95
	heliosupine	aerial, root	B 101
	viridiflorine	root	
	echinatine	root	
	acetylechimidine (or diastereoisomer)		
	7-acetyllycopsamine	aerial, root	B 31
	lycopsamine		A Huizing et al. (1981)
	intermedine		
	7-acetylintermedine		
	symlandine		C Roder et al. (1982a)
	myoscorpine	root	C Resch et al. (1982)
Symphytum orientale	anadoline	whole	B 102, 103, 104
	symphytine		
	echimidine		
Symphytum peregrinum Ledeb.	echimidine	root	A Gadella et al. (1983)
	symphytine		
Symphytum tuberosum L.	echimidine	whole	B 105
	anadoline		
	symlandine	root	A Gray et al. (1983)
Symphytum x uplandicum Nyman	lycopsamine	aerial	B 31, 17, 106
	intermedine		

Appendix II, Table 1 (contd).

Boraginaceae (contd)

Symphytum x uplandicum Nyman (contd)	uplandicine 7-acetyllycopsamine 7-acetylintermedine echimidine symphytine symlandine argelylechimidine (or diastereoisomer)	root	A Gadella et al. (1983)
Tournefortia sarmentosa Lam.	supinine	leaf, stem	B 107
Trichodesma africana	intermedine europine	aerial	B 419
Trichodesma incanum Alph. DC.	trichodesmine incanine	aerial seed, aerial, root	B 108, 109 B 110, 111, 112
Trichodesma zeylanicum (Burm. f) R. Br.	supinine	seed	B 113
Ulugbekia tschimganica	uluganine		B 114

Compositae

Adenostyles alliariae	platyphylline seneciphylline	root	B 115
Adenostyles glabra	seneciphylline		C Wiedenfeld et al. (1984)
Adenostyles rhombifolius (Willd.) M. Pimen. ssp. platyphylloides	platyphylline seneciphylline	aerial	B 116

Appendix II, Table 1 (contd).

Plant	Constituent alkaloids	Plant part	Reference[a]
Compositae (contd)			
Brachyglottis repanda Forst. et Forst.	senecionine senkirkine brachyglottine	aerial	B 117 B 118
Cacalia floridana (= Senecio floridanus Sch. Bip.)	otosenine florosenine floridanine floricaline	aerial	B 119
Cacalia hastata L. subsp. orientalis Kitamura	integerrimine	root	B 120
Cacalia yatabei Maxim	yamataimine	root	B 121
Conoclinium coelestinium (L.) DC	intermedine		C Herz et al. (1981)
Crassocephalum crepidioides	jacobine jacoline	aerial	A Asada et al. (1985)
Doronicum macrophyllum	otosenine floridanine doronine	root	B 122
Doronicum pardalianches Linn.	otosenine	root	A Rajagopalan & Negi (1985)
Echinacea angustifolia DC.	tussilagine isotussilagine	whole	C
Echinacea purpurea M.	tussilagine isotussilagine	whole	C

Appendix II, Table 1 (contd).

Compositae (contd)

Species	Alkaloids	Part	Reference
Emilia flammea Cass.	otosenine emiline	aerial, root	B 123, 124, 125
Erechtites hieracifolia (L.) Raf. ex DC.	senecionine seneciphylline	aerial	B 126, 127
Eupatorium altissimum L.	rinderine angelylheliotridine		C Herz et al. (1981)
Eupatorium cannabinum L.	echinatine supinine amabiline	aerial	B 128 A Luthy et al. (1984)
Eupatorium compositifolium Walt.	intermedine lycopsamine		C Herz et al. (1981)
Eupatorium maculatum L.	echinatine trachelanthamidine	root	B 129
Eupatorium purpureum	probably echinatine	aerial	B 130
Eupatorium serotinum Michx.	supinine rinderine	aerial	B 131
Eupatorium stoechadosmum Hance	lindelofine supinine	root	B 132
Farfugium japonicum Kitam.	senkirkine farfugine petasitenine	root, leaf whole whole	B 133 C Niwa et al. (1983) A Niwa et al. (1985)
Gynura scandens O. Hoffm.	gynuramine acetylgynuramine		C Wiedenfeld (1982)

Appendix II, Table 1 (contd).

Plant	Constituent alkaloids	Plant part	Reference[a]
Compositae (contd)			
Gynura segetum (Lour.) Merr.	senecionine	aerial	C Liang & Roder (1984)
Ligularia brachyphylla Hand.-Mazz.	clivorine ligularine ligudentine	aerial	B 134
Ligularia clivorum	clivorine	aerial	B 135, 136, 137, 138
Ligularia dentata (A. Gray) Hara	clivorine ligularine ligudentine ligularidine ligularinine ligularizine neoligularidine	aerial whole aerial, root	B 134 B 436 C Asada & Furuya (1984a)
Ligularia elegans (Cass.) [syn. Ligularia macrophylla (Ledeb. DC.)]	clivorine ligularine	aerial	B 134
Ligularia japonica	senecionine neopetasitenine platyphylline	root	A Asada et al. (1981)
Petasites albus L.	senkirkine	aerial	C Luthy et al. (1983)
Petasites hybridus L.	senecionine integerrimine senkirkine	aerial	C Luthy et al. (1983)

Appendix II, Table 1 (contd).

Compositae (contd)			
Petasites japonicus Maxim.	petasitenine (fukinotoxin)	aerial	B 20, 19, 139
	neopetasitenine		
	senkirkine	stem	B 140
	petasinine	aerial	B 141
	petasinoside		
Petasites laevigatus (Willd.) Reichenb. [syn. *Nardosmia laevigata* (Willd.) DC.]	platyphylline	aerial	B 142
	senkirkine (renardine)	aerial	B 143, 144
	senecionine		
Senecio abrotanifolius ssp. *abrotanifolius*	doronine	aerial	A Roder et al. (1984a)
	doronenine		
	bulgarsenine		
Senecio abrotanifolius ssp. *abrotanifolius* var. *tiroliensis*	doronine	aerial	A Roder et al. (1984a)
	doronenine		
	bulgarsenine		
Senecio aegypticus L.	senecionine	whole	B 145, 146
	otosenine	aerial	
	riddelliine		
	erucifoline		
Senecio alpinus (L.) Scop.	seneciphylline	aerial	B 147
	jacozine		
	jacobine	whole	A Luthy et al. (1981)
	integerrimine		
	jacoline		
	senecionine		
	jaconine		
Senecio ambrosioides (= *Senecio brasiliensis* Less.)	retrorsine	whole	B 148
	seneciphylline		
	senecionine	whole	B 420
	riddelliine		

Appendix II, Table 1 (contd).

Plant	Constituent alkaloids	Plant part	Reference[a]
Compositae (contd)			
Senecio ampullaceus Hook.	senecionine seneciphylline retrorsine	whole	B 149, 150, 151
Senecio anticuphorbium (L.) Sch. Bip.	integerrimine senkirkine	aerial	B 152
Senecio aquaticus Hill	seneciphylline	aerial	B 153, 154
Senecio aureus L.	senecionine otosenine floridanine florosenine	aerial aerial	B 155, 156 C Resch et al. (1983) C Roder et al. (1983)
Senecio auricola Bourg.	neosenkirkine	aerial	B 157
Senecio barbellatus DC.	swazine retrorsine		B 270
Senecio bipinnatisectus Belcher (syn. Erechtites atkinsoniae)	retrorsine	aerial, root	B 158
Senecio borysthenicus (= Senecio prealtus Berthol.)	seneciphylline	aerial, root	B 159, 160
Senecio brasiliensis DC. (syn. Senecio ambrosioides)	senecionine seneciphylline jacobine integerrimine retrorsine	leaf	B 161, 162 B 148, 163

Appendix II, Table 1 (contd).

Compositae (contd)				
Senecio brasiliensis Less var tripartitus	integerrimine		A	Nardi et al. (1980)
Senecio bupleuroides DC.	retrorsine	aerial	B	164
Senecio cacaliaster (Lam.)	sencalenine bulgarsenine 7-senecioylretronecine 7-senecioyl-9-sarracinoylretronecine		A	Roder et al. (1984b)
Senecio cannabifolius Less.	seneciphylline senecicannabine jacozine	aerial aerial, root	B C	165 Asada et al. (1982a)
Senecio carthamoides Greene	senecionine seneciphylline	whole	B	149, 151
Senecio caudatus DC.	7-senecioylretronecine 9-senecioylretronecine 7-senecioyl-9-sarracinylretronecine 7-senecioyl-9-(2-methyl-2,3-dihyroxy- butyryl)retronecine 7-senecioyl-9-(2-methy-2-hydroxy-3- acetoxybutyryl)retronecine retronecine 2-senecioyl-(-)-macronecine 9-senecioyl-(-)-macronecine senecicaudatine-9-senecioate senecidaudatine-9-isovalerate norsenecicaudatine-9-sencioate senecicaudatinal semiacetal	aerial	A	Bohlmann et al. (1986)
Senecio chrysanthemoides	seneciphylline		B	166

Appendix II, Table 1 (contd).

Plant	Constituent alkaloids	Plant part	Reference[a]
Compositae (contd)			
Senecio cineraria DC.	jacobine	aerial	B 153, 167
	senecionine	seed	B 168
	senecyphylline	aerial	B 169
	otosenine	aerial	B 170
	retrorsine	aerial	B 171
Senecio congestus (R. Br.) DC. [syn. Senecio palustris (L.) Hooker, Senecio tubicaulis Mansfeld]	senecionine neoplatyphylline platyphylline		C Roder et al. (1982b)
Senecio cruentus DC.	senecionine seneciphylline retrorsine riddelliine		C Asada et al. (1982b)
Senecio cymbaroides	senecionine seneciphylline riddelliine retrorsine	whole	B 420
Senecio desfonainei Druce	senecionine otosenine riddelliine seneciphylline retrorsine senkirkine angelylretronecine	aerial aerial	B 173, 145 B 146 A Rizk et al. (1983)
Senecio discolor DC.	retrorsine senecionine	leaf aerial	B 174 B 175

Appendix II, Table 1 (contd).

Compositae (contd)

Senecio dolichodoryius Cuatr.	15,20-dihyroxyerucifoline	aerial	A Bohlmann et al. (1986)
Senecio doronicum L.	bulgarsenine doronenine	leaf	B 176
Senecio douglasii DC.	retrorsine riddelliine seneciphylline senecionine	whole	B 149, 150, 151
Senecio durieui Gay	integerrimine	whole	B 157
Senecio cremophilus Richards	senecionine seneciphylline retrorsine riddelliine	aerial	B 149, 150, 151
Senecio erraticus Berthol.	senecionine otosenine floridanine	aerial	B 177
Senecio erraticus Berthol. subsp. barbaraeifolius Krock	senecionine otosenine erucifoline seneciphylline integerrimine	leaf aerial	B 178, 179, 180 B 181 B 182
Senecio erucifolius L.	senecionine seneciphylline erucifoline retrorsine	aerial aerial aerial	B 153 B 183, 180 B 184
Senecio faberi Hemsl.	integerrimine		C Wei et al. (1982)

Appendix II, Table 1 (contd).

Plant	Constituent alkaloids	Plant part	Reference[a]
Compositae (contd)			
Senecio filaginoides (H. et A.) DC.	senecionine retrorsine	root	A Pestchanker & Giordano (1986)
Senecio fistulosus Poepp. ex Less.	senecionine		A Gonzalez et al. (1986b)
Senecio fluviatilis Wallr.	seneciphylline otosenine florosenine	aerial	B 185
Senecio formosus	intgerrimine retrorsine	aerial	B 186
Senecio fremontii Torr. et A. Gray	seneciphylline senecionine	whole	B 148
Senecio gilliesiano	senecionine retrorsine	root	A Guidugli et al. (1986)
Senecio glabellus (Turcz.) DC.	senecionine	whole	B 187
Senecio glaberrimus DC.	retrorsine	aerial	B 153
Senecio glandulosus Don ex Hook. et Arn.	integerrimine retrorsine usaramine	root	A Pestchanker et al. (1985b)
Senecio graminifolius N.J. Jacq.	retrorsine graminifoline	aerial	B 188
Senecio grandifolia Jaqu.	platyphylline seneciphylline	root, leaf, stem	B 189

Appendix II, Table 1 (contd).

Compositae (contd)			
Senecio grandifolius Less.	senkirkine neosenkirkine	aerial	A Bohlmann et al. (1986)
Senecio griesbachii Baker	retrorsine	aerial	B 190
Senecio ilicifolius Thunb.	senecionine seneciphylline retrorsine	aerial	B 191, 192, 188, 127
Senecio illinitus Phill.	senkirkine O-acetylsenkirkine senecionine	aerial	A Gonzalez et al. (1986a)
Senecio inaequidens DC.	retrorsine senecionine senecivernine integerrimine	aerial	A Roder et al. (1981) A Bicchi et al. (1985)
Senecio incanus L. subsp. carniolicus (Willd.) Br.-Bl.	seneciphylline integerrimine	aerial	B 147
Senecio integerrimus Nutt.	integerrimine senecionine neoplatyphylline platyphylline	aerial whole	B 156 B 420
Senecio isatideus DC.	retrorsine	aerial	B 153, 193
Senecio jacobaea L.	seneciphylline senecionine jacobine jaconine jacozine	aerial	B 194, 195 B 196, 197 B 153, 197, 198 B 199, 200, 201
	otosenine senkirkine	aerial	B 51
	retrorsine	aerial	B 184

Appendix II, Table 1 (contd).

Plant	Constituent alkaloids	Plant part	Reference[a]
Compositae (contd)			
Senecio kirkii Hook. f. ex Kirk	senkirkine O-acetylsenkirkine	bark, leaf leaf	B 203 B 204
Senecio kleinia Sch. Bip.	integerrimine senkirkine	stem stem	B 205 B 206
Senecio krylovii	seneciphylline	aerial	B 207
Senecio kubensis Grossh.	seneciphylline	aerial	B 208
Senecio lampsanoides	seneciphylline	aerial, root	B 209, 210
Senecio laricifolius H.B.K.	senecionine seneciphylline senkirkine 19-hydroxysenkirkine 19-acetoxysenkirkine	aerial	A Bohlmann et al. (1986)
Senecio latifolius DC. (syn. Senecio sceleratus Schweikerdt)	retrorsine seneciphylline platyphylline sceleratine chlorodeoxysceleratine (merenskine)	aerial aerial aerial aerial	B 211, 212 B 213 B 260 B 261; C Bredenkamp et al. (1985)
Senecio leucostachys Baker	senecionine	root	A Pestchanker & Giordano (1986)
Senecio longilobus Benth.	seneciphylline retrorsine riddelliine	whole whole	B 156, 214, 150 B 151 B 149

Appendix II, Table 1 (contd).

Compositae (contd)			
Senecio magnificus F. Muell.	senecionine integerrimine	aerial	B 215 B 216
Senecio megaphyllus Green.	13,19-epoxysenecionphylline 13,19-epoxyspartioidine	aerial	A Bohlmann et al. (1986)
Senecio minimus Poir	seneciphylline	aerial	B 158
Senecio morrisonensis Hayata	integerrimine	whole	B 217
Senecio multilobatus	senecionine	aerial	A McCoy et al. (1983)
Senecio multivenius Benth. in Oerst.	seneciphylline senecionine	aerial	A Bohlmann et al. (1986)
Senecio nebrodensis L. var sicula	integerrimine senecionine	whole	B 218
Senecio nemorensis L. var bulgaricus (Vel) Stoj.	bulgarsenine retrosisosenine nemorensine	leaf	B 219
Senecio nemorensis L. ssp. fuchsii Gmel.	fuchsisenecionine senecionine		B 362, 363, 364, 365 B 423
Senecio nemorensis L. subdecurrens Griseb.	nemorensine retroisosenine bulgarsenine	aerial, root aerial	B 371 B 422
Senecio othonnae Bieb.	otosenine onetine seneciphylline floridanine doronine	aerial, root root aerial, root aerial	B 220 B 221 B 222 B 223

Appendix II, Table 1 (contd).

Plant	Constituent alkaloids	Plant part	Reference[a]
Compositae (contd)			
Senecio othonniformis Fourcade	bisline isoline	aerial	B 22, 225
Senecio palmatus Pall.	seneciphylline	root	B 226
Senecio paludosus L.	seneciphylline	root, aerial	B 153, 227, 228
Senecio pampeanus Cabrera	senecionine	aerial	B 229
Senecio pancicii Degen var arnautorum (Velen.) Stoj., Stef. et Kit.	senecionine seneciphylline	whole	C Jizba et al. (1982)
Senecio pancicii Degen var pancicii	senecionine	whole	C Jizba et al. (1982)
Senecio patagonicus Hook. and Arn.	senecionine seneciphylline	aerial	A Villaroel et al. (1985)
Senecio paucicalyculatus (Klatt.)	paucicaline retrorsine	whole	B 230
Senecio paucifolius S.G. Gmel.	seneciphylline		B 231
Senecio petasitis DC.	senecionine bisline (?)	leaf aerial	B 146 B 232
Senecio phillipicus Rogel et Koern.	retrorsine	aerial	A Gonzalez et al. (1986a)
Senecio pierotii	neosenkirkine senkirkine	aerial, root	C Asada & Furuya (1982)

Appendix II, Table 1 (contd).

Compositae (contd)			
Senecio pimpinellifolius H.B.K.	senecionine	aerial	A Bohlmann et al. (1986)
Senecio platyphylloides Somm. et Lev.	platyphylline seneciphylline	root	B 233, 234, 235
Senecio platyphyllus (Bieb.) DC.	platyphylline seneciphylline neoplatyphylline sarracine	root, aerial leaf root root	B 236, 237, 238 B 239 B 240, 241 B 242
Senecio pojarkovae	sarracine seneciphylline	root	B 243
Senecio procerus L. var procerus Stoj. Stef. et Kit.	senkirkine procerine	aerial, root	B 244
Senecio propinquus Ait.	seneciphylline	aerial, root	B 209, 245
Senecio pseudo-arnica Less.	senecionine	aerial	B 156
Senecio pterophorus	senecionine seneciphylline retrorsine rosmarinine acetylseneciphylline	aerial	B 192, 188, 127
Senecio quadridentatus Labill. (syn. Erechtites quadridentata DC.)	senecionine seneciphylline retrorsine	aerial	B 247
Senecio quebradensis Greenm.	senkirkine florosenine	aerial	A Bohlmann et al. (1986)
Senecio racemosus DC.	seneciphylline	root	B 248

Appendix II, Table 1 (contd).

Plant	Constituent alkaloids	Plant part	Reference[a]
Compositae (contd)			
Senecio renardii Winkl.	seneciphylline senkirkine (renardine) otosenine	aerial	B 249, 250
Senecio retrorsus DC.	retrorsine	aerial	B 194, 193
Senecio rhombifolius (Willd.) Sch. Bip.	sarracine platyphylline seneciphylline neoplatyphylline	root aerial, root	B 251, 233 B 208
Senecio riddellii Torr. et A. Gray	riddelliine retrorsine	aerial whole	B 156 B 252, 253 B 420
Senecio riddellii Torr. et A. Gray var. parksii (Cory)	retrorsine riddelliine	whole	B 149
Senecio ruderalis Harvey	retrorsine	aerial	B 254
Senecio ruwenzoriensis S. Moore	ruwenine ruzorine	whole	B 255
Senecio sandrasicus	senecionine		A Temizer et al. (1985)
Senecio scandens	senecionine seneciphylline	whole	B 256
Senecio seratophiloides Griseb.	senecionine senecivernine usaramine	root	A Pestchanker & Giordano (1986)

Appendix II, Table 1 (contd).

Compositae (contd)

Senecio seratophiloides Griseb. (contd)	retrorsine uspallatine		
Senecio spartioides Torr. et A. Gray	senecphylline senecionine spartiodine riddelliine retrorsine	aerial whole	B 156, 262 B 420
Senecio spathulatus A. Rich.	senecionine integerrimine seneciphylline	aerial, root	B 158 B 263
Senecio squalidus L.	senecionine integerrimine	aerial	B 153, 264, 265 B 205
Senecio stenocephalus Maxim.	seneciphylline	aerial	B 266
Senecio subalpinus C. Koch.	senecionine seneciphylline integerrimine	leaf aerial	B 267 B 147
Senecio subulatus Don ex Hook. et Arn. var. erectus	dihydroretrorsine retrorsine senecionine	root	A Pestchanker et al. (1985b)
Senecio swaziensis Compton	retrorsine swazine	aerial	B 268, 269, 270
Senecio tenuifolius Burm.	senecionine integerrimine senkirkine o-acetylsenkirkine	aerial	C Bhakuni & Gupta (1982)

Appendix II, Table 1 (contd).

Plant	Constituent alkaloids	Plant part	Reference[a]
Compositae (contd)			
Senecio tomentosus	senecionine otosenine (tomentosine)	aerial	B 271, 179
Senecio triangularis Hook.	senecionine integerrimine platyphylline rosmarinine retrorsine triangularine neotriangularine	aerial	B 272 C Roitman (1983b)
	7-angelylretronecine 7-senecioylretronecine 7-angelyl-9-sarracinylretronecine 7-senecioyl-9-sarracinylretronecine	whole	C Rueger & Benn (1973)
Senecio uintahensis	senkirkine senecionine	whole	B 420
Senecio umgeniensis Thell.	7-senecioyl-9-sarracinylretronecine	aerial	A Bohlmann et al. (1986)
Senecio usgorensis Cuatr.	13,19-epoxyseneciphylline	aerial	A Bohlmann et al. (1986)
Senecio uspallatensis	retrorsine uspallatine	root	A Pestchanker et al. (1985a)
Senecio variabilis Sch. Bip.	7-senecioylretronecine 9-senecioylretronecine 7-senecioyl-9-sarracinylretronecine	aerial, root	A Bohlmann et al. (1986)

Appendix II, Table 1 (contd).

Compositae (contd)

Senecio venosus Harvey	retrorsine	aerial	B 153
Senecio vernalis Walst. et Kit.	retrorsine senecionine senkirkine senecivernine integerrimine seneciphylline riddelline retronecine	aerial	B 273, 274 A Sener et al. (1986)
Senecio viscosus L.	senecionine integerrimine	aerial aerial	B 264, 153 B 182
Senecio vulgaris L.	senecionine seneciphylline retrorsine riddelline integerrimine spartioidine usaramine	aerial aerial whole	B 275, 264, 153 B 155, 276, 277 B 420 A Pieters & Vlietinck (1986)
Senecio werneriaefolius	senecionine retrorsine	whole	B 420
Syneilesis palmata Maxim.	syneilesine acetylsyneilesine senecionine	aerial, root	B 278, 279
Tussilago farfara L.	senkirkine	flower leaf, stem	B 18, 425 C Rosberger et al. (1981)
	senecionine tussilagine	leaf, stem	A Luthy et al. (1980) C Roder et al. (1981b)

Appendix II, Table 1 (contd).

Plant	Constituent alkaloids	Plant part	Reference[a]
Leguminosae			
Crotalaria aegyptica Benth.	crosemperine		B 419
	monocrotaline		
	7β-hydroxy-1-methylene-8α-pyrrolizidine		B 426
Crotalaria agatiflora Schweinf.	maduraensine	aerial	B 280
	anacrotine	aerial	B 281
	7-acetylmaduraensine		
	6-acetylanacrotine		
	7-acetyl-cis-maduraensine		
	6-acetyl-trans-anacrotine		
	crotaflorine		
	6-angelyl-trans-anacrotine		
Crotalaria assamica Benth.	monocrotaline		B 285, 286
Crotalaria axillaris Ait.	axillarine	seed	B 287, 288
	axillaridine		
Crotalaria barbata R. Graham ex R. Wight et Walk.-Arn.	crobarbatine	seed	B 289
Crotalaria berteroana DC. (syn. Crotalaria fulva Roxb.)	fulvine	aerial	B 297
Crotalaria brevidens Benth. var. intermedia (Kotschy) Polhill (syn. Crotalaria intermedia Kotschy)	integerrimine	seed	B 304
	usaramine		
Crotalaria breviflora DC.	integerrimine	seed	B 290, 291
	usaramine		

Appendix II, Table 1 (contd).

Leguminosae (contd)

Crotalaria burhia Ham. ex Benth.	crotalarine monocrotaline	aerial	B 292, 293
Crotalaria candicans W. and A.	crocandine isocorcandine isocromadurine crispatine turneforcidine cropodine	seed seed seed	B 380 C Suri et al. (1982) C Haksar et al. (1982)
Crotalaria cephalotes Steud. ex A. Rich	monocrotaline	seed	C Pilbeam et al. (1983)
Crotalaria crispata F. Muell. ex Benth.	monocrotaline fulvine crispatine	whole	B 294
Crotalaria cunninghamii R. Br.	monocrotaline	seed	C Pilbeam et al. (1983)
Crotalaria dura J.M. Wood et Evans	dicrotaline	aerial	B 295, 296
Crotalaria fulva Roxb. (see Crotalaria berteroana DC.)			
Crotalaria globifera E. Mey	dicrotaline globiferine grantianine grantaline trichodesmine	aerial seed	B 295, 296 C Brown et al. (1984)
Crotalaria grahamiana Wight & Arn.	monocrotaline grahamine monocrotalinine	seed seed whole	B 298 B 299 B 300

Appendix II, Table 1 (contd).

Plant	Constituent alkaloids	Plant part	Reference[a]
Leguminosae (contd)			
Crotalaria incana L.	integerrimine	seed	B 187
	anacrotine	aerial	B 302
	usaramine	seed	B 303
Crotalaria juncea L.	senecionine	seed	B 305, 306, 307
	seneciphylline		
	riddelliine		
	trichodesmine		
	junceine		
Crotalaria laburnifolia L.	anacrotine (crotalaburnine)	seed	B 308, 309, 310, 291, 311
Crotalaria laburnifolia L. subsp. eldomae	madurensine	aerial	B 312
	anacrotine		
	senkirkine		
	hydroxysenkirkine		
	crotafoline		
Crotalaria leschenaultii	monocrotaline	seed	B 313
Crotalaria leiloba Bartl. (syn. Crotalaria ferruginea Wall.)	monocrotaline	seed	B 314
Crotalaria madurensis R. Wight	madurensine	seed, flower, leaf	B 280
	crispatine	aerial	B 315
	fulvine		
	cromadurine	seed	B 62, 316
	isocromadurine	seed	B 317

Appendix II, Table 1 (contd).

Leguminosae (contd)			
Crotalaria micans Link. (syn. Crotalaria anagyroides Humb. et al.)	1-methylenepyrrolizidine senecionine anacrotine	seed seed	B 282, 283 B 284 B 280
Crotalaria mitchellii Benth.	monocrotaline retusamine	whole whole	B 318
Crotalaria mitchellii Benth. subsp. laevis A. Lee, published as "sp. aff. mitchellii")	retusamine	whole	B 318
Crotalaria mysorensis Roth.	monocrotaline	seed	B 319
Crotalaria nana Burm.	crotananine cronaburmine	seed seed	B 320 B 427
Crotalaria nitens Kunth.	monocrotaline		A Hoet et al. (1981)
Crotalaria novae-hollandiae DC. subsp. lasiophylla (Benth) A. Lee	monocrotaline retusamine	whole	B 318
Crotalaria novae-hollandiae DC. subsp. novae-hollandiae (syn. Crotalaria crassipes Hook.)	retusamine	seed	B 318
Crotalaria pallida Ait. (syn. Crotalaria mucronata, Crotalaria striata)	usaramine nilgirine crotastriatine	seed seed seed	B 291 B 322 B 323, 324
Crotalaria paniculata Willd.	fulvine	seed	B 325
Crotalaria paulina Schrank	monocrotaline	seed	C Pilbeam et al. (1983)
Crotalaria quinquefolia L.	monocrotaline	seed	C Pilbeam et al. (1983)

Appendix II, Table 1 (contd).

Plant	Constituent alkaloids	Plant part	Reference[a]
Leguminosae (contd)			
Crotalaria recta Steud. ex A. Rich	monocrotaline	aerial	B 326
Crotalaria retusa L.	monocrotaline retusine retusamine retronecine	seed seed, aerial	B 327 B 328
Crotalaria sagittalis L.	monocrotaline	seed	B 330
Crotalaria scassellatii Chiov.	axillaridine axillarine deoxyaxillarine	seed	A Wiedenfeld et al. (1985)
Crotalaria semperflorens Vent.	crosemperine	seed	B 331
Crotalaria spartioides DC.	retrorsine	aerial	B 334
Crotalaria spectabilis Roth. (syn. Crotalaria sericea Retz)	monocrotaline spectabiline	seed seed, whole	B 335, 227 B 336
Crotalaria stipularia Desv.	monocrotaline	seed	B 314
Crotalaria tetragona Roxb.	integerrimine trichodesmine	seed	B 314
Crotalaria verrucosa L.	anacrotine crotaverrine acetylcrotaverrine	seed seed	B 338 B 339

Appendix II, Table 1 (contd).

Leguminosae (contd)

Crotalaria virgulata subsp. grantiana (Harv.) Polhill (syn. Crotalaria grantiana Harv.)	grantianine grantaline 1-hydroxymethyl-1β,2β-epoxy-pyrrolizidine	seed	B 301, 259 C Smith & Culvenor (1984)
Crotalaria walkeri Arn.	crotaverrine acetylcrotaverrine	seed	B 340
Crotalaria wightiana Grah. ex Wight & Arn. (syn. Crotalaria rubiginosa Willd. var. wightiana J.G. Baker)	junceine trichodesmine	seed	B 329
Crotalaria zanzibarica Benth. (syn. Crotalaria usaramoensis E.G. Baker)	integerrimine usaramine senecionine retrorsine	seed seed	B 187 B 337

Ranunculaceae

Caltha biflora DC.	senecionine	aerial	B 341
Caltha leptosepala DC.	senecionine	aerial, root	B 341

Scrophulariaceae

Castilleja rhexifolia Rydb.	senecionine sarracine indicine or isomer		B 342 C Roby & Stermitz (1984)

a A = References in the reference list of this document.
   B = References in Smith & Culvenor (1981), J. nat. Prod., 44: 129-152 (with reference number).
   C = References in Mattocks (1986), Chemistry and toxicology of pyrrolizidine alkaloids.

APPENDIX II

Table 2. Plants containing known alkaloids that are non-hepatotoxic (aminoalcohols and esters)

Plant	Constituent alkaloids	Plant part	Reference[a]
A. Families in which hepatotoxic alkaloids also occur			
Apocynaceae			
Alafia multiflora	alafine	seed	B 343
Anodendron affine Druce	alloanodendrine anodendrin	aerial	B 344, 345
Boraginaceae			
Caccinea glauca Savi	7,9-dibenzoylretronecine	fl	B 346
Ehretia aspera Willd.	ehretinine	leaf	A Suri et al. (1980)
Heliotropium angiospermum Murray	1-hydroxymethyl-1β,2β-epoxy-pyrrolizidine	whole	C Birecka et al. (1983)
Heliotropium ovalifolium Forsk	heliofoline retronecine	whole	C Mohanraj et al. (1981)
Heliotropium spathulatum Rydb.	acetylcurassavine curassavine	aerial	A Birecka et al. (1980)
Heliotropium strigosum Willd.	strigosine	aerial	B 347
Lindelofia macrostyla (Bunge) M. Pop. (syn. Lindelofia anchusoides, Paracaryum heliocarpum Kern.)	lindelofine lindelofamine	aerial	B 348

Appendix II, Table 2 (contd).

Boraginaceae (contd)			
Lindelofia olgae (Regel et'Smirnov) Brand	viridiflorine	aerial	B 94
Lindelofia pterocarpa (Rupr.) M. Pop.	viridiflorine	aerial	B 93
Macrotomia echioides Boiss.	macrotomine	aerial	B 350
Paracaryum himalayense (Klotsch) C.B. Clark	viridiflorine	aerial	B 93
Tournefortia sibirica L.	turneforcine	aerial	B 351
Trachelanthus hissoricus Lipsky	viridiflorine trachelanthamine	leaf	B 352
Trachelanthus korolkovii (Lipsky) B. Fedtsch.	trachelanthamine	aerial	B 353, 354, 355, 94
Celastraceae			
Bhesa archboldiana (Merr. & Perry) Ding Hou [syn. Kurrimia archboldiana (Merr. & Perry)]	9-angelylretronecine	bk	B 356
Compositae			
Adenostyles rhombifolius (Willd.) M. Pimen. ssp. rhombifolia chemovar. sarracinifera	sarracine	aerial	B 116
Adenostyles rhombifolius (Willd.) M. Pimen. ssp. rhombifolia chemovar. platyphyllinifera	platyphylline	aerial	B 116
Cacalia hastata L.	hastacine	root	B 357, 241

Appendix II, Table 2 (contd).

Plant	Constituent alkaloids	Plant part	Reference[a]
Compositae (contd)			
Cacalia robusta	hastacine	aerial	B 358
Senecio amphibolus	macrophylline	aerial	B 359
Senecio angulatus L.	angularine rosmarinine	whole	B 360
Senecio aronicoides	hygrophylline	whole	B 420
Senecio brachypodus DC.	rosmarinine	aerial, root	B 361
Senecio francheti Winkl.	sarracine franchetine	aerial	B 352
Senecio glastifolius	sarracine		A Mortimer & White (1975)
Senecio hygrophyllus R.A. Dyer et C.A. Smith (syn. Senecio adnatus DC.)	platyphylline rosmarinine hygrophylline	aerial	B 366, 361
Senecio macrophyllus Bieb.	macrophylline	aerial	B 368
Senecio mikanioides Otto. ex Walp.	sarracine	aerial	B 155, 369, 370
Senecio nemorensis L. ssp. fuchsii var. nova (Zlatnik)	nemorensine	aerial	B 371
Senecio nemorensis L. ssp. jaquinianus (Rchb.) Durand	nemorensine	aerial	B 371
Senecio ovirensis ssp. gaudinii	angelylheliotridine		A Roder et al. (1980)

Appendix II, Table 2 (contd).

Compositae (contd)			
Senecio pauciligulatus Dyer et Sm.	rosmarinine	aerial	B 361
Senecio rivularis DC.	7-angelylheliotridine	aerial	B 182, 135
Senecio rosmarinifolius Linn.	rosmarinine	aerial	B 191, 192, 188
Senecio salignus DC.	7-angelylheliotridine	aerial	A Bohlmann et al. (1986)
Senecio sarracenius L.	sarracine	aerial	B 153, 372, 373
Senecio schvetsovii Korsh	macrophylline	aerial	B 231
Senecio sylvaticus L.	silvasenecine	aerial	B 362, 153
	sarracine	aerial	B 374
Senecio taiwanensis Hayata	rosmarinine	aerial	B 217
Senecio tournefortii Lap.	platyphylline	aerial	B 375
Leguminosae			
Crotalaria albida Heyne ex Roth. (syn. Crotalaria montana Roxb.)	croalbidine	aerial	B 376, 377
Crotalaria aridicola Domin.	1-methoxymethyl-1,2-dehydro-pyrrolizidine	aerial	B 378
	7β-hydroxy-1-methoxymethyl-1,2-dehydropyrrolizidine		
	7β-acetoxy-1-methoxymethyl-1,2-dehydropyrrolizidine	whole	B 379
Crotalaria damarensis Engl.	1-methylenepyrrolizidine	whole, seed	B 381, 283

Appendix II, Table 2 (contd).

Plant	Constituent alkaloids	Plant part	Reference[a]
Leguminosae (contd)			
Crotalaria goreensis Guill. et Perr.	7β-hydroxy-1-methylene-8β-pyrrolizidine 7β-hydroxy-1-methylene-8α-pyrrolizidine	aerial, seed	B 382
Crotalaria grandistipulata Harms	1-methylenepyrrolizidine	seed	B 434
Crotalaria lachnophora A. Rich	1-methylenepyrrolizidine	seed	B 434
Crotalaria maypurensis Humb et al.	7β-hydroxy-1-methylene-8β-pyrrolizidine 7β-hydroxy-1-methylene-8α-pyrrolizidine	aerial	B 383
Crotalaria medicaginea Lam.	1-methoxymethyl-1,2-dehydropyrrolizidine 7β-hydroxy-1-methoxymethyl-1,2-dehydropyrrolizidine	whole, seed	B 378
	1α-methoxymethyl-1β,2β-epoxypyrrolizidine	whole, seed	
	7α-hydroxy-1-methoxymethyl-1,2dehydropyrrolizidine	seed	
	1α-hydroxymethyl-1β,2β-epoxypyrrolizidine	aerial	
Crotalaria natalitia Meissner	1-methylenepyrrolizidine	seed	B 434
Crotalaria podocarpa DC.	7-hydroxy-1-methylenepyrrolizidine	seed	B 435
Crotalaria stolzii (Bak. f.) Milne-Redh. ex Polhill	1-methylenepyrrolizidine	seed	B 434
Ranunculacae laburnum L.	laburnine 1-hydroxymethyl-7-hydroxypyrrolizidine	seed	B 387, 388, 389 B 390

Appendix II, Table 2 (contd).

Scrophulariaceae			
Castilleja "rhexifolia aff. miniata"	sarracine 7-angelylplatynecine 8-angelylplatynecine		C Roby & Stermitz (1984)
B. Families in which hepatotoxic alkaloids are not known to occur			
Orchidaceae			
Chysis bractescens Lindl.	1α-methoxycarbonyl-8β- pyrrolizidine 1α-ethoxycarbonyl-8β- pyrrolizidine	whole	B 391, 392
Doritis pulcherrima (syn. Phalaenopsis esmerelda)	phalaenopsine La or T	whole	B 393
Hammarbya paludosa (L.) O.K.	paludosine hammarbine	whole whole	B 394 B 395
Kingiella taenialis (Lindl.) Rolfe	phalaenopsine La	whole	B 396
Liparis auriculata (Blume)	auriculine	whole	B 397
Liparis bicallosa Schltr.	laburnine malaxine	whole whole	B 397 B 398, 399
Liparis hachijoensis Nakai	laburnine malaxine	whole whole	B 397 B 398
Liparis keitaoensis Hay.	keitaoine keitine	whole	B 395
Liparis kumokiri F. Maekawa	kumokirine	whole	B 400, 398
Liparis loeselii (L.) L.C. Rich	auriculine	whole	B 394

Appendix II, Table 2 (contd).

Plant	Constituent alkaloids	Plant part	Reference[a]
Orchidaceae (contd)			
Liparis nervosa Lindl.	nervosine	whole	B 398, 401
Malaxis congesta comb. nov. (Rchb. f.)	malaxin	whole	B 402
Malaxis grandifolia Schltr.	grandifoline	whole	B 403
Phalaenopsis amabilis Bl.	phalaenopsine T	whole	B 404, 405, 393
Phalaenopsis amboinensis	phalaenopsine La	whole	B 393
Phalaenopsis aphrodite	phalaenopsine T	whole	B 393
Phalaenopsis cornu-cervi Rchb. f.	cornucervine	whole	B 404, 393
Phalaenopsis equestris Rchb. f.	phalaenopsin ls phalaenopsin T	whole	B 393
Phalaenopsis fimbriata	phalaenopsine T	whole	B 393
Phalaenopsis hieroglyfica	phalaenopsine T or La	whole	B 393
Phalaenopsis lueddemanniana	phalaenopsine T or La	whole	B 393
Phalaenopsis mannii Rchb. f.	phalaenopsine La	whole	B 393
Phalaenopsis sanderiana Rchb. f.	phalaenopsine La phalaenopsine T	whole	B 393
Phalaenopsis schilleriana	phalaenopsine La	whole	B 393
Phalaenopsis stuartiana Rchb. f.	phalaenopsine La phalaenopsine T	whole	B 393

Appendix II, Table 2 (contd).

Orchidaceae (contd)

Phalaenopsis sumatrana	phalaenopsis La	whole	B 393
Phalaenopsis violacea	phalaenopsis La or T	whole	B 393
Vanda cristata Lindl.	acetyllaburnine	whole	B 407
Vanda helvola Bl.	laburnine acetyllaburnine	whole	B 408
Vanda hindsii Lindl.	acetyllaburnine	whole	B 408
Vanda luzonica Loher	acetyllaburnine	whole	B 408
Vandopsis gigantea Pfitz.	laburnine lindelofidine acetyllaburnine acetyllindelofidine	whole	B 408
Vandopsis lissochiloides Pfitz.	laburnine lindelofidine acetyllaburnine acetyllindelofidine	whole	B 408

Rhizophoraceae

Cassipourea gummiflua Tulasne var. verticellata Lewis	cassipourine	stem, leaf bk	B 409 B 410

Appendix II, Table 2 (contd).

Plant	Constituent alkaloids	Plant part	Reference[a]
Santalaceae			
Thesium minkwitzianum B. Fedtsch.	thesine thesinine thesinicine	aerial	B 411 B 412, 413
	isoretronecanol	root	
Sapotaceae			
Mimusops elengi L.	1-hydroxymethylpyrrolizidine tiglate		B 414
Planchonella anteridifera (C.T. White et W.D. Francis) H.J. Lamb	planchonelline tiglyllaburnine benzoyllaburnine	leaf	B 415
Planchonella thyrsoidea C.T. White ex F.S. Walker	planchonelline tiglyllaburnine benzoyllaburnine	leaf	B 415
Planchonella sp. (NGF 24722)	trans-β-thioacryly1-(-)-isoretronecanol tiglylisoretronecanol	leaf	B 416

[a] A = References in the reference list of this document.
B = References in Smith & Culvenor (1981), J. nat. Prod., 44: 129-152 (with reference number).
C = References in Mattocks (1986), Chemistry and toxicology of pyrrolizidine alkaloids.

www.ingramcontent.com/pod-product-compliance
Ingram Content Group UK Ltd.
Pitfield, Milton Keynes, MK11 3LW, UK
UKHW021314180426
11947UKWH00015B/1229